FRACTURE MECHANICS

SOLID MECHANICS AND ITS APPLICATIONS
Volume 14

Series Editor: G.M.L. GLADWELL
Solid Mechanics Division, Faculty of Engineering
University of Waterloo
Waterloo, Ontario, Canada N2L 3G1

Aims and Scope of the Series

The fundamental questions arising in mechanics are: *Why?, How?,* and *How much?* The aim of this series is to provide lucid accounts written by authoritative researchers giving vision and insight in answering these questions on the subject of mechanics as it relates to solids.

The scope of the series covers the entire spectrum of solid mechanics. Thus it includes the foundation of mechanics; variational formulations; computational mechanics; statics, kinematics and dynamics of rigid and elastic bodies; vibrations of solids and structures; dynamical systems and chaos; the theories of elasticity, plasticity and viscoelasticity; composite materials; rods, beams, shells and membranes; structural control and stability; soils, rocks and geomechanics; fracture; tribology; experimental mechanics; biomechanics and machine design.

The median level of presentation is the first year graduate student. Some texts are monographs defining the current state of the field; others are accessible to final year undergraduates; but essentially the emphasis is on readability and clarity.

For a list of related mechanics titles, see final pages.

Fracture Mechanics

An Introduction

by

E. E. GDOUTOS

School of Engineering,
University of Thrace, Xanthi, Greece

Kluwer Academic Publishers

Dordrecht / Boston / London

Library of Congress Cataloging-in-Publication Data

Gdoutos, E. E., 1948-
Fracture mechanics : an introduction / by E.E. Gdoutos.
p. cm. -- (Solid mechanics and its applications ; v. 14)
Includes index.
ISBN 0-7923-1932-X (alk. paper)
1. Fracture mechanics. I. Title. II. Series.
TA409.G36 1993
620.1'126--dc20 92-27721

ISBN 0-7923-1932-X

Published by Kluwer Academic Publishers,
P.O. Box 17, 3300 AA Dordrecht, The Netherlands.

Kluwer Academic Publishers incorporates
the publishing programmes of
D. Reidel, Martinus Nijhoff, Dr W. Junk and MTP Press.

Sold and distributed in the U.S.A. and Canada
by Kluwer Academic Publishers,
101 Philip Drive, Norwell, MA 02061, U.S.A.

In all other countries, sold and distributed
by Kluwer Academic Publishers Group,
P.O. Box 322, 3300 AH Dordrecht, The Netherlands.

Printed on acid-free paper

Printed in the Netherlands

Contents

Conversion table

Length

1 m = 39.37 in — 1 in = 0.0254 m

1 ft = 0.3048 m — 1 m = 3.28 ft

Force

1 N = 0.102 Kgf — 1 Kgf = 9.807 N

1 N = 0.2248 lb — 1 lb = 4.448 N

1 dyne = 10^{-5} N

1 kip = 4.448 kN — 1 kN = 0.2248 kip

Stress

1 Pa = 1 N/m^2

1 lb/in^2 = 6.895 kPa — 1 kPa = 0.145 lb/in^2

1 ksi = 6.895 MPa — 1 MPa = 0.145 ksi

Stress intensity factor

1 MPa$\sqrt{m}$ = 0.910 ksi$\sqrt{in}$ — 1 ksi$\sqrt{in}$ = 1.099 MPa$\sqrt{m}$

Preface

Traditional failure criteria cannot adequately explain many structural failures that occur at stress levels considerably lower than the ultimate strength of the material. Example problems include bridges, tanks, pipes, weapons, ships, railways and aerospace structures. On the other hand, experiments performed by Griffith in 1921 on glass fibers led to the conclusion that the strength of real materials is much smaller, typically by two orders of magnitude, than their theoretical strength. In an effort to explain these phenomena the discipline of fracture mechanics has been created. It is based on the realistic assumption that all materials contain crack-like defects which constitute the nuclei of failure initiation. A major objective of fracture mechanics is to study the load-carrying capacity of structures in the presence of initial defects, where a dominant crack is assumed to exist.

A new design philosophy is therefore introduced by fracture mechanics as opposed to the use of the traditional fracture criteria. Since structures that have no defects cannot be constructed on the grounds of practicality, the safe design of structures should proceed along two lines: either the safe operating load should be determined when a crack of a prescribed size is assumed to exist in the structure; or, given the operating load, the size of the crack that is created in the structure should be determined.

Design by fracture mechanics necessitates knowledge of a critical crack size and a parameter which characterizes the propensity of a crack to extend. Such a parameter should be able to relate laboratory test results to structural performance, so that the response of a structure with cracks can be predicted from laboratory test data. This is determined as a function of material behavior, crack size, structural geometry and loading conditions. On the other hand, the critical value of this parameter – known as fracture toughness, a property of the material – is determined from laboratory tests. Fracture toughness expresses the ability of the material to resist fracture in the presence of cracks. By equating this parameter to its critical value we obtain a relation between applied load, crack size and structure geometry which gives the necessary information for structural design. Fracture toughness is used to rank the ability of a material to resist fracture within the framework of fracture mechanics, in the same way that yield or ultimate strength is used to rank the resistance of the material to yield or fracture in the conventional design criteria. In selecting materials for structural applications we must choose between materials with a high yield strength, but comparatively low fracture toughness, or those with a lower yield strength, but higher fracture toughness.

This book has been prepared to meet the continuing demand for a text designed to present a clear, consistent, straightforward and unified interpretation of the basic concepts and underlying principles of the discipline of fracture mechanics. A general survey of the field would serve no purpose other than give a collection of references and outline equations and results. A realistic application of fracture mechanics could not be made without a sound understanding of the fundamentals. The book is self-contained; the presentations are concise and each topic can be understood by advanced undergraduates in material science and continuum mechanics. Each chapter contains illustrative example problems and homework problems. A total of about fifty example problems and more than two hundred unsolved problems are included in the book.

The book is divided into ten chapters. The first, introductory, chapter gives a brief account of some characteristic failures that could not be explained by the traditional failure criteria, and of Griffith's experiments which gave impetus to the development of a new philosophy in engineering design based on fracture mechanics. The next two chapters deal with the determination of the stress and deformation fields in cracked bodies, and provide the necessary prerequisite for the development of the criteria of fracture mechanics. More specifically, Chapter 2 covers the Westergaard method for determining the linear elastic stress field in cracked bodies, with particular emphasis on the local behavior around the crack tip, and Chapter 3 is devoted to the determination of the elastic-plastic stress and displacement distribution around cracks for time-independent plasticity. Addressed in the fourth chapter is the theory of crack growth based on the global energy balance of the entire system. The fifth chapter is devoted to the critical stress intensity factor fracture criterion. The sixth chapter deals with the theoretical foundation of the path-independent J-integral and its use as a fracture criterion. Furthermore, a brief presentation of the crack opening displacement fracture criterion is given. Chapter 7 studies the underlying principles of the strain energy density theory and demonstrates its usefulness and versatility in solving a host of two- and three-dimensional problems of mixed-mode crack growth in brittle and ductile fracture. Chapter 8 presents in a concise form the basic concepts and the salient points of dynamic fracture mechanics. Addressed in Chapter 9 is the phenomenon of fatigue and environment-assisted crack growth which takes place within the framework of the macroscopic scale level. Finally, Chapter 10 briefly outlines the basic mechanisms of fracture which take place in metals at the microscopic scale level and presents a concise description of the more widely used nondestructive testing methods for defect detection.

Most of the material of the theoretical presentation of the various chapters of the book is contained in the previous book by the author *Fracture Mechanics Criteria and Applications*, published by Kluwer Academic Publishers. That book contains a more detailed description of the various aspects of fracture mechanics than the present book and includes an extensive list of references for further study. The present book was especially written as a potential textbook for fracture mechanics courses at undergraduate and postgraduate level. The instructive character of the book is enhanced by many illustrative example problems and homework problems included in each chapter.

The author wishes to express his gratitude to Professor G.C. Sih for very stimulating discussions and his comments and suggestions during the writing of the book. Thanks are also extended to my secretary Mrs L. Adamidou for typing the manuscript and to my student Mr N. Prassos for the preparation of illustrations. Finally, I wish to express my profound gratitude to my wife, Maria, and my children, Eleftherios and Alexandra-Kalliope for their understanding and patience during the writing of the book.

Xanthi, Greece, 1993 EMMANUEL E. GDOUTOS

2 types < ductile / brittle

Chapter 1

Introduction

1.1. Conventional failure criteria

The mechanical design of engineering structures usually involves an analysis of the stress and displacement fields in conjunction with a postulate predicting the event of failure itself. Sophisticated methods for determining stress distributions in loaded structures are available today. Detailed theoretical analyses based on simplifying assumptions regarding material behavior and structural geometry are undertaken to obtain an accurate knowledge of the stress state. For complicated structure or loading situations, experimental or numerical methods are preferable. Having performed the stress analysis, we select a suitable failure criterion for an assessment of the strength and integrity of the structural component.

Conventional failure criteria have been developed to explain strength failures of load-bearing structures which can be classified roughly as ductile at one extreme and brittle at another. In the first case, breakage of a structure is preceded by large deformation which occurs over a relatively long time period and may be associated with yielding or plastic flow. The brittle failure, on the other hand, is preceded by small deformation, and is usually sudden. Defects play a major role in the mechanism of both these types of failure; those associated with ductile failure differ significantly from those influencing brittle fracture. For ductile failures, which are dominated by yielding before breakage, the important defects (dislocations, grain boundary spacings, interstitial and out-of-size substitutional atoms, precipitates) tend to distort and warp the crystal lattice planes. Brittle fracture, however, which takes place before any appreciable plastic flow occurs, initiates at larger defects such as inclusions, sharp notches, surface scratches or cracks.

For a uniaxial test specimen failure by yielding or fracture takes place when

$$\sigma = \sigma_Y \quad \text{or} \quad \sigma = \sigma_u \tag{1.1}$$

where σ is the applied stress and σ_Y or σ_u is the yield or breakage stress of the material in tension.

Materials that fail in a ductile manner undergo yielding before they ultimately fracture. Postulates for determining those macroscopic stress combinations that result in initial yielding of ductile materials have been developed and are known as yield criteria. At this point we should make it clear that a material may behave in a ductile or brittle manner, depending on the temperature, rate of loading and other variables present. Thus, when we speak about ductile or brittle materials we actually mean the ductile or brittle states of materials. Although the onset of yielding is influenced by factors such as temperature, time and size effects, there is a wide range of circumstances where yielding is mainly determined by the stress state itself. Under such conditions, for isotropic materials, there is extensive evidence that yielding is a result of distortion and is mainly influenced by shear stresses. Hydrostatic stress states, however, play a minor role in the initial yielding of metals. Following these reasonings Tresca and von Mises developed their yield criteria.

The Tresca criterion states that a material element under a multiaxial stress state enters a state of yielding when the maximum shear stress becomes equal to the critical shear stress in a pure shear test at the point of yielding. The latter is a material parameter. Mathematically speaking, this criterion is expressed by [1.1]

$$\frac{|\sigma_1 - \sigma_3|}{2} = k \,, \quad \sigma_1 > \sigma_2 > \sigma_3 \,, \tag{1.2}$$

where σ_1, σ_2, σ_3 are the principal stresses and k is the yield stress in a pure shear test.

The von Mises criterion is based on the distortional energy, and states that a material element initially yields when it absorbs a critical amount of distortional strain energy which is equal to the distortional energy in uniaxial tension at the point of yield. The yield condition is written in the form [1.1]

$$(\sigma_1 - \sigma_2)^2 + (\sigma_2 - \sigma_3)^2 + (\sigma_3 - \sigma_1)^2 = 2\sigma_Y^2 \,, \tag{1.3}$$

where σ_y is the yield stress in uniaxial tension.

However, for porous or granular materials, as well as for some polymers, it has been established that the yield condition is sensitive to hydrostatic stress states. For such materials, the yield stress in simple tension is not equal in general to the yield stress in simple compression. A number of pressure-dependent yield criteria have been proposed in the literature.

On the other hand, brittle materials – or, more strictly, materials in the brittle state – experience fracture without appreciable plastic deformation. For such cases the maximum tensile stress and the Coulomb-Mohr [1.1] criterion are popular. The maximum tensile stress criterion assumes that rupture of a material occurs when the maximum tensile stress exceeds a specific stress which is a material parameter. The Coulomb-Mohr criterion, which is used mainly in rock and soil mechanics states that fracture occurs when the shear stress τ on a given plane becomes equal to a critical value which depends on the normal stress σ on that plane. The fracture condition can be written as

$$|\tau| = F(\sigma) \,, \tag{1.4}$$

where the curve $\tau = F(\sigma)$ on the $\sigma - \tau$ plane is determined experimentally and is considered as a material parameter.

The simplest form of the curve $\tau = F(\sigma)$ is the straight line, which is expressed by

$$\tau = c - \mu\sigma \,. \tag{1.5}$$

Under such conditions the Coulomb-Mohr fracture criterion is expressed by

$$\left(\frac{1 + \sin\,\omega}{2c\,\cos\,\omega}\right)\sigma_1 - \left(\frac{1 - \sin\,\omega}{2c\,\cos\,\omega}\right)\sigma_3 = 1\,, \tag{1.6}$$

where $\tan\,\omega = \mu$ and $\sigma_1 > \sigma_2 > \sigma_3$.

Equation (1.6) suggests that fracture is independent of the intermediate principal stress σ_2. Modifications to the Coulomb-Mohr criterion have been introduced to account for the influence of the intermediate principal stress on the fracture of pressure-dependent materials.

These macroscopic failure criteria describe the onset of yield in materials with ductile behavior, or fracture in materials with brittle behavior; they have been used extensively in the design of engineering structures. In order to take into account uncertainties in the analysis of service loads, material or fabrication defects and high local or residual stresses, a safety factor is employed to limit the calculated critical equivalent yield or fracture stress to a portion of the nominal yield or fracture stress of the material. The latter quantities are determined experimentally. This design procedure has been successful for the majority of structures for many years.

However, it was early realized that there is a broad class of structures, especially those made of high-strength materials, whose failure could not be adequately explained by the conventional design criteria. Griffith [1.2, 1.3], from a series of experiments run on glass fibers, came to the conclusion that the strength of real materials is much smaller, typically by two orders of magnitude, than their theoretical strength. The theoretical strength is determined by the properties of the internal structure of the material, and is defined as the highest stress level that the material can sustain. In the following two sections we shall give a brief account of some characteristic failures which could not be explained by the traditional failure criteria, and describe some of Griffith's experiments. These were the major events that gave impetus to the development of a new philosophy in structural design based on fracture mechanics.

1.2. Characteristic brittle failures

The phenomenon of brittle fracture is frequently encountered in many aspects of everyday life. It is involved, for example, in splitting logs with wedges, in the art of sculpture, in cleaving layers in mica, in machining materials, and in many manufacturing and constructional processes. On the other hand, many catastrophic structural failures involving loss of life have occurred as a result of sudden, unexpected brittle fracture. The history of technology is full of such incidents. We do not intend to overwhelm the reader with the vast number of disasters involving failures of bridges, tanks, pipes, weapons, ships, railways and aerospace structures, but rather to present a few characteristic cases which substantially influenced the development of fracture mechanics.

Although brittle fractures have occurred in many structures over the centuries, the problem arose in acute form with the introduction of all-welded designs. In riveted structures, for example, fracture usually stopped at the riveted joints and did not propagate into adjoining plates. A welded structure, however, appears to be continuous, and a crack growth may propagate from one plate to the next through the welds, resulting in global structural failure. Furthermore, welds may have defects of various kinds, including cracks, and usually introduce high-tensile residual stresses.

The most extensive and widely known massive failures are those that occurred in tankers and cargo ships that were built, mainly in the U.S.A., under the emergency shipbuilding programs of the Second World War [1.4–1.8]. Shortly after these ships were commissioned, several serious fractures appeared in some of them. The fractures were usually sudden and were accompanied by a loud noise. Of approximately 5000 merchant ships built in U.S.A., more than one-fifth developed cracks before April 1946. Most of the ships were less than three years old. In the period between November 1942 and December 1952 more than 200 ships experienced serious failures. Ten tankers and three Liberty ships broke completely in two, while about 25 ships suffered complete fractures of the deck and bottom plating. The ships experienced more failures in heavy seas than in calm seas and a number of failures took place at stresses that were well below the yield stress of the material. A characteristic brittle fracture concerns the tanker *Schenectady*, which suddenly broke in two while in the harbor in cool weather after she had completed successful sea trials. The fracture occurred without warning, extended across the deck just aft of the bridge about midship, down both sides and around the bilges. It did not cross the bottom plating [1.9].

Extensive brittle fractures have also occurred in a variety of large steel structures. Shank [1.10], in a report published in 1954, covers over 60 major structural failures including bridges, pressure vessels, tanks and pipelines. According to Shank, the earliest structural brittle failure on record is a riveted standpipe 250 ft high in Long Island that failed in 1886 during a hydrostatic acceptance test. After water had been pumped to a height of 227 ft, a 20 ft long vertical crack appeared in the bottom, accompanied by a sharp rending sound, and the tower collapsed. In 1938 a welded bridge of the Vierendeel truss type built across the Albert Canal in Belgium with a span of 245 ft collapsed into the canal in quite cold weather. Failure was accompanied by a sound like a shot, and a crack appeared in the lower cord. The bridge was only one year old. In 1940 two similar bridges over the Albert Canal suffered major structural failures. In 1962 the one-year-old King's Bridge in Melbourne, Australia, fractured after a span collapsed as a result of cracks that developed in a welded girder [1.11]. A spherical hydrogen welded tank of 38.5 ft diameter and 0.66 in thickness in Schenectady, New York, failed in 1943 under an internal pressure of about 50 lb/in^2

and at ambient temperature of 10°F [1.10]. The tank burst catastrophically into 20 fragments with a total of 650 ft of herringboned brittle tears. In one of the early aircraft failures, two British de Havilland jet-propelled airliners known as Comets (the first jet airplane designed for commercial service) crashed near Elba and Naples in the Mediterranean in 1954 [1.12]. After these accidents, the entire fleet of these passenger aircraft was grounded. In order to shed light into the cause of the accident a water tank was built at Farnborough into which was placed a complete Comet aircraft. The fuselage was subjected to a cyclic pressurization, and the wings to air loads that simulated the corresponding loads during flight. The plane tested had already flown for 3500 hours. After tests with a total lifetime equivalent to about 2.25 times the former flying time, the fuselage burst in a catastrophic manner after a fatigue crack appeared at a rivet hole attaching reinforcement around the forward escape hatch. For a survey and analysis of extensive brittle failures the interested reader is referred to reference [1.13] for large rotating machinery, to [1.14] for pressure vessels and piping, to [1.15] for ordnance structures and to [1.16] for airflight vehicles.

From a comprehensive investigation and analysis of the above structural failures, we can draw the following general conclusions.

(a) Most fractures were mainly brittle in the sense that they were accompanied by very little plastic deformation, although the structures were made of materials with ductile behavior at ambient temperatures.
(b) Most brittle failures occurred in low temperatures.
(c) Usually, the nominal stress in the structure was well below the yield stress of the material at the moment of failure.
(d) Most failures originated from structural discontinuities including holes, notches, re-entrant corners, etc.
(e) The origin of most failures, excluding those due to poor design, was pre-existing defects and flaws, such as cracks accidentally introduced into the structure. In many cases the flaws that triggered fracture were clearly identified.
(f) The structures that were susceptible to brittle fracture were mostly made of high-strength materials which have low notch or crack toughness (ability of the material to resist loads in the presence of notches or cracks).
(g) Fracture usually propagated at high speeds which, for steel structures, were in the order of 1000 m/s. The observed crack speeds were a fraction of the longitudinal sound waves in the medium.

These findings were essential for the development of a new philosophy in structural design based on fracture mechanics.

1.3. Griffith's work

Long before 1921, when Griffith published his monumental theory on the rupture of solids, a number of pioneering results had appeared which gave evidence of the existence of a size effect on the strength of solids. These findings, which could be considered as a prelude to the Griffith theory, will now be briefly described.

Leonardo da Vinci (1452–1519) ran tests to determine the strength of iron wires [1.17]. He found an inverse relationship between the strength and the length, for wires of constant diameter. We quote from an authoritative translation of da Vinci's sketch book [1.18]:

> Observe what the weight was that broke the wire, and in what part the wire broke . . . Then shorten this wire, at first by half, and see how much more weight it supports; and then make it one quarter of its original length, and so on, making various lengths and noting the weight that breaks each one and the place in which it breaks.

Todhunter and Pearson [1.19] refer to two experimental results analogous to those of da Vinci. According to [1.19], Lloyd (about 1830) found that the average strength of short iron bars is higher than that of long iron bars and Le Blanc (1839) established long iron wires to be weaker than short wires of the same diameter. Stanton and Batson [1.20] reported the results of tests conducted on notched-bar specimens at the National Physical Laboratory, Teddington, after the First World War. From a series of tests it was obtained that the work of fracture per unit volume was decreased as the specimen dimensions were increased. Analogous results were obtained by Docherty [1.21, 1.22] who found that the increase of the plastic work at fracture with the specimen size was smaller than that obtained from geometrical similarity of the strain patterns. This means that the specimens behaved in a more brittle fracture as their size was increased.

All these early results gave indication of the so-called size effect of the strength of solids, which is expressed by an increase in strength as the dimensions of the testpiece decrease. Results at the U.S. Naval Research Laboratory on the strength of glass fibers [1.23] corroborated the early findings of Leonardo da Vinci. Figure 1.1, taken from reference [1.23], shows a decrease of the logarithm of the average strength of glass as a function of the logarithm of the specimen length. The upper line refers to fibers for which precautions have been taken to prevent damage in handling; the lower line was obtained for fibers that came in a loose skein and had a number of flaws.

A plausible explanation of these results is that all structural materials contain flaws which have a deteriorating effect on the strength of the material. The larger the volume of the material tested, the higher the possibility that large cracks exist which, as will be seen, reduce the material strength in a square root relation to their dimensions. However, the first systematic study of the size effect was made by Griffith [1.2, 1.3] who, with his key ideas about the strength of solids, laid down the foundation of the present theory of fracture.

Griffith was motivated in his work by the study of the effect of surface treatment on the strength of metallic machine parts. Early results by Kommers [1.24] indicated that the strength of polished specimens was about 45–50 percent higher than the strength of turned specimens. Other results indicated an increase of the order of 20 percent. Furthermore, the strength was increased by decreasing the size of the scratches. Following the Inglis solution [1.25] of the stress field in an infinite plate

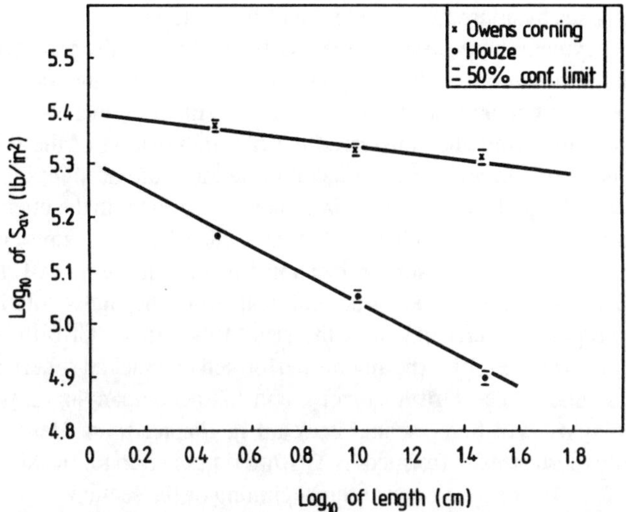

Fig. 1.1. Logarithm of average tensile strength versus logarithm of specimen length for carefully protected glass fibers (×) and fibers damaged by rough handling (•) [1.23].

weakened by an elliptic hole, Griffith observed that tensile stresses – higher than the applied stress by almost an order of magnitude – appeared near the holes according to their shape. Furthermore, he noticed that these maximum stresses were independent of the absolute size of the hole and depended only on the ratio of the axes of the elliptic hole. Indeed, according to [1.25] the maximum stress in the plate, $\sigma_{\max}$, occurs at the end point of the major axis of the ellipse and is given by

$$\sigma_{\max} = \sigma\left(1 + \frac{2a}{b}\right) \simeq 2\sigma\sqrt{\frac{a}{\rho}}\,, \tag{1.7}$$

where σ is the applied stress at infinity in a direction normal to the major axis of the hole, $2a$ and $2b$ are the lengths of the major and minor axes of the ellipse and ρ is the radius of curvature at the ends of the major axis of the ellipse.

These results were in conflict with experiments. Indeed, first, the strength of scratched plates depends on the size and not only on the shape of the scratch; second, higher stresses could be sustained by a scratched plate than those observed in an ordinary tensile test. In experiments performed on cracked circular tubes made of glass, Griffith observed that the maximum tensile stress in the tube was of the magnitude of 344 kip*/in^2 (2372 MPa), while the tensile strength of glass was 24.9 kip/in^2 (172 MPa). These results led him to raise the following questions (we quote from reference [1.2]):

If the strength of this glass, as ordinarily interpreted, is not constant, on what does it depend? What is the greatest possible strength, and can this strength be made for

* kip = 1000 lb.

technical purposes by appropriate treatment of the material?

In order to explain these discrepancies, Griffith attacked the problem of rupture of elastic solids from a different standpoint. He extended the theorem of minimum potential energy to enable it to be applied to the critical moment at which rupture of the solid occurs. Thus, he considered the rupture position of the solid to be an equilibrium position. In applying the theorem he took into account the increase of potential energy due to the formation of new material surfaces in the interior of solids. Using the Inglis solution, Griffith obtained the critical breaking stress of a cracked plate, and found it to be inversely proportional to the square root of the length of the crack. Thus he resolved the paradox arising from the Inglis solution, that the strength of the plate is independent of the size of the crack. Griffith corroborated his theoretical predictions by experiments performed on cracked spherical bulbs and circular glass tubes. The Griffith theory, and his accompanying experiments on cracked specimens, will be presented in detail in chapter four. Here, we describe some further experiments performed by Griffith on the strength of thin glass; these relate to the size effects mentioned at the beginning of the section.

Glass fibers of various diameters were prepared and tested in tension until they broke. The fibers were drawn by hand as quickly as possible from glass bead heated to about 1400–1500°C. For a few seconds after preparation, the strength of the fibers was found to be very high. Values of tensile strength in the range 220–900 kip/in^2 (1500–6200 MPa) for fibers of about 0.02 in diameter were observed. These values were obtained by bending the fibers to fracture and measuring the critical radius of curvature. It was found that the fibers remained almost perfectly elastic until breakage. The strength of the fibers decreased for a few hours until a steady state was reached in which the strength depended upon the diameter only. These fibers were then tested in order to obtain a relation between the strength and the diameter. The fiber diameter ranged from 0.13×10^{-3} to 4.2×10^{-3} in and the fibers were left for about 40 hours before being tested. The specimens had a constant length of about 0.05 in and were obtained after breaking the long fibers several times. Thus, the probability of material defects along the entire specimen length was low, and this was the same for all specimens. The results of the tests are shown in Table 1.1, taken from reference [1.2]. Note that the strength increases as the fiber diameter decreases. The strength tends to that of bulk glass for large thicknesses. The limit as the diameter decreases was obtained by Griffith by plotting the reciprocals of the strength and extrapolating to zero diameter. The maximum strength of glass was found to be 1600 kip/in^2 (11000 MPa), and this value agreed with that obtained from experiments on cracked plates in conjunction with the Griffith theory.

Analogous results on the maximum strength of other materials had been obtained long before Griffith's results. Based on the molecular theory of matter, it had been established that the tensile strength of an isotropic solid or liquid is of the same order as, and always less than, its intrinsic pressure. The latter quantity can be determined using the Van der Waals equation or by measuring the heat that is required to vaporize the substance. According to Griffith [1.2], Traube [1.26] gives values of the intrinsic pressure of various metals including nickel, iron, copper, silver, antimony, zinc, tin

TABLE 1.1. Strength of glass fibers according to Griffith's experiments.

Diameter (10^{-3} in)	Breaking stress (lb/in^2)	Diameter (10^{-3} in)	Breaking stress (lb/in^2)
40.00	24 900	0.95	117 000
4.20	42 300	0.75	134 000
2.78	50 800	0.70	164 000
2.25	64 100	0.60	185 000
2.00	79 600	0.56	154 000
1.85	88 500	0.50	195 000
1.75	82 600	0.38	232 000
1.40	85 200	0.26	332 000
1.32	99 500	0.165	498 000
1.15	88 700	0.130	491 000

and lead which are from 20 to 100 times the tensile strength of the metals. Based on these results, Griffith concluded that the actual strength is always a small fraction of that estimated by molecular theory.

Long before Griffith established the dependence of the strength of glass fibers on the fiber diameter, Karmarsch [1.27] in 1858 gave the following expression for the tensile strength of metal wires,

$$\sigma_{\max} = A + \frac{B}{d}, \tag{1.8}$$

where d is the diameter of the wire and A and B are constants. Griffith's results of Table 1.1 can be represented by the expression

$$\sigma_{\max} = 22\,400\,\frac{4.4 + d}{0.06 + d}, \tag{1.9}$$

where $\sigma_{\max}$ is in lb/in^2 and d in thousandths of an inch. For the range of diameters available to Karmarsch, Equation (1.9) differs little from

$$\sigma_{\max} = 22\,400 + \frac{98\,600}{d}, \tag{1.10}$$

which is of the same form as Equation (1.8).

Griffith's experiments on glass fibers established the 'size effect' in solids and gave an explanation of his observations that 'the maximum tensile stress in the corners of the crack is more than ten times as great as the tensile strength of the material, as measured in an ordinary test' [1.2]. The maximum tensile stress in a cracked plate was estimated from the Inglis solution by measuring the radius of curvature ρ at the ends of the crack. The latter quantity was measured by Griffith by inspection of the interference colors there. He inferred that the width of the crack at its end is about one-quarter of the shortest wavelength of visible light. He found that $\rho = 2 \times 10^{-6}$ in, so that Equation (1.7) gives $\sigma_{max} = 350$ kip/in^2 (2 400 MPa), which is almost one-fifth of the theoretical strength of glass. Thus, near the crack ends, the stresses could approach the theoretical strength of the material. For such small distances, however, Griffith raised the question of appropriateness of the continuum theory. We quote from reference [1.2]: 'The theory of isotropic homogeneous solids may break down if applied to metals in cases where the smallest linear dimension involved is not many times the length of a crystal'. The consequences of this observation will be discussed later.

1.4. Fracture mechanics

Griffith attributed the observed low strength of glass tension test specimens, of the order of 24.9 kip/in^2 (172 MPa), as compared to the maximum stress observed in cracked bodies of the order of 344 kip/in^2 (2372 MPa) and to the theoretical strength of glass of the order of 1600 kip/in^2 (11 000 MPa), to the presence of discontinuities or flaws. For the tension specimen he calculated that flaws of length 2×10^{-4} in should exist.

By his flaw hypothesis Griffith gave a solid explanation of the size effect and laid down the foundations of a new theory of fracture of solids. This theory received no further consideration until almost after the Second World War, when the massive failures of tankers and cargo ships and other catastrophic fractures occurred, as reported in Section 1.2. These failures could not be explained by the conventional design criteria of that time. Attempts were made to use Griffith's ideas in the formulation of a new philosophy for structural design. These efforts led to the development of a new discipline, which is known as *fracture mechanics*.

Before discussing the basic concepts of the discipline of fracture mechanics further, it is appropriate for us to pay attention to the phenomenon of the fracture of solids. During the fracture process in solids, new material surfaces are formed in the medium in a thermodynamically irreversible manner. The fracture may roughly be classified from the macroscopic point of view as brittle and ductile. Brittle fracture is associated with low energy, and for unstable loading conditions it usually takes place under high fracture velocities. Ductile fracture is associated with large deformations, high energy dissipation rates and slow fracture velocities. The phenomenon of the fracture of solids is complicated and depends on a wide variety of factors, including the macroscopic effects, the microscopic phenomena which occur at the locations where the fracture nucleates or grows, and the composition of the material. The study of the fracture process depends on the scale level at which it is considered. At one extreme there is a rupture of cohesive bonds in the solid, and the associated phenomena take place within distances of the order of 10^{-7} cm. For such studies the principles of quantum mechanics should be used. At the other extreme the material is considered as a homogeneous continuum, and the phenomenon of fracture is studied within the framework of continuum mechanics and classical thermodynamics. Fracture studies which take place at scale levels between these two extremes concern

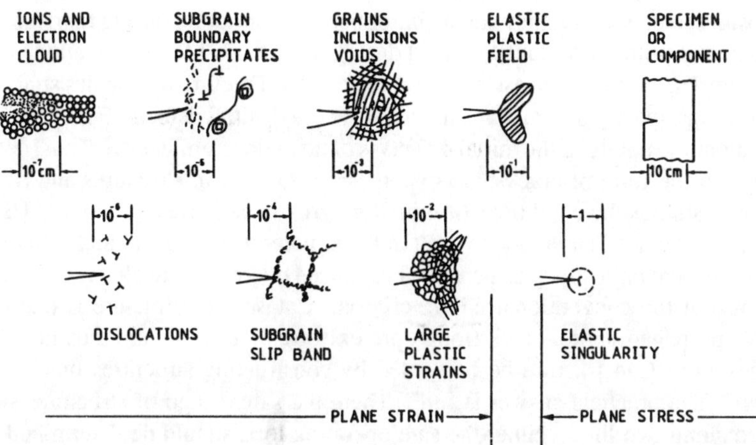

Fig. 1.2. Fracture mechanisms at different scale levels. (After McClintock and Irwin, in *Fracture Toughness Testing and its Applications*, ASTM STP 381, p. 84, 1962, with permission.)

movement of dislocations, formation of subgrain boundary precipitates and slip bands, grain inclusions and voids. The size range of significant events involved in the process of crack extension is shown in Figure 1.2. An understanding of the phenomenon of fracture depends to a large extent on the successful integration of continuum mechanics with materials science, metallurgy, physics and chemistry. Due to the insurmountable difficulties encountered in an interdisciplinary approach, the phenomenon of fracture is usually studied within one of the three scale levels: the atomic, the microscopic or the continuum. Attempts have been made to bridge the gap between these three approaches.

Two key factors gave impetus to the development of fracture mechanics: the size effect, and the inadequacy of traditional failure criteria. The first was demonstrated by Griffith and later by other investigators. It is that the strength of a material measured from a laboratory specimen is many times lower than that predicted from calculations. The traditional failure criteria were inadequate because they could not explain failures which occur at a nominal stress level considerably lower than the ultimate strength of the material. Fracture mechanics is based on the principle that all materials contain initial defects in the form of cracks, voids or inclusions which can affect the load carrying capacity of engineering structures. This is revealed experimentally. Near the defects, high stresses prevail that are often responsible for lowering the strength of the material. One of the objectives of fracture mechanics, as applied to engineering design, is the determination of the critical load of a structure by accounting for the size and location of initial defects. Thus, the problems of initiation, growth and arrest of cracks play a major role in the understanding of the mechanism of failure of structural components.

There are at least three ways in which defects can appear in a structure: first, they can exist in a material due to its composition, as second-phase particles, debonds in composities, etc.; second, they can be introduced in a structure during fabrication, as in welds; and third, they can be created during the service life of a component, like fatigue cracks, environment assisted or creep cracks. Fracture mechanics studies the load-bearing capacity of structures in the presence of initial defects. For engineering applications the nature of the initial defects is of no major significance. Thus, defects, basically in the form of cracks, are hypothesized to exist in structures and fracture mechanics studies the conditions of initiation, growth and arrest of cracks. Usually one dominant crack is assumed to exist in the structural component under study.

A new design philosophy is therefore introduced by fracture mechanics as opposed to the use of the conventional failure criteria. Catastrophic fracture is due to the unstable propagation of a crack from a pre-existing defect. We are thus faced with the question: 'Can fracture be prevented by constructing structures that have no defects?' The practical answer is 'no'. Then, the safe design of structures should proceed along two lines: either the safe operating load should be determined for a crack of a prescribed size, assumed to exist in the structure; or, given the operating load, the size of the crack that is created in the structure should be determined. In this case the structure should be inspected periodically to ensure that the actual crack size is smaller than the crack size the material can sustain safely. Then the following questions arise:

(a) What is the maximum crack size that a material can sustain safely?
(b) What is the strength of a structure as a function of crack size?
(c) How does the crack size relate to the applied loads?
(d) What is the critical load required to extend a crack of known size, and is the crack extension stable or unstable?
(e) How does the crack size increase as a function of time?

In answering these questions fracture mechanics is searching for parameters which characterize the propensity of a crack to extend. Such a parameter should be able to relate laboratory test results to structural performance, so that the response of a structure with cracks can be predicted from laboratory test data. If we call such a parameter the *crack driving force* we should be able to determine that force as a function of material behavior, crack size, structural geometry and loading conditions. On the other hand, the critical value of this parameter, which is taken as a property of the material, should be determined from laboratory tests. The critical value of the crack driving force, known as the *fracture toughness*, expresses the ability of the material to resist fracture in the presence of cracks. By equating the crack driving force to the fracture toughness, we obtain a relation between applied load, crack size and structure geometry which gives the necessary information for structural design.

An additional material parameter, the fracture toughness, is therefore introduced into structural design by the methodology of fracture mechanics. This parameter is used to rank the ability of material to resist fracture within the framework of fracture mechanics, in the same way that yield or ultimate strength ranks the resistance of a material to yield or fracture in the conventional design criteria. In selecting materials for structural applications we must choose between materials with a high yield strength, but comparatively low fracture toughness, on the one hand, or with lower

Von-Mises

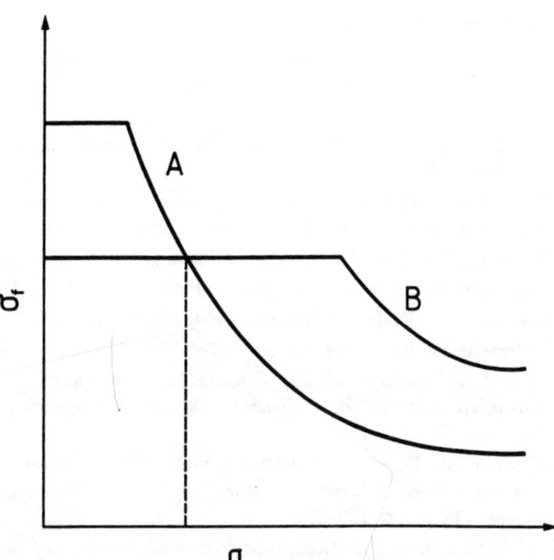

Fig. 1.3. Failure strength versus crack size for two different materials A and B.

yield strength, but higher fracture toughness on the other. As Griffith discovered, the fracture strength is inversely proportional to the square root of the crack size for brittle fracture behavior. Failure by general yield, however, intervenes at some point. Figure 1.3 presents the variation of the strength of a structure versus crack size for two materials A and B different in yield strength and fracture toughness. Material A has higher yield strength, but lower fracture toughness, than material B. The two horizontal lines in the figure represent the failure strength governed by the general yield, while the two downward sloping curves depict the failure strength according to linear elastic fracture mechanics. It is observed that for crack sizes smaller than the crack size corresponding to the intersection of the curves, the strength of the structure is higher for the lower toughness material. Thus, for a structural design in situations where small cracks are anticipated to exist, a material with a higher yield strength should be used, whereas for larger crack sizes a material with a higher fracture toughness would be preferable.

References

1.1. Nadai, A. (1950) *Theory of Flow and Fracture of Solids*, McGraw-Hill, New York.

1.2. Griffith, A.A. (1921) 'The phenomena of rupture and flow in solids', *Philosophical Transactions of the Royal Society of London* **A221**, 163–198.

1.3. Griffith, A.A. (1924) 'The theory of rupture', *Proceedings of First International Congress of Applied Mechanics*, Delft, pp. 55–63.

1.4. Finally report of the board to investigate 'The design and methods of construction of welded steel merchant vessels', July 15 (1946), Government Printing Office, Washington (1947); reprinted in part in *Welding Journal* **26**, 569–619 (1947).

1.5. Technical progress report of the ship structure committee (1948) *Welding Journal* **27**, 337s.

1.6. Second technical report of the ship structure committee, July 1 (1950); reprinted in *Welding Journal* **30**, 169s–181s (1951).

1.7. Williams, M.L. and Ellinger, G.A. (1953) 'Investigation of structural failures of welded ships', *Welding Journal* **32**, 498s–527s.

1.8. Boyd, G.M. (1969) 'Fracture design practices for ship structures', in *Fracture–An Advance Treatise*, Vol. V, *Fracture Design of Structures* (ed. H. Liebowitz), Pergamon Press, pp. 383–470.

1.9. Parker, E.R. (1957) *Brittle Behavior of Engineering Structures*, Wiley, New York.

1.10. Shank, M.E. (1954) 'Brittle failure of steel structures–a brief history', *Metal Progress* **66**, 83–88.

1.11. Fractured girders of the King's Bridge, Melbourne (1964) *Engineering* **217**, 520–522.

1.12. Bishop, T. (1955) 'Fatigue and the Comet disasters', *Metal Progress* **67**, 79–85.

1.13. Yukawa, S., Timo, D.P., and Rudio, A. (1969) 'Fracture design practices for rotating equipment', in *Fracture–An Advanced Treatise*, Vol. V, *Fracture Design of Structures* (ed. H. Liebowitz), Pergamon Press, pp. 65–157.

1.14. Duffy, A.R., McClure, G.M., Eiber, R.J. and Maxey, W.A. (1969) 'Fracture design practices for pressure piping', in *Fracture–An Advanced Treatise*, Vol. V, *Fracture Design of Structures* (ed. H. Liebowitz), Pergamon Press, pp. 159–232.

1.15. Adachi, J. (1969) 'Fracture design practices for ordnance structures', in *Fracture–An Advanced Treatise*, Vol. V, *Fracture Design of Structures* (ed. H. Liebowitz), Pergamon Press, pp. 285–381.

1.16. Kuhn, P. (1969) 'Fracture design analysis for airflight vehicles', in *Fracture–An Advanced Treatise*, Vol. V, *Fracture Design of Structures* (ed. H. Liebowitz), Pergamon Press, pp. 471–500.

1.17. Timoshenko, S.P. (1953) *History of the Strength of Materials*, McGraw-Hill, New York.

1.18. Irwin, G.R. and Wells, A.A. (1965) 'A continuum-mechanics view of crack propagation', *Metallurgical Reviews* **10**, 223–270.

1.19. Todhunter, I. and Pearson, K. (1986) *History of the Theory of Elasticity and of the Strength of Materials*, Sections 1503 and 936, Cambridge Univ. Press.

1.20. Stanton, T.E. and Batson, R.G.C. (1921) *Proceedings of the Institute of Civil Engineers* **211**, 67–100.

1.21. Docherty, J.G. (1932) 'Bending tests on geometrically similar notched bar specimens', *Engineering* **133**, 645–647.

1.22. Docherty, J.G. (1935) 'Slow bending tests on large notched bars', *Engineering* **139**, 211–213.

1.23. Irwin, G.R. (1964) 'Structural aspects of brittle fracture', *Applied Materials Research* **3**, 65–81.

1.24. Kommers, J.B. (1912) *International Association for Testing Materials* **4A, 4B**.

1.25. Inglis, C.E. (1913) 'Stresses in a plate due to the presence of cracks and sharp corners', *Transactions of the Institute of Naval Architects* **55**, 219–241.

1.26. Traube, I. (1903) 'Die physikalischen Eigenschaften der Elemente vom Standpunkte der Zustandsgleichung von van der Waals', *Zeitschrift für Anorganische Chemie* **XXXIV**, 413–426.

1.27. Karmarsch, I. (1858) *Mitteilungen des gew. Ver für Hannover*, pp. 138–155.

Chapter 2

Linear Elastic Stress Field in Cracked Bodies

2.1. Introduction

Fracture mechanics is based on the assumption that all engineering materials contain cracks from which failure starts. The estimation of the remaining life of machine or structural components requires knowledge of the redistribution of stresses caused by the introduction of cracks in conjunction with a crack growth condition. Cracks lead to high stresses near the crack tip; this point should receive particular attention since it is here that further crack growth takes place. Loading of a cracked body is usually accompanied by inelastic deformation and other nonlinear effects near the crack tip, except for ideally brittle materials. There are, however, situations where the extent of inelastic deformation and the nonlinear effects are very small compared to the crack size and any other characteristic length of the body. In such cases the linear theory is adequate to address the problem of stress distribution in the cracked body. Situations where the extent of inelastic deformation is pronounced necessitate the use of nonlinear theories and will be dealt with in the next chapter.

It is the objective of this chapter to study the linear elastic stress field in cracked bodies, with emphasis on the problem of a single crack in an infinite plate. The Westergaard semi-inverse method is used for this purpose. Particular attention is given on the stress intensity factor which governs the linear elastic stress field near the crack tip.

2.2. Crack deformation modes and basic concepts

Consider a plane crack extending through the thickness of a flat plate. Let the crack plane occupy the plane xz and the crack front be parallel to the z-axis. Place the origin of the system $Oxyz$ at the midpoint of the crack front. There are three independent kinematic movements of the upper and lower crack surfaces with respect to each other. These three basic modes of deformation are illustrated in Figure 2.1, which presents the displacements of the crack surfaces of a local element containing the

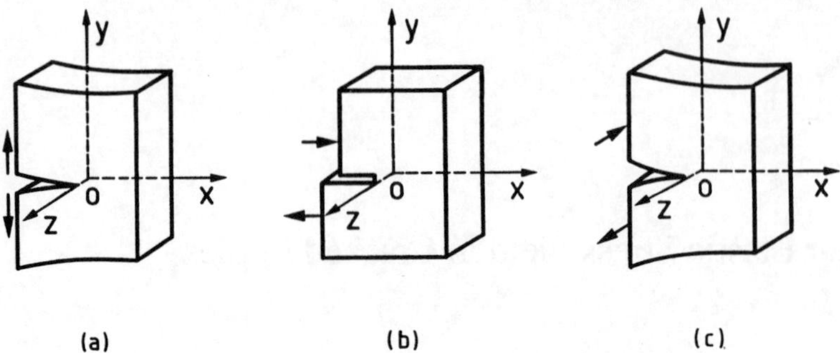

Fig. 2.1. The three basic modes of crack extension. (a) Opening mode, I, (b) Sliding mode, II, and (c) Tearing (or antiplane) mode, III.

crack front. Any deformation of the crack surfaces can be viewed as a superposition of these basic deformation modes, which are defined as follows:

(a) *Opening mode*, I. The crack surfaces separate symmetrically with respect to the planes xy and xz.

(b) *Sliding mode*, II. The crack surfaces slide relative to each other symmetrically with respect to the plane xy and skew-symmetrically with respect to the plane xz.

(c) *Tearing mode*, III. The crack surfaces slide relative to each other skew-symmetrically with respect to both planes xy and xz.

The stress and deformation fields associated with each of these three deformation modes will be determined in the sequel for the cases of plane strain and generalized plane stress. A body is said to be in a state of plane strain parallel to the plane xy if

$$u = u(x, y)\ , \quad v = v(x, y)\ , \quad w = 0 \tag{2.1}$$

where u, v and w denote the displacement components along the axes x, y and z. Thus, the strains and stresses depend only on the variables x and y. Plane strain conditions are realized in long cylindrical bodies which are subjected to loads normal to the cylinder axis and uniform in the z-direction. In crack problems, plane strain conditions are approximated in plates with large thickness relative to the crack length.

A generalized plane stress state parallel to the xy plane is defined by

$$\begin{aligned} &\sigma_z = \tau_{zx} = \tau_{zy} = 0 \\ &\sigma_x = \sigma_x(x, y)\ , \quad \sigma_y = (x, y)\ , \quad \tau_{xy} = \tau_{xy}(x, y) \end{aligned} \tag{2.2}$$

where σ_x, σ_y, σ_z and τ_{xy}, τ_{zx}, τ_{zy} denote the normal and shear stresses associated with the system xyz. Generalized plane stress conditions are realized in thin flat plates with traction-free surfaces. In crack problems, the generalized plane stress conditions are approximated in plates with crack lengths that are large in relation to the plate thickness. We recall from the theory of elasticity that a plane strain problem may be solved as a generalized plane stress problem by replacing the value of Poisson's ratio ν by the value $\nu/(1+\nu)$.

2.3. Westergaard method

(a) Description of the method

The Westergaard semi-inverse method [2.1, 2.2] constitutes a simple and versatile tool for solving a certain class of plane elasticity problems. It uses the Airy stress function representation, in which the solution of a plane elasticity problem is reduced to finding a function U which satisfies the biharmonic equation

$$\nabla^2\nabla^2 U = \frac{\partial^4 U}{\partial x^4} + 2\frac{\partial^4 U}{\partial x^2\,\partial y^2} + \frac{\partial^4 U}{\partial y^4} = 0 \qquad (2.3)$$

and the appropriate boundary conditions.

The stress components are given by

$$\sigma_x = \frac{\partial^2 U}{\partial y^2}\,, \quad \sigma_y = \frac{\partial^2 U}{\partial x^2}\,, \quad \tau_{xy} = -\frac{\partial^2 U}{\partial x\,\partial y}\,. \qquad (2.4)$$

If we choose the function U in the form

$$U = \psi_1 + x\psi_2 + y\psi_3\,, \qquad (2.5)$$

where the functions ψ_i $(i = 1, 2, 3)$ are harmonic, that is,

$$\nabla^2\psi_i = \frac{\partial^2\psi_i}{\partial x^2} + \frac{\partial^2\psi_i}{\partial y^2} = 0\,, \qquad (2.6)$$

U will automatically satisfy Equation (2.3).

Following the Cauchy–Riemann conditions for an analytic function of the form

$$Z(z) = \text{Re } Z + i \text{ Im } Z \qquad (2.7)$$

we have

$$\text{Re}\,\frac{dZ}{dz} = \frac{\partial \text{ Re } Z}{\partial x} = \frac{\partial \text{ Im } Z}{\partial y}$$

$$\text{Im}\,\frac{dZ}{dz} = \frac{\partial \text{ Im } Z}{\partial x} = -\frac{\partial \text{ Re } Z}{\partial y} \qquad (2.8)$$

and, therefore,

$$\nabla^2 \text{ Re } Z = \nabla^2 \text{ Im } Z = 0\,. \qquad (2.9)$$

Thus, the functions ψ_i $(i = 1, 2, 3)$ in Equation (2.5) can be considered as the real or imaginary part of an analytic function of the complex variable z.

Introducing the notation

$$\tilde{Z} = \frac{\mathrm{d}\tilde{\tilde{Z}}}{\mathrm{d}z}\,, \quad Z = \frac{\mathrm{d}\tilde{Z}}{\mathrm{d}z}\,, \quad Z' = \frac{\mathrm{d}Z}{\mathrm{d}z}\,. \tag{2.10}$$

Westergaard defined an Airy function U_{I} for symmetric problems by

$$U_{\mathrm{I}} = \mathrm{Re}\ \tilde{\tilde{Z}}_{\mathrm{I}} + y\ \mathrm{Im}\ \tilde{Z}_{\mathrm{I}}\,. \tag{2.11}$$

U_{I} automatically satisfies Equation (2.3). Using Equations (2.4) we find the stresses from U_{I} to be

$$\begin{aligned}
\sigma_x &= \mathrm{Re}\ Z_{\mathrm{I}} - y\ \mathrm{Im}\ Z'_{\mathrm{I}} \\
\sigma_y &= \mathrm{Re}\ Z_{\mathrm{I}} + y\ \mathrm{Im}\ Z'_{\mathrm{I}} \\
\tau_{xy} &= -y\ \mathrm{Re}\ Z'_{\mathrm{I}}\,.
\end{aligned} \tag{2.12}$$

Using Hooke's law and the strain-displacement equations we obtain the displacement components

$$\begin{aligned}
2\mu u &= \frac{\kappa - 1}{2}\ \mathrm{Re}\ \tilde{Z}_{\mathrm{I}} - y\ \mathrm{Im}\ Z_{\mathrm{I}} \\
2\mu v &= \frac{\kappa + 1}{2}\ \mathrm{Im}\ \tilde{Z}_{\mathrm{I}} - y\ \mathrm{Re}\ Z_{\mathrm{I}}
\end{aligned} \tag{2.13}$$

where $\kappa = 3 - 4\nu$ for plane strain and $\kappa = (3 - \nu)/(1 + \nu)$ for generalized plane stress.

For skew-symmetric problems with respect to the x-axis the Airy function U_{II} is defined by

$$U_{\mathrm{II}} = y\ \mathrm{Im}\ \tilde{Z}_{\mathrm{II}} \tag{2.14}$$

and the stresses and displacements by

$$\begin{aligned}
\sigma_x &= 2\ \mathrm{Re}\ Z_{\mathrm{II}} - y\ \mathrm{Im}\ Z'_{\mathrm{II}} \\
\sigma_y &= y\ \mathrm{Im}\ Z'_{\mathrm{II}} \\
\tau_{xy} &= -\ \mathrm{Im}\ Z_{\mathrm{II}} - y\ \mathrm{Re}\ Z'_{\mathrm{II}}
\end{aligned} \tag{2.15}$$

and

$$\begin{aligned}
2\mu u &= \frac{\kappa + 1}{2}\ \mathrm{Re}\ \tilde{Z}_{\mathrm{II}} - y\ \mathrm{Im}\ Z_{\mathrm{II}} \\
2\mu v &= \frac{\kappa - 1}{2}\ \mathrm{Im}\ \tilde{Z}_{\mathrm{II}} - y\ \mathrm{Re}\ Z_{\mathrm{II}}\,.
\end{aligned} \tag{2.16}$$

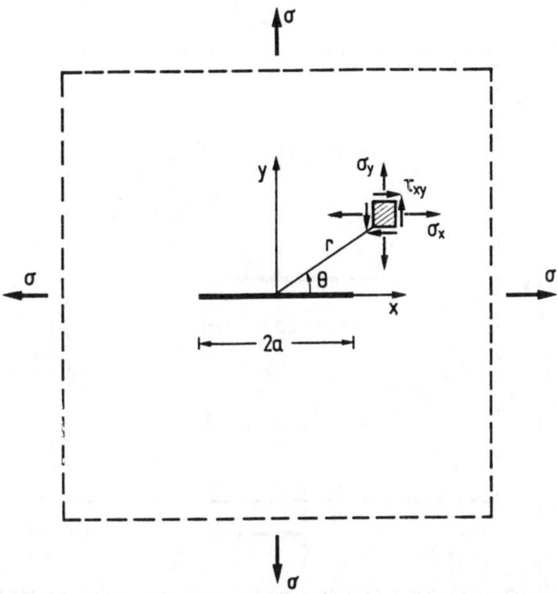

Fig. 2.2. A crack of length $2a$ in an infinite plate subjected to a uniform stress σ at infinity.

(b) Crack problems

Consider a crack of length $2a$ which occupies the segment $-a \leq x \leq a$ along the x-axis in an infinite plate subjected to uniform equal stresses σ along the y and x directions at infinity (Figure 2.2). The boundary conditions of the problem may be stated as follows:

$$\sigma_y + i\tau_{xy} = 0 \quad \text{for} \quad y = 0\,, \quad -a < x < a \tag{2.17}$$

and

$$\sigma_x = 0\,, \quad \sigma_y = \sigma\,, \quad \tau_{xy} = 0 \quad \text{for} \quad (x^2 + y^2)^{1/2} \to \infty\,. \tag{2.18}$$

The function defined by

$$Z_{\mathrm{I}} = \frac{\sigma z}{(z^2 - a^2)^{1/2}} \tag{2.19}$$

satisfies the boundary conditions (2.17) and (2.18) and therefore is the Westergaard function for the problem shown in Figure 2.2.

For the problem of a crack of length $2a$ which occupies the segment $-a \leq x \leq a$ along the x-axis in an infinite plate subjected to uniform in-plane shear stresses τ at infinity (Figure 2.3), the boundary conditions of the problem may be stated as

$$\sigma_y + i\tau_{xy} = 0 \quad \text{for} \quad y = 0\,, \quad -a < x < a$$

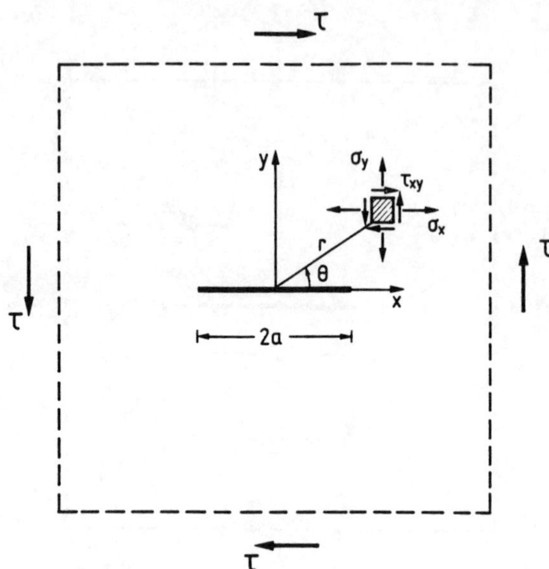

Fig. 2.3. A crack of length $2a$ in an infinite plate subjected to uniform in-plane shear stresses τ at infinity.

$$\sigma_x = \sigma_y = 0\,, \quad \tau_{xy} = \tau \quad \text{for} \quad (x^2 + y^2)^{1/2} \to \infty\,. \tag{2.20}$$

and the Westergaard function of the problem is

$$Z_{\mathrm{II}} = -\frac{i\tau z}{(z^2 - a^2)^{1/2}}\,. \tag{2.21}$$

2.4. Singular stress and displacement fields

The study of stress and displacement fields near the crack tip is very important, because these fields govern the fracture process that takes place at the crack tip. In this section we shall make a thorough study of the stresses and displacements near the crack tip for the three deformation modes.

(a) Opening mode

The Westergaard function for an infinite plate with a crack of length $2a$ subjected to equal stresses σ at infinity (Figure 2.2) is given by Equation (2.19). If we place the origin of the coordinate system at the crack tip $z = a$ through the transformation

$$\zeta = z - a\,, \tag{2.22}$$

Equation (2.19) takes the form

$$Z_{\rm I} = \frac{\sigma(\zeta + a)}{[\zeta(\zeta + 2a)]^{1/2}} \,. \tag{2.23}$$

Expanding Equation (2.23) we obtain

$$Z_{\rm I} = \frac{\sigma(\zeta + a)}{(2a\zeta)^{1/2}} \left[1 - \frac{1}{2}\frac{\zeta}{2a} + \frac{1 \cdot 3}{2 \cdot 4}\left(\frac{\zeta}{2a}\right)^2 - \frac{1 \cdot 3 \cdot 5}{2 \cdot 4 \cdot 6}\left(\frac{\zeta}{2a}\right)^3 + \ldots \right]. \tag{2.24}$$

For small $|\zeta|$, ($|\zeta| \to 0$), that is near the crack tip at $x = a$, Equation (2.24) may be written

$$Z_{\rm I} = \frac{K_{\rm I}}{(2\pi\zeta)^{1/2}} \,, \tag{2.25}$$

where

$$K_{\rm I} = \sigma\sqrt{\pi a} \,. \tag{2.26}$$

Using polar coordinates, r, θ we have

$$\zeta = r\, e^{i\theta} \tag{2.27}$$

and the stresses near the crack tip are

$$\begin{aligned} \sigma_x &= \frac{K_{\rm I}}{\sqrt{2\pi r}} \cos\frac{\theta}{2}\left(1 - \sin\frac{\theta}{2}\sin\frac{3\theta}{2}\right) \\ \sigma_y &= \frac{K_{\rm I}}{\sqrt{2\pi r}} \cos\frac{\theta}{2}\left(1 + \sin\frac{\theta}{2}\sin\frac{3\theta}{2}\right) \\ \tau_{xy} &= \frac{K_{\rm I}}{\sqrt{2\pi r}} \cos\frac{\theta}{2}\sin\frac{\theta}{2}\cos\frac{3\theta}{2} \,. \end{aligned} \tag{2.28}$$

Now suppose that the cracked plate is subjected to uniform stresses σ and $k\sigma$ along the y and x directions, respectively, at infinity. The stress field may be obtained by superimposing the stress field in Equation (2.28) and the uniform field $\sigma_x = (k-1)\,\sigma$. Thus

$$\sigma_x = \frac{K_{\rm I}}{\sqrt{2\pi r}} \cos\frac{\theta}{2}\left(1 - \sin\frac{\theta}{2}\sin\frac{3\theta}{2}\right) - (1-k)\,\sigma \tag{2.29}$$

while the stresses σ_y and τ_{xy} are the same as in Equation (2.28).

The quantity $K_{\rm I}$ is the opening-mode stress intensity factor and expresses the strength of the singular elastic stress field. As was shown by Irwin [2.3], Equation (2.28) applies to all crack-tip stress fields independently of crack/body geometry and the loading conditions. The constant term entering σ_x in Equation (2.29) takes different values depending on the applied loads and the geometry of the cracked plate. The stress intensity factor depends linearly on the applied load and is a function of the crack length and the geometrical configuration of the cracked body. Results for stress intensity factors for some crack problems of practical importance are presented in Appendix 2.1.

Introducing the value of the Westergaard function Z_{I} from Equation (2.25) into Equations (2.13) we obtain the displacements

$$u = \frac{K_{\mathrm{I}}}{2\mu}\sqrt{\frac{r}{2\pi}}\cos\frac{\theta}{2}\,(\kappa - \cos\theta)$$

$$v = \frac{K_{\mathrm{I}}}{2\mu}\sqrt{\frac{r}{2\pi}}\sin\frac{\theta}{2}\,(\kappa - \cos\theta)\,. \tag{2.30}$$

Equations (2.28) and (2.30) express the stress and displacement fields for plane strain and plane stress conditions near the crack tip. It is observed that the stresses and strains always have an inverse square root singularity at the crack tip. For plane strain conditions the stress $\sigma_z (= \nu(\sigma_x + \sigma_y))$ normal to the plane of the plate is given by

$$\sigma_z = \frac{2\nu K_{\mathrm{I}}}{\sqrt{2\pi r}}\cos\frac{\theta}{2}\,. \tag{2.31}$$

When the Westergaard function Z_{I} of a crack problem is known it can always, as in the previous case, be put in the form

$$Z_{\mathrm{I}} = \frac{f(\zeta)}{\zeta^{1/2}} \tag{2.32}$$

where the function $f(\zeta)$ is well behaved for small $|\zeta|$. Thus, for $|\zeta| \to 0$ Equation (2.32) takes the form

$$Z_{\mathrm{I}} = \frac{f(0)}{\zeta^{1/2}}\,. \tag{2.33}$$

If we compare the stress σ_y along the x-axis computed from Equation (2.33) with that given by Equations (2.28), we obtain

$$K_{\mathrm{I}} = \lim_{|\zeta|\to 0} \sqrt{2\pi\zeta}\, Z_{\mathrm{I}}\,. \tag{2.34}$$

Equation (2.34) can be used to determine the K_{I} stress intensity factor when the function Z_{I} is known.

From Equations (2.28) we obtain the following expressions for the singular polar components of stress and displacement

$$\sigma_r = \frac{K_{\mathrm{I}}}{\sqrt{2\pi r}}\left(\frac{5}{4}\cos\frac{\theta}{2} - \frac{1}{4}\cos\frac{3\theta}{2}\right)$$

$$\sigma_\theta = \frac{K_{\mathrm{I}}}{\sqrt{2\pi r}}\left(\frac{3}{4}\cos\frac{\theta}{2} + \frac{1}{4}\cos\frac{3\theta}{2}\right) \tag{2.35}$$

$$\tau_{r\theta} = \frac{K_{\mathrm{I}}}{\sqrt{2\pi r}}\left(\frac{1}{4}\sin\frac{\theta}{2} + \frac{1}{4}\sin\frac{3\theta}{2}\right)$$

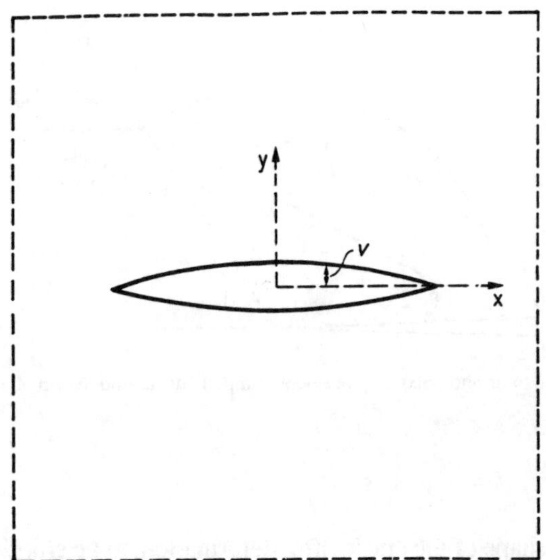

Fig. 2.4. Crack deformation shape for mode-I loading.

and

$$u_r = \frac{K_I}{4\mu}\sqrt{\frac{r}{2\pi}}\left[(2\kappa - 1)\cos\frac{\theta}{2} - \cos\frac{3\theta}{2}\right]$$

$$u_\theta = \frac{K_I}{4\mu}\sqrt{\frac{r}{2\pi}}\left[-(2\kappa + 1)\sin\frac{\theta}{2} + \sin\frac{3\theta}{2}\right] \tag{2.36}$$

and for the principal singular stresses

$$\sigma_1 = \frac{K_I}{\sqrt{2\pi r}}\cos\frac{\theta}{2}\left(1 + \sin\frac{\theta}{2}\right)$$

$$\sigma_2 = \frac{K_I}{\sqrt{2\pi r}}\cos\frac{\theta}{2}\left(1 - \sin\frac{\theta}{2}\right). \tag{2.37}$$

It is important to determine the displacement v (Figure 2.4) of the crack faces. From Equation (2.13), this is obtained as

$$2\mu v = \frac{\kappa + 1}{2}\ \text{Im}\ \tilde{Z}_I . \tag{2.38}$$

For a crack of length $2a$ in an infinite plate subjected to equal stresses σ at infinity the function Z_I is given by Equation (2.19). We obtain

$$v = \frac{\kappa + 1}{4\mu}\sigma\sqrt{a^2 - x^2}\ , \quad -a \le x \le a\ . \tag{2.39}$$

Equation (2.39) can be put in the form

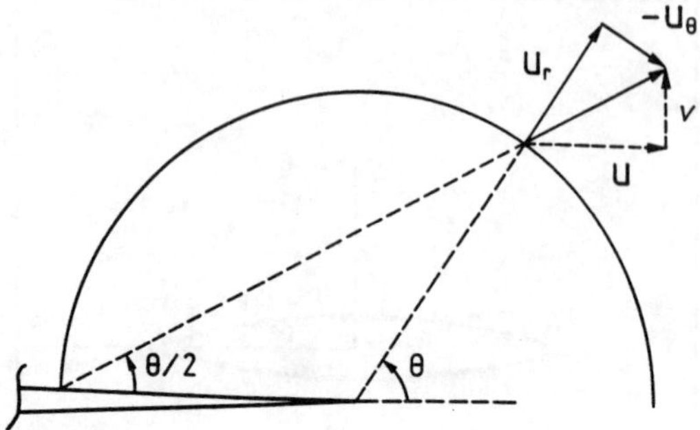

Fig. 2.5. Rectangular and polar displacement components around the tip of a mode-I crack.

$$\frac{v^2}{t^2} + x^2 = a^2 \, , \quad t = \frac{\kappa + 1}{4\mu} \, \sigma \tag{2.40}$$

which shows the shape of the crack, after deformation, to be elliptic.

The vertical displacement v becomes maximum at the center of the crack. This is obtained from Equation (2.39):

$$v_{\max} = \frac{\kappa + 1}{4\mu} \, \sigma a \, . \tag{2.41}$$

From Equations (2.30) and (2.36) which express the polar and rectangular components of the displacement in the vicinity of the crack tip we obtain

$$u_r = u \, , \quad u_\theta = -v \, . \tag{2.42}$$

This property of the displacement components is shown in Figure 2.5.

(b) Sliding mode

The Westergaard function $Z_{\rm II}$ for a crack of length $2a$ in an infinite plate subjected to uniform in-plane shear stress τ at infinity (Figure 2.3) is given by Equation (2.21). The stresses and displacements are obtained from Equations (2.15) and (2.16). Following the same procedure as in the previous case, and recognizing the general applicability of the singular solution for all sliding-mode crack problems, we obtain the following equations for the stresses and displacements

$$\begin{aligned} \sigma_x &= -\frac{K_{\rm II}}{\sqrt{2\pi r}} \, \sin\frac{\theta}{2} \left(2 + \cos\frac{\theta}{2} \, \cos\frac{3\theta}{2}\right) \\ \sigma_y &= \frac{K_{\rm II}}{\sqrt{2\pi r}} \, \sin\frac{\theta}{2} \, \cos\frac{\theta}{2} \, \cos\frac{3\theta}{2} \\ \tau_{xy} &= \frac{K_{\rm II}}{\sqrt{2\pi r}} \, \cos\frac{\theta}{2} \left(1 - \sin\frac{\theta}{2} \, \sin\frac{3\theta}{2}\right) \end{aligned} \tag{2.43}$$

and

$$u = \frac{K_{\text{II}}}{2\mu} \sqrt{\frac{r}{2\pi}} \sin \frac{\theta}{2} (2 + \kappa + \cos \theta)$$

$$v = \frac{K_{\text{II}}}{2\mu} \sqrt{\frac{r}{2\pi}} \cos \frac{\theta}{2} (2 - \kappa - \cos \theta) . \tag{2.44}$$

The quantity K_{II} is the sliding-mode stress intensity factor and, as for the opening mode, it expresses the strength of the singular elastic stress field. When the Westergaard function Z_{II} is known, K_{II} is determined, following the same procedure as previously, by

$$K_{\text{II}} = \lim_{|\zeta| \to 0} i \sqrt{2\pi\zeta}\, Z_{\text{II}} . \tag{2.45}$$

For a crack of length $2a$ in an infinite plate subjected to in-plane shear stress τ at infinity, we obtain from Equations (2.45) and (2.21)

$$K_{\text{II}} = \tau \sqrt{\pi a} . \tag{2.46}$$

(c) Tearing mode

For the tearing (or antiplane) mode of crack deformation the in-plane displacements u and v are zero, while the displacement w is a function of the in-plane coordinates x and y, that is

$$u = v = 0 , \quad w = w(x, y) . \tag{2.47}$$

Equation (2.47) suggests that the movement of the crack surfaces can be related to the warping action of noncircular cylinders subjected to torsion. Equation (2.47) renders

$$\varepsilon_x = \varepsilon_y = \varepsilon_z = \gamma_{xy} = 0$$

and from Hooke's law we have

$$\gamma_{xz} = \frac{\partial w}{\partial x} , \quad \gamma_{yz} = \frac{\partial w}{\partial y} \tag{2.48}$$

$$\sigma_x = \sigma_y = \sigma_z = \tau_{xy} = 0$$

$$\tau_{xz} = \mu \frac{\partial w}{\partial x} , \quad \tau_{yz} = \mu \frac{\partial w}{\partial y} . \tag{2.49}$$

Substituting Equation (2.49) into the non-self-satisfied equilibrium equation

$$\frac{\partial \tau_{xz}}{\partial x} + \frac{\partial \tau_{yz}}{\partial y} = 0 \tag{2.50}$$

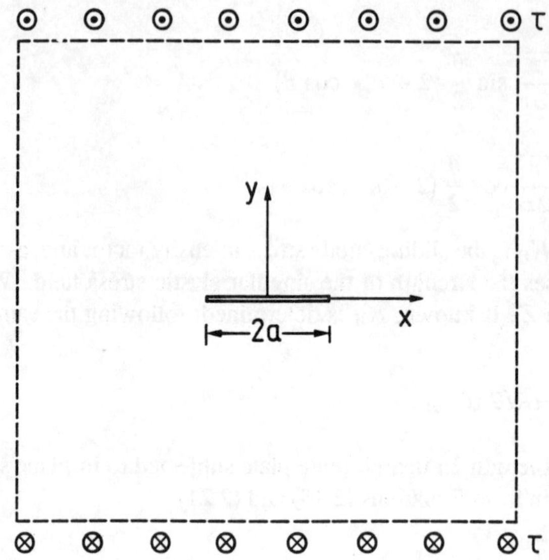

Fig. 2.6. A crack of length $2a$ in an infinite plate subjected to uniform out-of-plane shear stress τ at infinity.

we obtain for w

$$\nabla^2 w = 0 . \tag{2.51}$$

Since w satisfies the Laplace equation it can be put in the form

$$w = \frac{1}{\mu} \text{ Im } Z_{\text{III}} \tag{2.52}$$

where Z_{III} is an analytic function. From Equations (2.49) we obtain

$$\tau_{xz} = \text{ Im } Z'_{\text{III}} \, , \quad \tau_{yz} = \text{ Re } Z'_{\text{III}} \, . \tag{2.53}$$

Consider a crack of length $2a$ which occupies the segment $-a \leq x \leq a$ along the x-axis in an infinite plate subjected to uniform out-of-plane shear stress τ at infinity (Figure 2.6). The boundary conditions of the problem may be stated as

$$\tau_{yz} = 0 \quad \text{for} \quad y = 0 \, , \quad -a < x < a$$

$$\tau_{xz} = 0 \, , \quad \tau_{yz} = \tau \quad \text{for} \quad (x^2 + y^2)^{1/2} \to \infty \, . \tag{2.54}$$

Following the same procedure as in the case of the opening mode we introduce the function Z'_{III}, defined by

$$Z'_{\text{III}} = \frac{\tau z}{(z^2 - a^2)^{1/2}} \, , \tag{2.55}$$

This satisfies the boundary conditions (2.54).

Near the crack tip, we obtain for the stresses τ_{xz}, τ_{yz} and the displacement w

$$\tau_{xz} = -\frac{K_{\mathrm{III}}}{\sqrt{2\pi r}} \sin \frac{\theta}{2}\,, \quad \tau_{yz} = \frac{K_{\mathrm{III}}}{\sqrt{2\pi r}} \cos \frac{\theta}{2}\,,$$

$$w = \frac{2K_{\mathrm{III}}}{\mu} \sqrt{\frac{r}{2\pi}} \sin \frac{\theta}{2} \tag{2.56}$$

where

$$K_{\mathrm{III}} = \tau \sqrt{\pi a}\,. \tag{2.57}$$

Equation (2.56) applies to all tearing-mode crack problems near the crack tip. The quantity K_{III} is the tearing-mode stress intensity factor and expresses the strength of the singular elastic stress field. When the function Z'_{III} is given, K_{III} is determined by

$$K_{\mathrm{III}} = \lim_{|\zeta| \to 0} \sqrt{2\pi\zeta}\, Z'_{\mathrm{III}}\,. \tag{2.58}$$

In cases where two or all three deformation modes exist in a crack problem, the singular elastic stress field in the neighborhood of the crack tip is obtained by superimposing the solutions corresponding to each of the three deformation modes and it is characterized by the respective stress intensity factors.

2.5. Stress intensity factor solutions

The stress intensity factor is a fundamental quantity that governs the stress field near the crack tip. As will be seen in chapter five, it can be used to predict the failure of a cracked plate. The stress intensity factor depends on both the geometrical configuration and the loading conditions of the body. A number of methods have been used for the determination of stress intensity factors. They may be classified as

1. Theoretical (Westergaard semi-inverse method and method of complex potentials).
2. Numerical (Green's function, weight functions, boundary collocation, alternating method, integral transforms, continuous dislocations and finite elements methods).
3. Experimental (photoelasticity, moiré, holography, caustics, and combinations of these methods).

Theoretical methods are generally restricted to plates of infinite extent with simple geometrical configurations of cracks and boundary conditions. For more complicated situations one must resort to numerical or experimental methods. Description of these methods is outside the scope of the present book. For a brief outline of these methods the reader is referred to [2.4]. To obtain an idea of stress intensity factor solutions

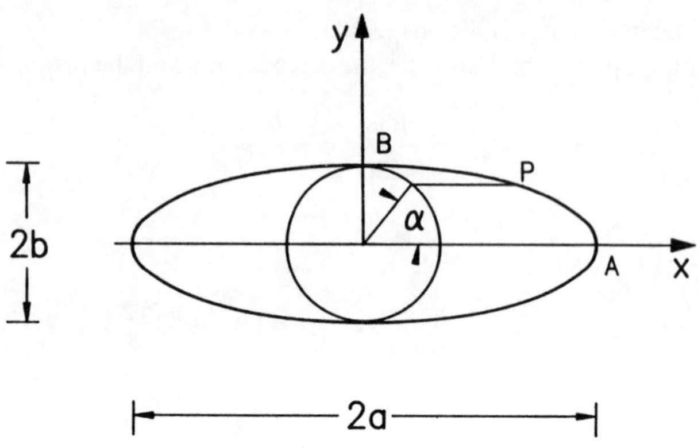

Fig. 2.7. Elliptical crack with axes $2a$ and $2b$.

we present results for some cases of practical importance in Appendix 2.1. For a compilation of stress intensity factor solutions for crack problems the reader is referred to the handbooks of references [2.5, 2.6].

2.6. Three-dimensional cracks

So far, attention has been paid only to the problem of a plane crack extending through the thickness of a flat plate. However, many embedded cracks or flaws in engineering structures have irregular shapes. These flaws are three-dimensional and, for purposes of analysis, are usually idealized as planes of discontinuities bounded by smooth curves. The basic shapes that are most suitable for analysis are the circular or penny-shaped and the elliptical embedded cracks. In the elliptical crack, various degrees of crack-border curvedness may be obtained by varying the ellipticity.

Three-dimensional surface and embedded cracks are frequently encountered in engineering structures. Thus, surface cracks are usually initiated from the interior of pressure vessels and pipelines used in the nuclear industry. Because of their importance in the design of a variety of structures, three-dimensional cracks have attracted the interest of engineers and researchers. A great amount of effort has been spent on the determination of the stress distribution in bodies with three-dimensional cracks. The stress field in the neighborhood of a point of the border of an elliptical crack is a combination of the opening-mode, sliding-mode and tearing-mode, as for a through crack in a plate, and it is governed by the values of the corresponding stress intensity factors, K_{I}, K_{II} and K_{III}. As in the two-dimensional case, these factors are independent of the coordinate variables r, θ and depend only on the position of the point P at the crack front, the nature of loading and the crack geometry.

Consider an elliptical crack with semi-axes a and b embedded in an infinite body which is subjected to a uniform uniaxial stress σ normal to the crack plane (Figure 2.7). The opening-mode stress intensity factor for this problem is

$$K_{\mathrm{I}} = \frac{\sigma\sqrt{\pi b}}{E(k)}\left(\sin^2\alpha + \frac{b^2}{a^2}\cos^2\alpha\right)^{1/4} \tag{2.59}$$

where the angle α is defined in Figure 2.7 and $E(k)$ is the complete elliptic integral of second kind. It is defined as

$$E(k) = \int_0^{\pi/2} (1 - k^2 \sin^2\varphi)^{1/2}\,\mathrm{d}\varphi\,. \tag{2.60}$$

with

$$k = \sqrt{1 - \frac{b^2}{a^2}}\,. \tag{2.61}$$

Equation (2.59) gives the variation of the opening-mode stress intensity factor K_{I} along the border of the elliptical crack. We find that K_{I} takes its maximum value at point $B(\alpha = 90^\circ)$ of the minor crack axis and its minimum value at point $A(\alpha = 0)$ of the major crack axis (Figure 2.7). These values are given by

$$K_{\mathrm{I}A} = \frac{\sigma\sqrt{\pi b}}{E(k)}\sqrt{\frac{b}{a}}\,,\quad K_{\mathrm{I}B} = \frac{\sigma\sqrt{\pi b}}{E(k)}\,. \tag{2.62}$$

For a circular crack of radius a in an infinite solid, Equation (2.59) gives

$$K_{\mathrm{I}} = \frac{2}{\pi}\sigma\sqrt{\pi a}\,. \tag{2.63}$$

In the limit as $b/a \to 0$, $k \to 1$ and $E(k) \to 1$, Equation (2.59) for $\alpha = 90^\circ$ gives

$$K_{\mathrm{I}} = \sigma\sqrt{\pi b} \tag{2.64}$$

which is the stress intensity factor at the tip of a through-the-thickness crack of length $2b$.

These formulas for the stress intensity factor refer to an embedded elliptical crack in an infinite solid. Results for semi-elliptical and quarter-elliptical cracks in plates of finite width have been obtained using numerical methods and can be found in references [2.5, 2.6].

Examples

Example 2.1.

For a semi-infinite crack subjected to tearing (antiplane) mode of deformation assume that the displacement w has the form

$$w = r^{\lambda} f(\theta)\,. \tag{1}$$

Based on this expression for w determine the singular stress and displacement components.

Solution: Substituting Equation (1) into Equation (2.51) we obtain

$$\frac{\mathrm{d}^2 f(\theta)}{\mathrm{d}\theta^2} + \lambda^2 f(\theta) = 0 \tag{2}$$

which has the solution

$$f(\theta) = A \sin \lambda\theta + B \cos \lambda\theta\,. \tag{3}$$

The term $B \cos \lambda\theta$ should be excluded from the solution due to the antisymmetry of the problem with regard to $\theta = 0$. Then Equation (3) becomes

$$f(\theta) = A \sin \lambda\theta\,. \tag{4}$$

The boundary condition along the crack faces is

$$\tau_{\theta z} = 0\,, \quad \text{for} \quad \theta = \pm\pi \tag{5}$$

and implies, with Hooke's law

$$\tau_{\theta z} = \frac{\mu}{r}\frac{\partial w}{\partial \theta} \tag{6}$$

that

$$\cos \lambda\pi = 0\,. \tag{7}$$

Equation (7) gives the eigenvalues

$$\lambda = \pm\frac{n}{2}\,, \quad n = 1, 3, 5, \ldots\,. \tag{8}$$

Negative values of n are ignored because they produce infinite displacements w at the crack tip ($r = 0$). For $n = 1$ the singular stress and displacement components are obtained. We obtain for w

$$w = Ar^{1/2} \sin \frac{\theta}{2}\,. \tag{9}$$

The non-zero stresses, τ_{rz} and $\tau_{\theta z}$ are calculated by

$$\tau_{r\theta} = \mu\frac{\partial w}{\partial r}\,, \quad \tau_{\theta z} = \frac{\mu}{r}\frac{\partial w}{\partial \theta} \tag{10}$$

and take the form

$$\tau_{rz} = \frac{\mu}{2} Ar^{-1/2} \sin\frac{\theta}{2}\,, \quad \tau_{\theta z} = \frac{\mu}{2} Ar^{-1/2} \cos\frac{\theta}{2}\,. \tag{11}$$

Putting

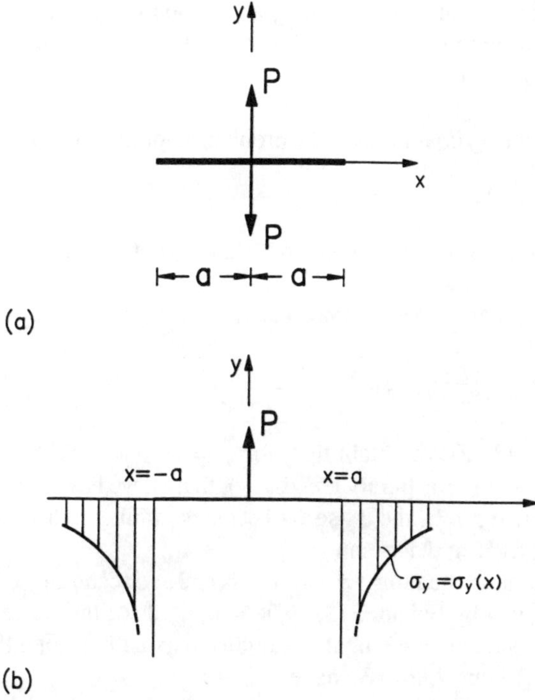

Fig. 2.8. An infinite plate with a crack of length $2a$ (a) opened by a pair of splitting forces P and (b) stress distribution along the x-axis of the half-plane $y > 0$.

$$A = \frac{K_{\mathrm{III}}}{\mu}\sqrt{\frac{2}{\pi}} \tag{12}$$

we obtain

$$w = \frac{2K_{\mathrm{III}}}{\mu}\sqrt{\frac{r}{2\pi}}\sin\frac{\theta}{2} \tag{13}$$

and

$$\tau_{rz} = \frac{K_{\mathrm{III}}}{\sqrt{2\pi r}}\sin\frac{\theta}{2}\,, \quad \tau_{\theta z} = \frac{K_{\mathrm{III}}}{\sqrt{2\pi r}}\cos\frac{\theta}{2}\,. \tag{14}$$

Equations (13) and (14) express the displacement and the singular stress components, and coincide with the earlier results in Equation (2.56) (Problem 2.17).

Example 2.2.

Verify that the Westergaard function

$$Z_{\mathrm{I}} = \frac{Pa}{\pi z(z^2 - a^2)^{1/2}} \tag{1}$$

corresponds to the case of a crack of length $2a$ along the x-axis in an infinite plate opened by a pair of splitting forces P acting at $x = 0$, $y = 0$ (Figure 2.8a). Determine the stress intensity factor K_{I}.

Solution: The boundary conditions of the problem at infinity may be stated as follows:

$$\sigma_x = 0\,, \quad \sigma_y = 0\,, \quad \tau_{xy} = 0 \quad \text{for} \quad |z| \to \infty\,. \tag{2}$$

Substituting Equation (1) into Equation (2.12), we obtain that all stresses at infinity are zero.

The function Z_{I} along the x-axis becomes

$$Z_{\mathrm{I}}(x,0) = \frac{Pa}{\pi x (x^2 - a^2)^{1/2}}\,. \tag{3}$$

From Equation (2.12) we obtain that, for $y = 0$, $|x| < a$, the σ_y-stress is zero (the quantity $(x^2 - a^2)^{1/2}$ is purely imaginary), that is, we have a traction-free crack surface. However, at $x = 0$ the σ_y-stress becomes infinite, indicating the existence of a concentrated force at that point.

For $y = 0$, $x > a$, the quantity $(x^2 - a^2)^{1/2}$ is real and according to Equation (2.12) the σ_y-stress is given by Equation (3). The magnitude of the concentrated force at $x = 0$, $y = 0$ is obtained by taking the equilibrium equation along the x-axis of the half-plane $y > 0$ (Figure 2.8b). We have

$$2 \int_a^{\infty} \sigma_y \, \mathrm{d}x + P_y = 0 \tag{4}$$

or

$$2 \frac{Pa}{\pi} \int_a^{\infty} \frac{\mathrm{d}x}{x(x^2 - a^2)^{1/2}} + P_y = 0$$

or

$$P_y = -P\,. \tag{5}$$

Equation (5) shows that a concentrated force of magnitude P along the $+y$ direction acts at $x = 0$, $y = 0$. From symmetry, we find that a similar force acts at $x = 0$, $y = 0$ along the $-y$ direction.

The stress intensity factor K_{I} is calculated from Equation (2.34) as

$$K_{\mathrm{I}} = \lim_{|\zeta| \to 0} \sqrt{2\pi\zeta}\, Z_{\mathrm{I}} \tag{6}$$

where

$$\zeta = z - a\,. \tag{7}$$

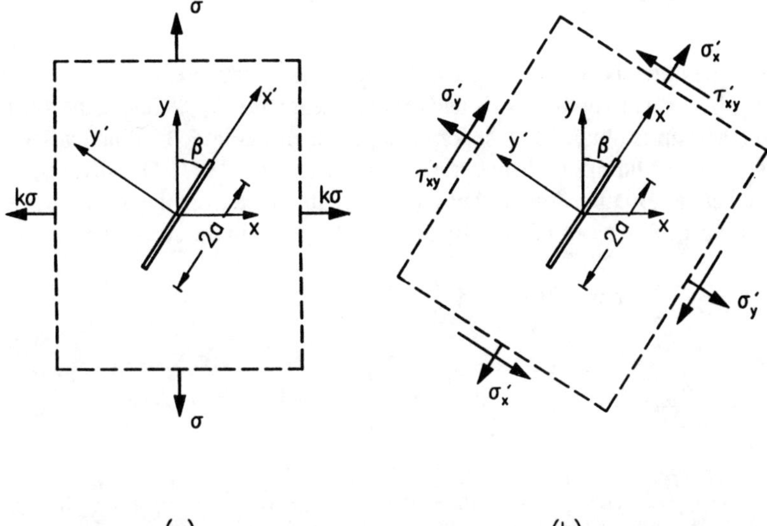

Fig. 2.9. An inclined crack (a) in a biaxial stress field and (b) stress transformation along and perpendicular to the crack plane.

We have

$$Z_{\mathrm{I}} = \frac{Pa}{\pi(a+\zeta)\,[\zeta(\zeta+2a)]^{1/2}} \tag{8}$$

implying that

$$K_{\mathrm{I}} = \frac{P}{(\pi a)^{1/2}}\,. \tag{9}$$

Example 2.3.

Consider a crack of length $2a$ that makes an angle β with the y direction in an infinite plate subjected to stresses σ and $k\sigma$ along the y and x direction, respectively, at infinity (Figure 2.9a). Derive the expressions of the singular stress components σ_x, σ_y and τ_{xy}. For the stress σ_x obtain also the constant term.

Solution: By stress transformation we obtain the following stresses σ'_x, σ'_y, τ'_{xy} in the system $x'y'$ (Figure 2.9b)

$$\sigma'_x = \frac{k+1}{2}\,\sigma - \frac{k-1}{2}\,\sigma\,\cos 2\beta$$

$$\sigma'_y = \frac{k+1}{2}\,\sigma + \frac{k-1}{2}\,\sigma\,\cos 2\beta \tag{1}$$

$$\tau'_{xy} = -\frac{k-1}{2}\,\sigma\,\sin 2\beta\,.$$

Thus the crack is subjected at infinity: (a) to a biaxial stress σ'_y, (b) to a normal stress $(\sigma'_x - \sigma'_y)$ along the x-axis and (c) to a shear stress $\tau_{x'y'}$. Thus, the stress field at the crack tip is obtained by superposing an opening-mode loading caused by the stress σ'_y and a sliding-mode loading caused by the stress τ'_{xy}. The stress $(\sigma'_y - \sigma'_x)$ does not create singular stresses but should be subtracted from the σ'_x stress along the x'-axis. From Equations (1), (2.28) and (2.43) we obtain the stresses $\sigma_{x'}, \sigma_{y'}, \tau_{xy'}$.

$$\sigma_{x'} = \frac{K_{\mathrm{I}}}{\sqrt{2\pi r}}\cos\frac{\theta}{2}\left(1-\sin\frac{\theta}{2}\sin\frac{3\theta}{2}\right) - \frac{K_{\mathrm{II}}}{\sqrt{2\pi r}}\sin\frac{\theta}{2}\left(2+\cos\frac{\theta}{2}\cos\frac{3\theta}{2}\right) - (k-1)\,\sigma\,\cos 2\beta$$

$$\sigma_{y'} = \frac{K_{\mathrm{I}}}{\sqrt{2\pi r}}\cos\frac{\theta}{2}\left(1+\sin\frac{\theta}{2}\sin\frac{3\theta}{2}\right) + \frac{K_{\mathrm{II}}}{\sqrt{2\pi r}}\sin\frac{\theta}{2}\cos\frac{\theta}{2}\cos\frac{3\theta}{2} \tag{2}$$

$$\tau_{x'y'} = \frac{K_{\mathrm{I}}}{\sqrt{2\pi r}}\cos\frac{\theta}{2}\sin\frac{\theta}{2}\cos\frac{3\theta}{2} + \frac{K_{\mathrm{II}}}{\sqrt{2\pi r}}\cos\frac{\theta}{2}\left(1-\sin\frac{\theta}{2}\sin\frac{3\theta}{2}\right)$$

where

$$K_{\mathrm{I}} = \frac{1}{2}\,[k+1+(k-1)\cos 2\beta]\,\sigma\sqrt{\pi a}$$

$$K_{\mathrm{II}} = -\frac{k-1}{2}\sin 2\beta\,\sigma\sqrt{\pi a}\,. \tag{3}$$

Example 2.4.

Consider a short crack of length a emanating from a circular hole along the x-axis in a plate subjected to uniaxial tension σ along the y-axis (Figure 2.10). Determine the stress intensity factor. Then consider another crack of length a emanating from the circular hole along the y-axis and determine the stress intensity factor. Finally, determine the stress intensity factor for the above two cracks when the plate is subjected to an additional stress $k\sigma$ along the x-axis.

Note that when a plate with a circular hole is subjected to stress σ normal to the x-axis the hoop stresses along the x- and y-axes are 3σ and $-\sigma$ respectively.

Solution: Consider a material element at the rim of the hole along the x-axis, when there is no crack. Due to stress concentration the element is subjected to a tensile stress 3σ along the y-axis, while the other two stresses are zero. By the superposition shown in Figure 2.11 and for a small crack of length a (see case 2 of Appendix 2.1) we have

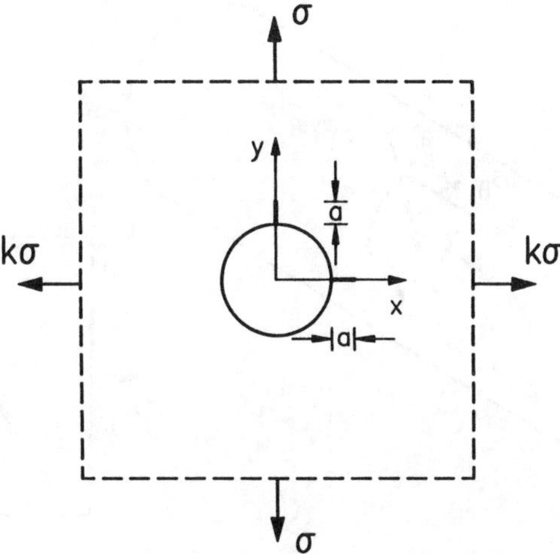

Fig. 2.10. A short crack emanating from a circular hole in an infinite plate subjected to a biaxial stress field.

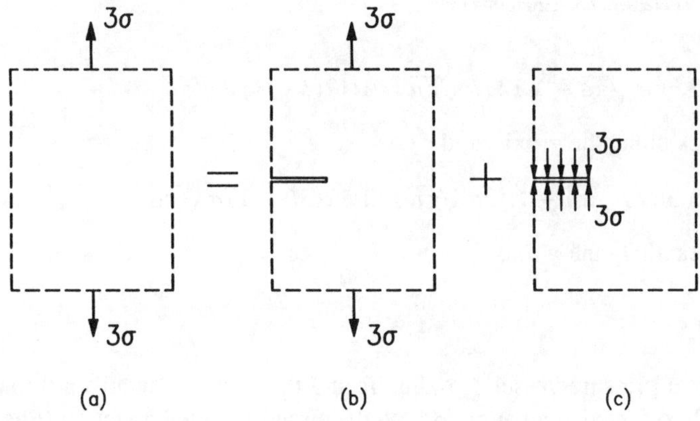

Fig. 2.11. Superposition of stresses.

$$K_{\rm I} = 1.12(3\sigma)\sqrt{\pi a} = 3.36\sigma\sqrt{\pi a}\,. \tag{1}$$

For a small crack of length a along the y-axis emanating from the hole we have in a similar manner

$$K_{\rm I} = -1.12\sigma\sqrt{\pi a}\,. \tag{2}$$

When the plate is subjected to an additional stress $k\sigma$ along the x-axis we can superpose the previous results and obtain

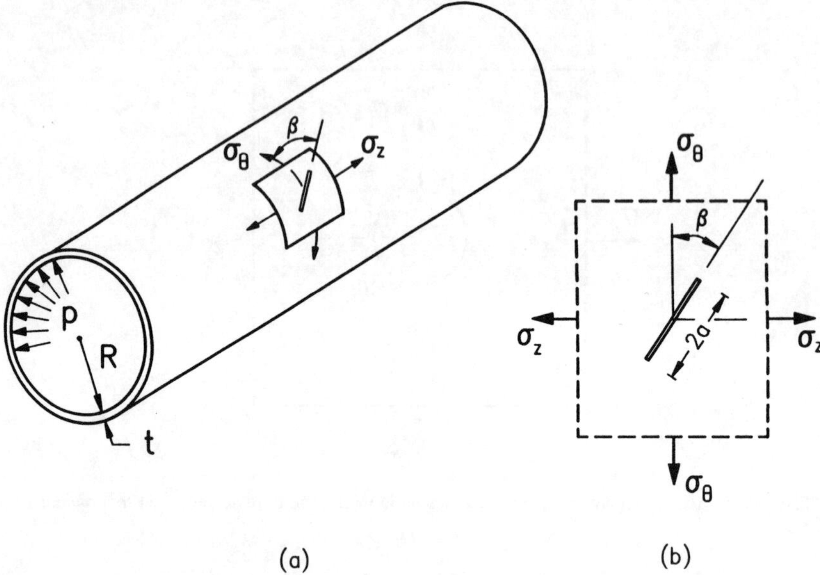

Fig. 2.12. (a) A cylindrical pressure vessel with an inclined through the thickness crack and (b) stresses acting in a local element containing the crack.

$$K_{\mathrm{I}} = 3.36\sigma\sqrt{\pi a} - 1.12\sigma k\sqrt{\pi a} = 1.12(3-k)\,\sigma\sqrt{\pi a} \tag{3}$$

for the crack along the x-axis, and

$$K_{\mathrm{I}} = 3.36k\sigma\sqrt{\pi a} - 1.12\sigma\sqrt{\pi a} = 1.12(3k-1)\,\sigma\sqrt{\pi a} \tag{4}$$

for the crack along the y-axis.

Example 2.5.

A cylindrical pressure vessel of radius R and thickness t contains a through crack of length $2a$ oriented at an angle β with the circumferential direction (Figure 2.12). When the vessel is subjected to an internal pressure p, determine the stress intensity factors at the crack tip.

Solution: The hoop σ_θ and longitudinal σ_z stresses in the vessel are obtained by equilibrium. Equilibrium along the longitudinal axis of the vessel (Figure 2.13a) gives

$$(2\pi R)\,t\sigma_z = \pi R^2 p$$

or

$$\sigma_z = \frac{pR}{2t}\,.$$

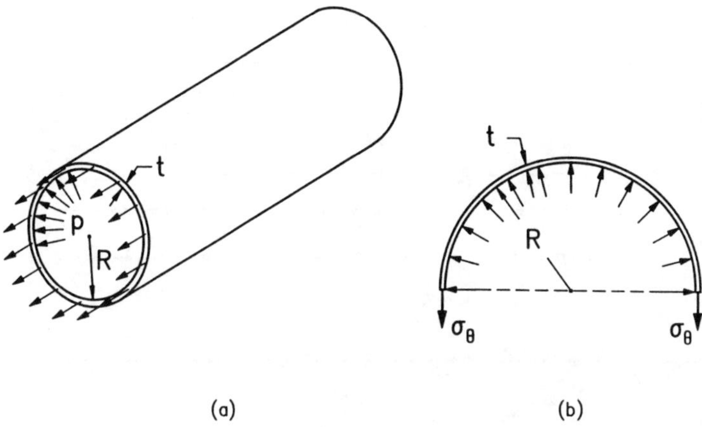

Fig. 2.13. Stress equilibrium along (a) the longitudinal and (b) hoop directions of the cylindrical vessel of Figure 2.12.

Equilibrium along the hoop direction (Figure 2.13b) gives

$$2t\sigma_\theta = 2Rp$$

or

$$\sigma_\theta = \frac{pR}{t} .$$

From Equation (3) of Example 2.3 and the same reasoning as in Example 2.4 we obtain

$$K_{\mathrm{I}} = \frac{pR}{2t} \sqrt{\pi a}\,(1 + \sin^2 \beta) , \quad K_{\mathrm{II}} = \frac{pR}{2t} \sqrt{\pi a} \sin \beta \cos \beta .$$

Problems

2.1. In William's eigenfunction expansion method the Airy stress function for a semi-infinite crack in an infinite plate subjected to general loading is assumed in the form

$$U = r^{\lambda+1} f(\theta)$$

where r, θ are polar coordinates centered at the crack tip and λ is real.

Using the boundary conditions along the crack faces, determine the function U and find the expressions for the singular stress and displacement components for opening mode and sliding mode loading.

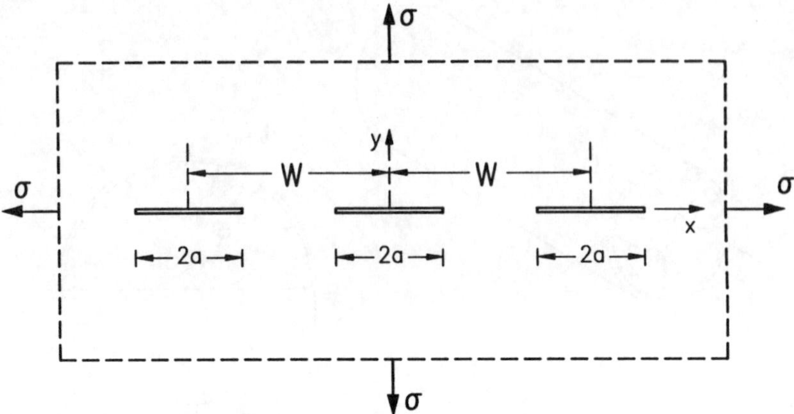

Fig. 2.14. An infinite periodic array of equally spaced cracks in an infinite plate subjected to equal uniform stresses σ at infinity.

Observe that negative values of λ are ignored since they produce infinite displacements at the crack tip. Furthermore, use the result that the total strain energy contained in any circular region surrounding the crack tip is bounded to show that the value $\lambda = 0$ should also be excluded from the solution.

2.2. Show that the Airy stress function

$$U = Cr^{3/2}\left(\cos\frac{\theta}{2} + \frac{1}{3}\cos\frac{3\theta}{2}\right)$$

corresponds to a semi-infinite mode I crack with the crack faces at $\theta = \pm\pi$ being unloaded.

2.3. Consider an infinite periodic array of equally spaced cracks along the x-axis in an infinite plate subjected to equal uniform stresses σ along the x- and y-axes at infinity (Figure 2.14). Verify that the Westergaard function is

$$Z_{\mathrm{I}} = \frac{\sigma\sin\left(\frac{\pi z}{W}\right)}{\left[\sin^2\left(\frac{\pi z}{W}\right) - \sin^2\left(\frac{\pi a}{W}\right)\right]^{1/2}}.$$

Then show that the stress intensity factor is given by

$$K_{\mathrm{I}} = \sigma(\pi a)^{1/2}\left(\frac{W}{\pi a}\tan\frac{\pi a}{W}\right)^{1/2}.$$

2.4. Consider an infinite periodic array of equally spaced cracks along the x-axis with each crack subjected to a pair of concentrated forces at the center of the crack

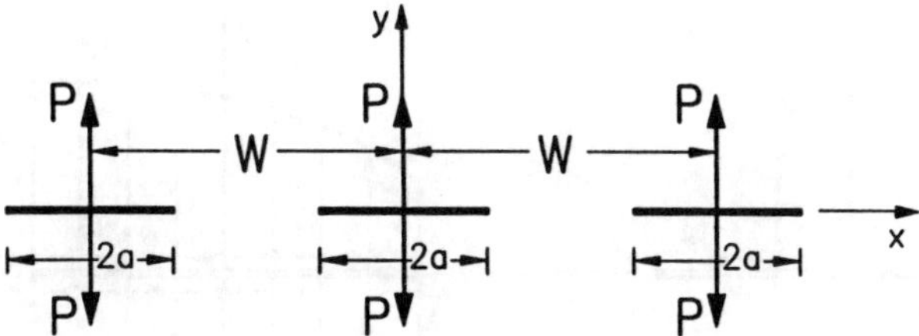

Fig. 2.15. An infinite periodic array of equally spaced cracks subjected to a pair of concentrated forces P at their center in an infinite plate.

(Figure 2.15). Verify that the Westergaard function is

$$Z_{\mathrm{I}} = \frac{P \sin(\pi a/W)}{W(\sin(\pi z/W))^2}\left[1 - \left(\frac{\sin(\pi a/W)}{\sin(\pi z/W)}\right)^2\right]^{-1/2}.$$

Then show that the stress intensity factor is given by

$$K_{\mathrm{I}} = \frac{P}{\left(\dfrac{W}{2}\sin\dfrac{\pi a}{W}\right)^{1/2}}.$$

2.5. The Westergaard function Z for the concentrated forces P and Q applied at the point $x = b$ $(b < a)$ of a crack AB of length $2a$ in an infinite plate (Figure 2.16a) is

$$Z = \frac{Q + iP}{2\pi}\left[\left(\frac{\kappa - 1}{\kappa + 1}\right)\frac{1}{\sqrt{z^2 - a^2}} + \frac{1}{b - z}\left(\sqrt{\frac{b^2 - a^2}{z^2 - a^2}} + 1\right)\right]$$

Show that the complex stress intensity factor $K = K_{\mathrm{I}} - iK_{\mathrm{II}}$ at the tip B of the crack is

$$K = \frac{Q + iP}{2\sqrt{\pi a}}\left(\frac{\kappa - 1}{\kappa + 1} + \sqrt{\frac{a + b}{b - a}}\right).$$

Then show that for equal and opposite distributed forces $\sigma_y(x, 0)$ and $\tau_{xy}(x, 0)$ on the upper and lower crack faces (Figure 2.16b) K_{I} and K_{II} are given by

$$K_{\mathrm{I}} = \frac{1}{\sqrt{\pi a}}\int_{-a}^{a}\sigma_y(x, 0)\sqrt{\frac{a + x}{a - x}}\,\mathrm{d}x$$

$$K_{\mathrm{II}} = \frac{1}{\sqrt{\pi a}}\int_{-a}^{a}\tau_{xy}(x, 0)\sqrt{\frac{a + x}{a - x}}\,\mathrm{d}x.$$

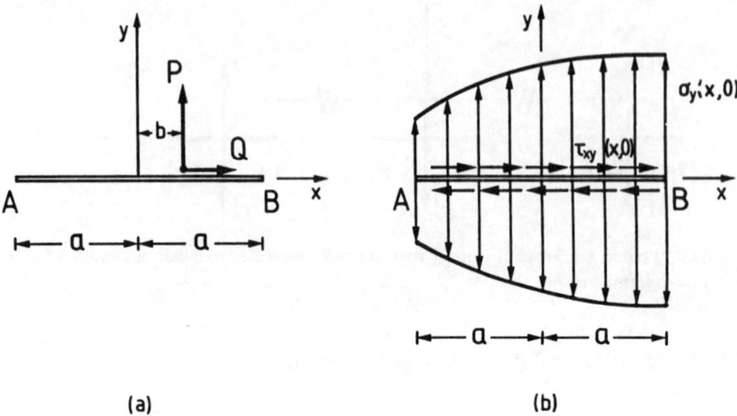

Fig. 2.16. A crack of length $2a$ subjected (a) to concentrated forces P and Q and (b) to distributed forces $\sigma_y(x,0)$ and $\tau_{xy}(x,0)$ along the crack faces.

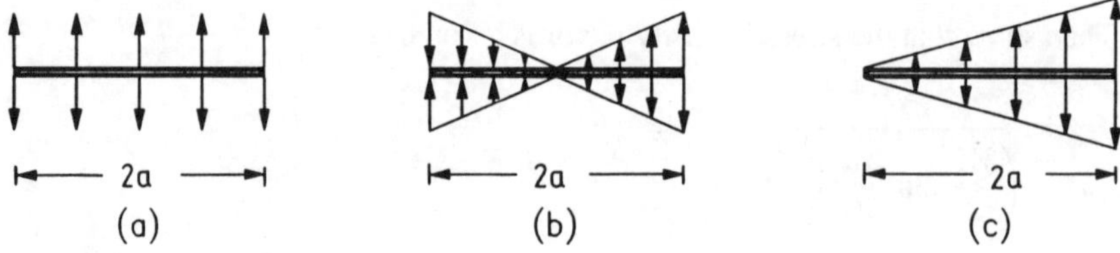

Fig. 2.17. A crack of length $2a$ in an infinite plate subjected to (a) a uniform and (b,c) triangular opposite forces on the upper and lower crack faces.

2.6. Using Problem 2.5 determine the values of K_{I} for uniform (Figure 2.17a) and triangular (Figure 2.17b,c) equal and opposite distributed forces on the upper and lower faces of a crack of length $2a$ in an infinite plate.

2.7. Verify that the Westergaard function for an infinite plate with a crack of length $2a$ subjected to a pair of forces at $x = b$ (Figure 2.18a) is

$$Z_{\mathrm{I}} = \frac{P}{\pi(z-b)}\left(\frac{a^2-b^2}{z^2-a^2}\right)^{1/2}.$$

Then show that the stress intensity factor of the tip $x = a$ is given by

$$K_{\mathrm{I}} = \frac{P}{(\pi a)^{1/2}}\left(\frac{a+b}{a-b}\right)^{1/2}.$$

Use these results to show that for an additional pair of forces at $x = -b$ (2.18b) the Westergaard function is

$$Z_{\mathrm{I}} = \frac{2Pz}{\pi(z^2-b^2)}\left(\frac{a^2-b^2}{z^2-a^2}\right)^{1/2}$$

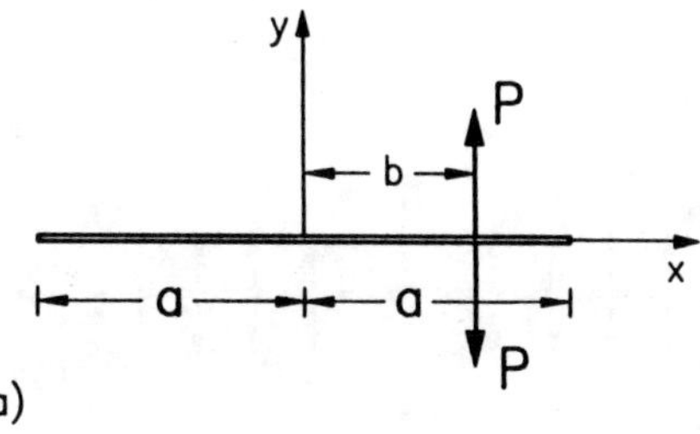

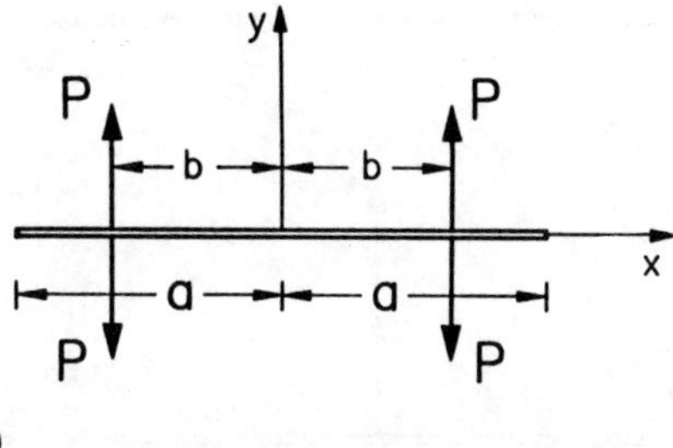

Fig. 2.18. An infinite plate with a crack of length $2a$ subjected (a) to a pair of forces P at $x = b$ and (b) to two pair of forces at $x = \pm b$.

and the stress intensity factor is

$$K_{\mathrm{I}} = \frac{2P}{(a^2 - b^2)^{1/2}} \left(\frac{a}{\pi}\right)^{1/2} .$$

2.8. Use Problem 2.7 to show that the Westergaard function for the configuration of Figure 2.19 is

$$Z_{\mathrm{I}} = \frac{2\sigma}{\pi} \left[\frac{z}{(z^2 - a^2)^{1/2}} \arccos\left(\frac{b}{a}\right) - \operatorname{arccot}\left[\frac{b}{z}\left(\frac{z^2 - a^2}{a^2 - b^2}\right)^{1/2}\right]\right]$$

and the stress intensity factor is

$$K_{\mathrm{I}} = 2\sigma \sqrt{\frac{a}{\pi}} \arcsin\left(\frac{b}{a}\right) .$$

2.9. Consider a crack of length $2a$ in an infinite plate subjected to the concentrated forces P at a distance y_0 from the crack (Figure 2.20). Verify that the Westergaard function is

$$Z_{\mathrm{I}} = \frac{P}{\pi} \left[1 - \frac{y_0}{2(1-\nu)} \frac{\partial}{\partial y_0}\right] f(z, y_0, a)$$

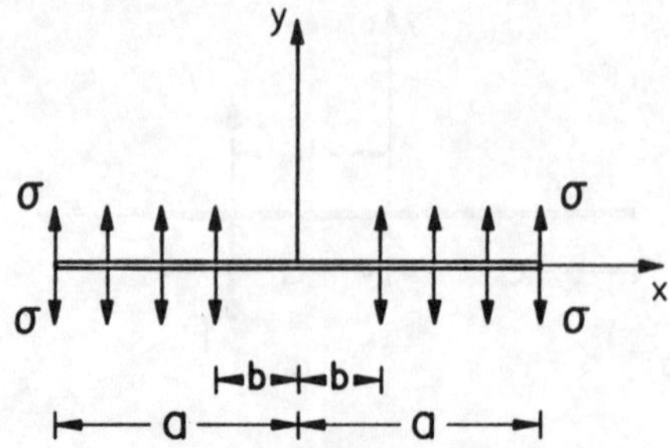

Fig. 2.19. A crack of length $2a$ in an infinite plate subjected to a uniform stress distribution σ along the interval $b \leq |x| \leq a$.

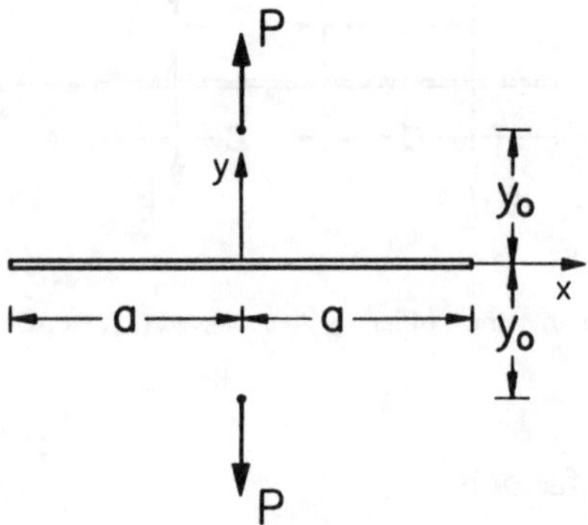

Fig. 2.20. A crack of length $2a$ in an infinite plate subjected to concentrated forces P.

where

$$f(z, y_0, a) = \frac{(a^2 + y_0^2)^{1/2}}{z^2 + y_0^2} \frac{z}{(z^2 - a^2)^{1/2}} \, .$$

Determine the stress intensity factor $K_{\rm I}$.

2.10. Verify that the Westergaard function for the plate of Figure 2.14 subjected to a uniform in-plane shear stress τ at infinity is

$$Z_{\rm II} = \frac{-i\tau \sin\left(\dfrac{\pi z}{W}\right)}{\left[\sin^2\left(\dfrac{\pi z}{W}\right) - \sin^2\left(\dfrac{\pi a}{W}\right)\right]^{1/2}}$$

and the stress intensity factor is

$$K_{\rm II} = \tau(\pi a)^{1/2} \left(\frac{W}{\pi a} \tan \frac{\pi a}{W}\right)^{1/2} .$$

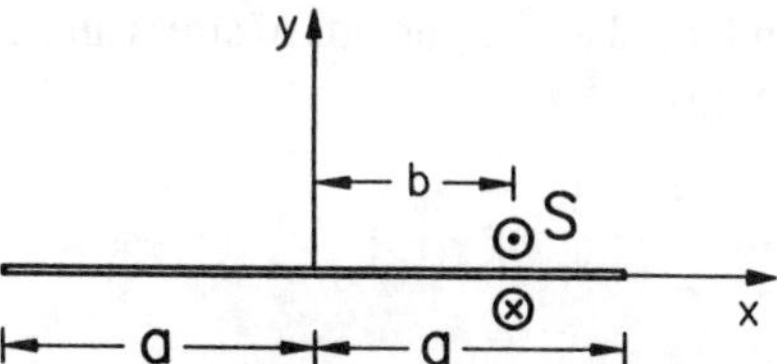

Fig. 2.21. An infinite plate with a crack of length $2a$ subjected to a pair of shear forces S at $x = b$.

2.11. Consider the Westergaard stress function

$$Z_{\mathrm{I}} = \frac{\sigma(2z^2 - a^2)}{2a(z^2 - a^2)^{1/2}}$$

and find the loading that represents in an infinite plate with a crack of length $2a$ along the x-axis. Determine the stress intensity factor.

2.12. Verify that the function Z'_{III} for an infinite plate with a crack of length $2a$ subjected to a pair of shear forces S at $x = b$ (Figure 2.21) is

$$Z'_{\mathrm{III}} = -\frac{S}{\pi(z - b)} \left(\frac{a^2 - b^2}{z^2 - a^2}\right)^{1/2}.$$

Determine the stress intensity factor.

2.13. The Westergaard function for the stress field near the tip of an opening-mode crack is put in the form

$$Z_{\mathrm{I}} = \frac{K_{\mathrm{I}}}{(2\pi\zeta)^{1/2}} \left[1 + \beta(\zeta/a)\right]$$

where the parameter β models the effect of near field boundaries and boundary loading.

Determine the singular stresses σ_x, σ_y and τ_{xy} from Z_{I}. According to photoelastic law, the isochromatic fringe order N is related to the maximum shear stress τ_m by

$$2\tau_m = \frac{Nf}{t}$$

where f is the stress-optical constant and t is the plate thickness.

Show that this equation can be used to determine K_{I} from the isochromatic fringe pattern in the neighborhood of the crack tip.

2.14. Use Equation (2.24) to find Taylor series for the stresses σ_x, σ_y, τ_{xy} for a crack of length $2a$ in an infinite plate subjected to stresses σ and $k\sigma$ along the y and x-axis, respectively, at infinity (Figure 2.2).

2.15. Show that the singular polar components of stress and displacement for sliding-mode crack problems are given by

$$\sigma_r = \frac{K_{\text{II}}}{\sqrt{2\pi r}} \left(-\frac{5}{4} \sin \frac{\theta}{2} + \frac{3}{4} \sin \frac{3\theta}{2} \right)$$

$$\sigma_\theta = \frac{K_{\text{II}}}{\sqrt{2\pi r}} \left(-\frac{3}{4} \sin \frac{\theta}{2} - \frac{3}{4} \sin \frac{3\theta}{2} \right)$$

$$\tau_{r\theta} = \frac{K_{\text{II}}}{\sqrt{2\pi r}} \left(\frac{1}{4} \cos \frac{\theta}{2} + \frac{3}{4} \cos \frac{3\theta}{2} \right)$$

and

$$u_r = \frac{K_{\text{II}}}{4\mu} \sqrt{\frac{r}{2\pi}} \left[-(2\kappa - 1) \sin \frac{\theta}{2} + 3 \sin \frac{3\theta}{2} \right]$$

$$u_\theta = \frac{K_{\text{II}}}{4\mu} \sqrt{\frac{r}{2\pi}} \left[-(2\kappa + 1) \cos \frac{\theta}{2} + 3 \cos \frac{3\theta}{2} \right] .$$

2.16. Derive the expressions for the singular principal stresses for sliding-mode loading.

2.17. Use Equation (2.56) to show that the singular polar components of stress for tearing mode crack problems are given by

$$\tau_{rz} = \frac{K_{\text{III}}}{\sqrt{2\pi r}} \sin \frac{\theta}{2} , \quad \tau_{\theta z} = \frac{K_{\text{III}}}{\sqrt{2\pi r}} \cos \frac{\theta}{2} .$$

2.18. Consider a crack in a mixed-mode stress field governed by the values of the opening-mode K_{I} and sliding-mode K_{II} stress intensity factors. Obtain the singular stress components and subtract the constant term σ_{0x} from the stress σ_x to account for distant field stresses. Determine the isochromatic fringe order N from equation

$$2\tau_m = \frac{Nf}{t}$$

where τ_m is the maximum in-plane shear stress, f is the stress-optical constant and t is the plate thickness. Obtain an expression for N. Consider the opening-mode. If r_m and θ_m are the polar coordinates of the point on an isochromatic loop, furthest from the crack tip (Figure 2.22), show that

$$K_{\text{I}} = \frac{Nf\sqrt{2\pi r_m}}{t \sin \theta_m} \left[1 + \left(\frac{2}{3 \tan \theta_m} \right)^2 \right]^{1/2} \left(1 + \frac{2 \tan (3\theta_m/2)}{3 \tan \theta_m} \right)$$

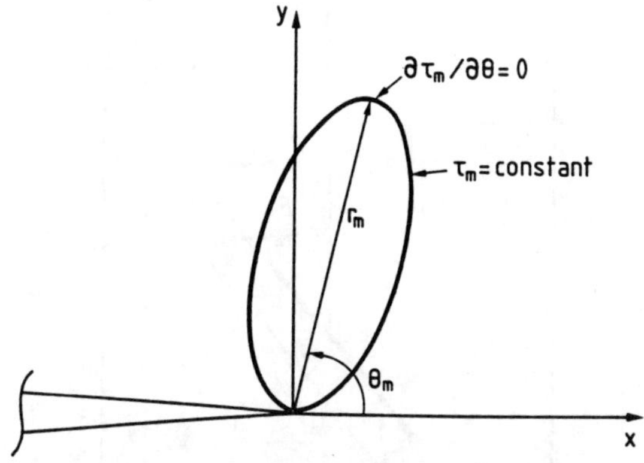

Fig. 2.22. A crack-tip isochromatic fringe loop.

$$\sigma_{0x} = -\frac{Nf}{t} \frac{\cos\theta_m}{\cos(3\theta_m/2)\,(\cos^2\theta_m + \frac{9}{4}\sin^2\theta_m)^{1/2}}\,.$$

2.19. Consider a crack in a mixed-mode stress field governed by the values of the opening-mode K_{I} and sliding-mode K_{II} stress intensity factors. If only the singular stresses are considered, show that the maximum in-plane shear stress τ_m is given by

$$\tau_m = \frac{1}{2\sqrt{2\pi r}} [\sin^2\theta\, K_{\mathrm{I}}^2 + 2\sin 2\theta\, K_{\mathrm{I}} K_{\mathrm{II}} + (4 - 3\sin^2\theta)\, K_{\mathrm{II}}^2]^{1/2}\,.$$

Then show that the polar angle θ_m of the point furthest from the crack tip on the curve $\tau_{\max}$ = constant (Figure 2.22) satisfies the following equation

$$\left(\frac{K_{\mathrm{II}}}{K_{\mathrm{I}}}\right)^2 - \frac{4}{3}\left(\frac{K_{\mathrm{II}}}{K_{\mathrm{I}}}\right) \cot 2\theta_m - \frac{1}{3} = 0\,.$$

2.20. Show that the strain energy contained in a small circle of radius r_0 surrounding the crack tip is given by

$$W_{\mathrm{I}} = \frac{(2\kappa - 1)}{16\mu} r_0 K_{\mathrm{I}}^2$$

$$W_{\mathrm{II}} = \frac{(2\kappa + 3)}{16\mu} r_0 K_{\mathrm{II}}^2$$

$$W_{\mathrm{III}} = \frac{r_0 K_{\mathrm{III}}^2}{2\mu}$$

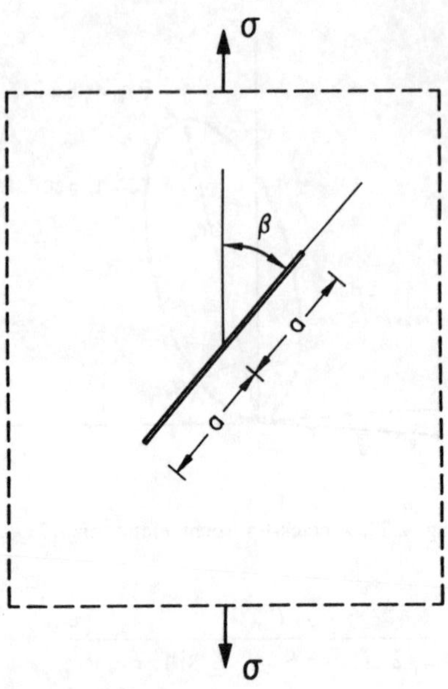

Fig. 2.23. An inclined crack in an infinite plate.

for the deformation modes I, II, III respectively.

2.21. Use Problem 2.3 to show that the stress intensity factor for a strip of width $2b$ containing a central crack of length $2a$ can be approximated by

$$K_{\mathrm{I}} = \sigma(\pi a)^{1/2} \left(\frac{2b}{\pi a} \tan \frac{\pi a}{2b} \right)^{1/2} .$$

Compare the values of K_{I} with those of case 1 of Appendix 2.1.

2.22. Consider a crack of length $2a$ in an infinite plate subjected to a uniform stress σ at infinity that makes an angle β with the crack axis (Figure 2.23). Show that the stress intensity factors K_{I} and K_{II} are given by

$$K_{\mathrm{I}} = \sigma \sqrt{\pi a} \sin^2 \beta$$

$$K_{\mathrm{II}} = \sigma \sqrt{\pi a} \sin \beta \cos \beta .$$

2.23. Consider a crack AB of length $2a$ in an infinite plate subjected to a linear stress distribution at infinity (Figure 2.24). Using the principle of superposition and the results of Problem 2.5, show that the stress intensity factor is given by

$$K_{\mathrm{I}} = \frac{\sigma_0 \sqrt{\pi a}}{2} .$$

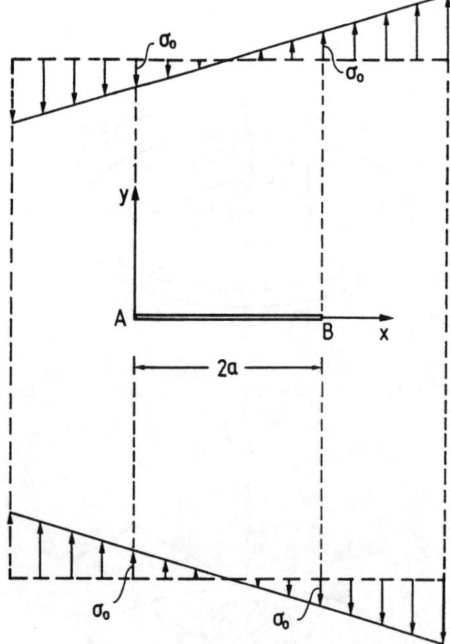

Fig. 2.24. A crack of length $2a$ in an infinite plate subjected to a linear stress distribution at infinity.

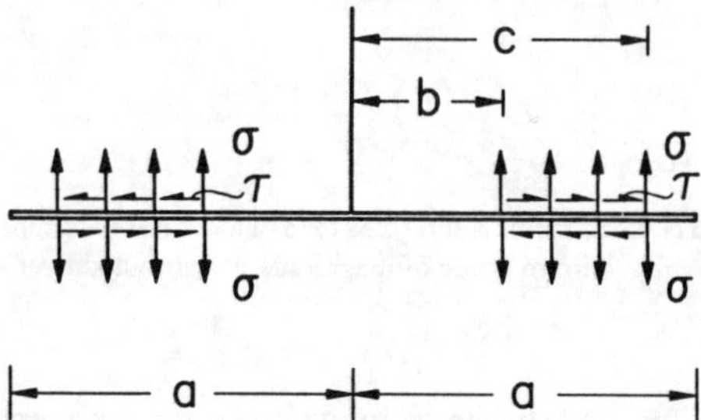

Fig. 2.25. A crack of length $2a$ in an infinite plate subjected to a uniform normal stress σ and shear stress τ from $x = \pm b$ to $x = \pm a$.

2.24. Use Problem 2.5 to determine the stress intensity factor for a crack of length $2a$ in an infinite plate subjected to uniform normal stress σ and shear stress τ along the upper crack surface from $x = b$ to $x = c$. Then determine the stress intensity factor when the same stresses apply to the lower crack surface. Finally use Problem 2.7 to determine the stress intensity factor when additional normal and shear stresses σ and τ apply along the upper and lower crack surface from $x = -c$ to $x = -b$ (Figure 2.25).

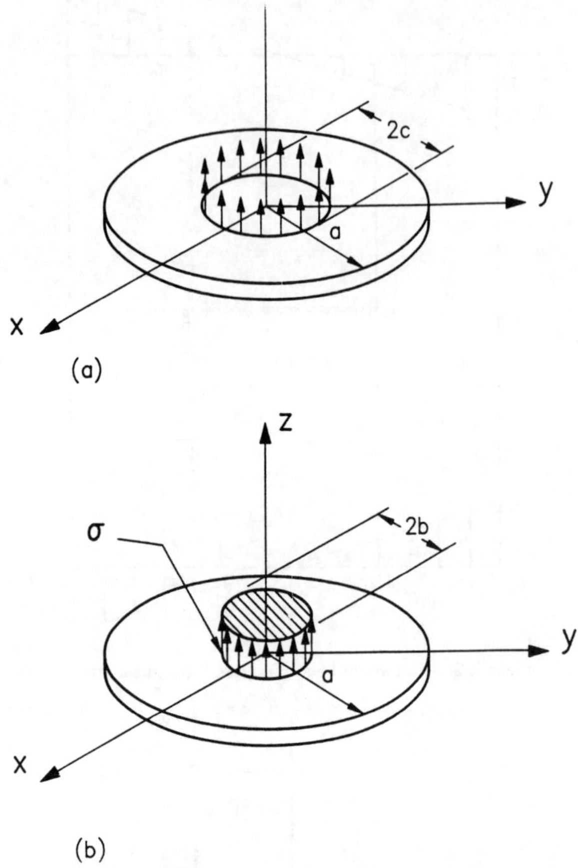

Fig. 2.26. A penny-shaped crack of radius a subjected to (a) a load stress P acting along the circumference of a circle of radius c and (b) a uniform stress of magnitude σ distributed over a concentric circular area of radius b.

2.25. Use Problem 2.7 to find the stress intensity factor for a crack of length $2a$ in an infinite plate subjected to self-balanced normal stresses acting along the crack faces.

2.26. The stress intensity factor for a penny-shaped crack of radius a subjected to a load P acting along the circumference of a circle of radius c (Figure 2.26a) is given by

$$K_{\mathrm{I}} = \frac{P}{(\pi a)^{3/2}\,[1 - \left(\frac{c}{a}\right)^2]^{1/2}} .$$

Determine the stress intensity factor for this crack when it is subjected to a uniform stress of magnitude σ distributed over a concentric circular area of radius $b (b < a)$ (Figure 2.26b).

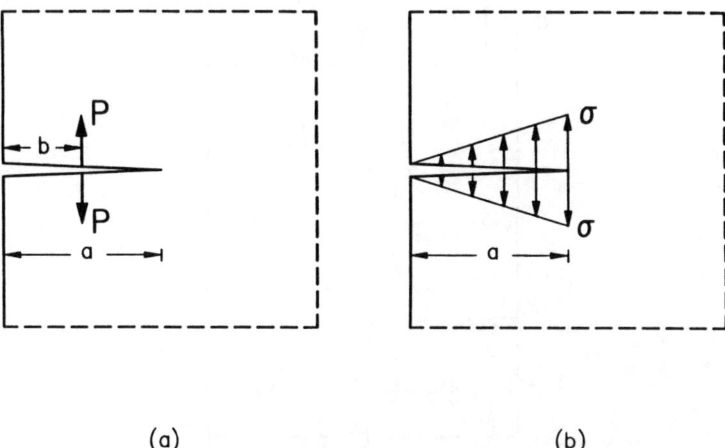

Fig. 2.27. An edge crack in an infinite plate subjected to (a) a pair of concentrated forces at a distance b from the plate edge and (b) a self-balanced linear tensile stress distribution acting along the crack faces.

2.27. Show that the stress intensity factor for an edge crack of length a in a semi-infinite solid subjected to a uniform out-of-plane shear stress τ at infinity is the same as for a crack of length $2a$ in a full solid.

2.28. The stress intensity factor for an edge crack of length a in a semi-infinite plate subjected to a pair of equal and opposite concentrated forces at a distance b from the plate edge (Figure 2.27a) is given by

$$K_{\mathrm{I}} = \frac{2}{\sqrt{\pi}} \frac{1 + F(b/a)}{(a^2 - b^2)^{1/2}} P a^{1/2}$$

where

$$F(b/a) = [1 - (b/a)^2]\,[0.2945 - 0.3912(b/a)^2 + 0.7685(b/a)^4 - 0.9942(b/a)^2 + 0.5094(b/a)^8]\,.$$

Using this result show that the stress intensity factor for this crack subjected to a self-balanced linear tensile stress distribution acting along the crack faces (Figure 2.27b) is

$$K_{\mathrm{I}} = 0.683\sigma\sqrt{\pi a}\,.$$

2.29. Determine the stress intensity factor for a strip with an edge crack loaded as in Figure 2.28 for $a/b = 0.4$. Consult Appendix 2.1.

2.30. A cylindrical pressure vessel with internal radius $R = 600$ mm and thickness $B = 20$ mm contains a longitudinal crack of length $L = 100$ mm and depth $a = 2$ mm

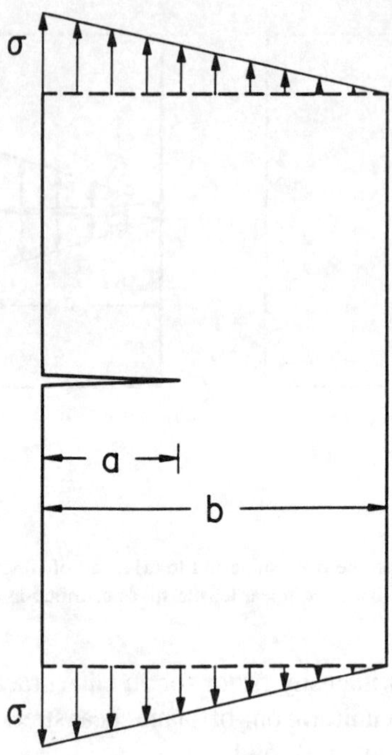

Fig. 2.28. A strip with an edge crack subjected to a triangular stress distribution perpendicular to the crack along its upper and lower boundaries.

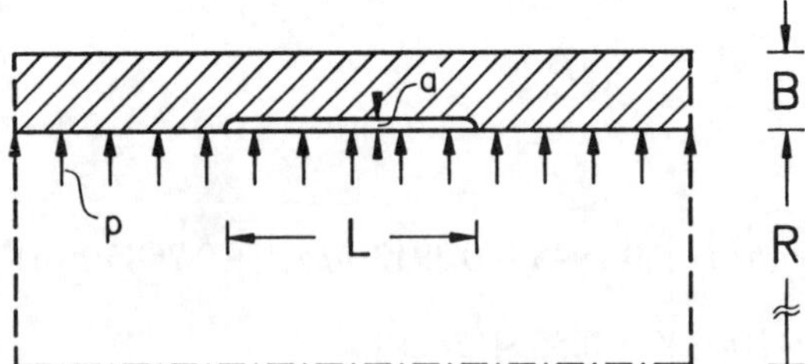

Fig. 2.29. A cylindrical pressure vessel with a longitudinal crack of length L and depth a subjected to internal pressure p.

(Figure 2.29). When the vessel is subjected to internal pressure $p = 1$ MPa determine the stress intensity factor at the crack tip.

2.31. A three-point bend specimen of thickness 1 mm (Figure 2.30) contains a crack of length $a = 2$ mm. Determine the stress intensity factor K_{I} without resorting to case 4 of Appendix 2.1. Then determine K_{I} from Appendix 2.1 and compare the results.

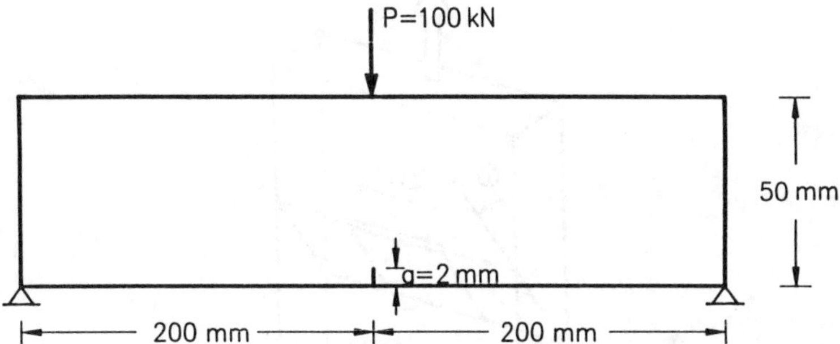

Fig. 2.30. A cracked three-point bend specimen.

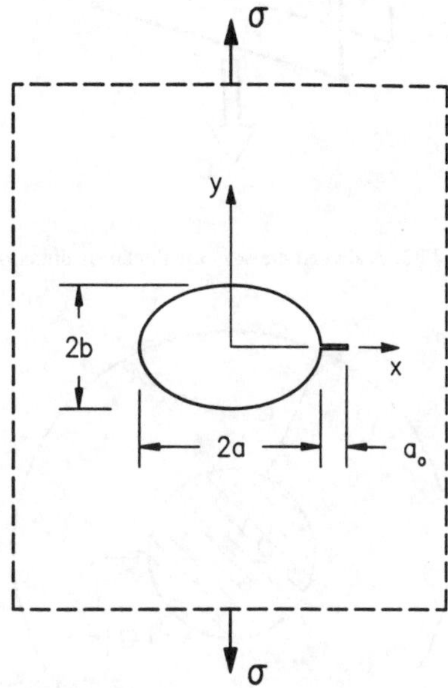

Fig. 2.31. A short crack emanating from an elliptical hole in an infinite plate subjected to a stress σ perpendicular to the major axis of the hole.

2.32. Consider a short crack of length a_0 emanating from an elliptical hole with axes $2a$ and $2b$ in an infinite plate subjected to a stress σ perpendicular to the major axis of the hole (Figure 2.31). Determine the stress intensity factor. Note that the hoop stress at the end point of the major axis of the ellipse is given by Equation (1.7).

2.33. Consider a center crack of length $2a$ in a plane specimen titled at an angle ω with the specimen surface and subjected to a uniaxial stress σ (Figure 2.32). Ignoring the influence of the plate surface, determine the stress intensity factors at the crack front.

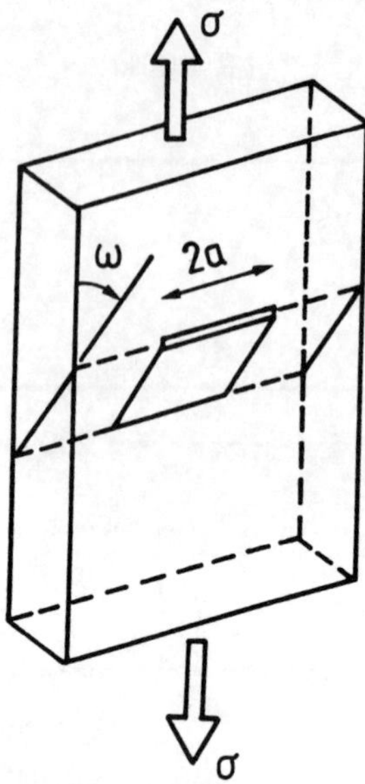

Fig. 2.32. A slanted crack in the thickness direction.

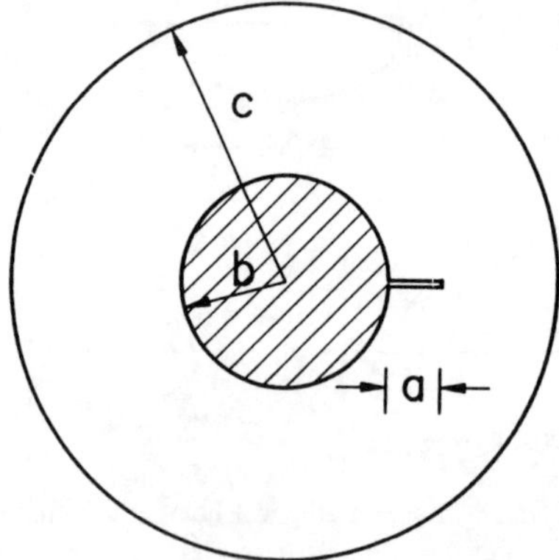

Fig. 2.33. A shaft of radius b press fitted into a wheel of outside radius c containing a small crack of length a.

2.34. A spherical vessel of radius R and thickness t contains a crack of length $2a$ oriented at an angle ω with the meridional direction. When the vessel is subjected to an internal pressure p determine the stress intensity factors at the crack tip.

2.35. A shaft of radius b is press fitted into a wheel of outside radius c with a radial interference δ (Figure 2.33). A small crack of length a emanates from the inner boundary of the wheel. Determine the stress intensity factor at the crack tip.

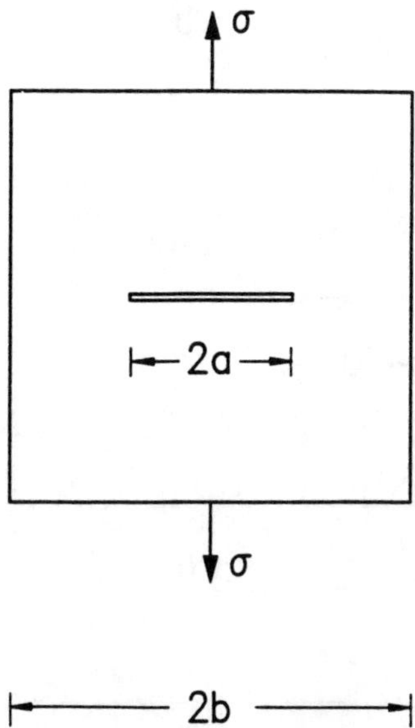

Fig. 2.1a. A center-cracked plate under uniform tension.

2.36. A cylindrical pipe with inner radius b and outer radius c is subjected to a temperature difference ΔT across the wall. Assume a radial crack of length a emanating from the inner bore. Determine the stress intensity factor at the crack tip.

Appendix 2.1.

1. Center-Cracked Plate Under Uniform Tension (Figure 2.1a)

$$K_\mathrm{I} = \sigma\sqrt{\pi a}\left[1.0 + 0.128\left(\frac{a}{b}\right) - 0.288\left(\frac{a}{b}\right)^2 + 1.523\left(\frac{a}{b}\right)^3\right]$$

$$0 < \frac{a}{b} < 0.7.$$

2. Single-Edge-Cracked Plate Under Uniform Tension (Figure 2.1b)

$$K_\mathrm{I} = \sigma\sqrt{\pi a}\left[1.12 - 0.23\left(\frac{a}{b}\right) + 10.55\left(\frac{a}{b}\right)^2 - 21.72\left(\frac{a}{b}\right)^3 + 30.39\left(\frac{a}{b}\right)^4\right]$$

$$\frac{a}{b} < 0.6.$$

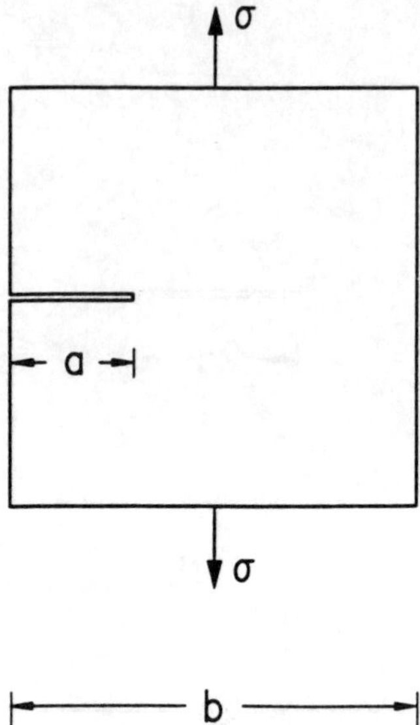

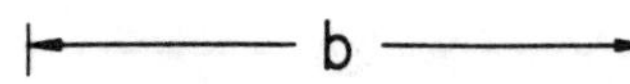

Fig. 2.1b. A single-edge-cracked plate under uniform tension.

3. Double-Edge-Cracked Plate Under Uniform Tension (Figure 2.1c)

$$K_{\mathrm{I}} = \sigma\sqrt{\pi a}\left[1.12 - 0.20\left(\frac{a}{b}\right) - 1.20\left(\frac{a}{b}\right)^2 + 1.93\left(\frac{a}{b}\right)^3\right]$$

$$0 < \frac{a}{b} < 0.7\,.$$

4. Single-Edge-Cracked Three-Point Bend Specimen (Figure 2.1d)

$$K_{\mathrm{I}} = \frac{PS}{BW^{3/2}}\left[2.9\left(\frac{a}{W}\right)^{1/2} - 4.6\left(\frac{a}{W}\right)^{3/2} + 21.8\left(\frac{a}{W}\right)^{5/2} - \right.$$

$$\left. -37.6\left(\frac{a}{W}\right)^{7/2} + 38.7\left(\frac{a}{W}\right)^{9/2}\right].$$

5. Finite Width Strip with Edge Crack Under Bending (Figure 2.1e)

$$K_{\mathrm{I}} = \frac{6M}{BW^2}\sqrt{\pi a}\left[1.12 - 1.40\left(\frac{a}{W}\right) + 7.33\left(\frac{a}{W}\right)^2 - \right.$$

$$\left. -13.08\left(\frac{a}{W}\right)^3 + 14.0\left(\frac{a}{W}\right)^4\right].$$

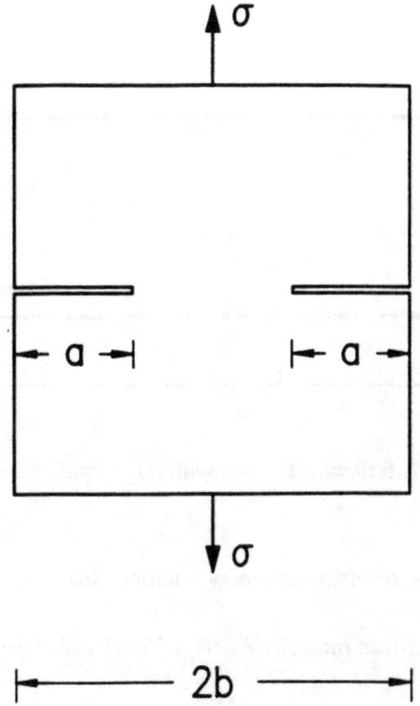

Fig. 2.1c. A double-edge-cracked plate under uniform tension.

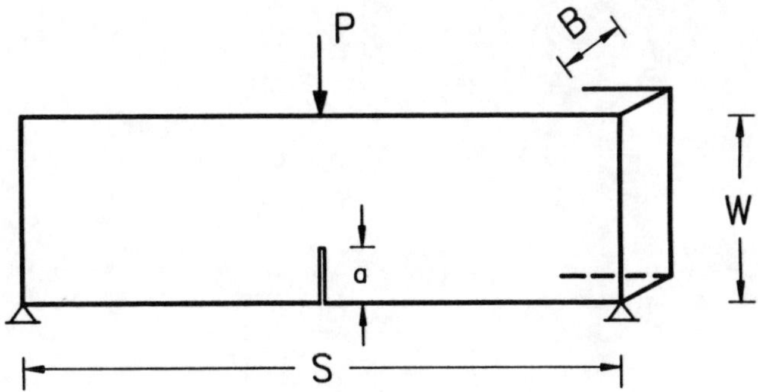

Fig. 2.1d. A single-edge-cracked three-point bend specimen.

References

2.1. Westergaard, H.M. (1934) 'Stresses at a crack, size of the crack and the bending of reinforced concrete', *Proceedings of the American Concrete Institute* **30**, 93–102.

2.2. Westergaard, H.M. (1937) 'Bearing pressures and cracks', *Journal of Applied Mechanics, Trans. ASME* **6**, A.49–A.53.

2.3. Irwin, G.R. (1958) 'Fracture', in *Encyclopedia of Physics*, Vol. VI, *Elasticity and Plasticity* (ed. S. Flügge), Springer-Verlag, pp. 551–590.

2.4. Gdoutos, E.E. (1990) *Fracture Mechanics Criteria and Applications*, Kluwer Academic Publishers.

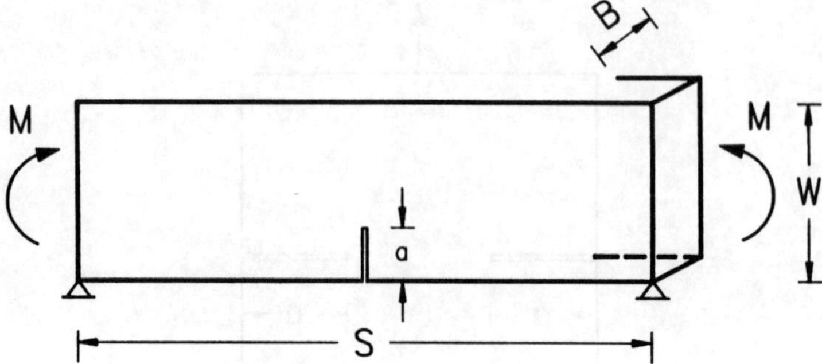

Fig. 2.1e. A finite width strip with edge crack under bending.

2.5. Sih, G.C. (1973) 'Handbook of Stress-Intensity Factors', Institute of Fracture and Solid Mechanics, Lehigh University.

2.6. Murakami, Y. (ed.) (1987) *Stress Intensity Factors Handbook*, Pergamon Press.

Chapter 3

Elastic-Plastic Stress Field in Cracked Bodies

3.1. Introduction

The linear elastic analysis of the stress field in cracked bodies, dealt with in the preceding chapter, applies, strictly speaking, only to ideal brittle materials for which the amount of inelastic deformation near the crack tip is negligible. In most cases, however, there is some inelasticity, in the form of plasticity, creep or phase change in the neighborhood of the crack tip. A study of the local stress fields for the three modes of loading showed that they have general applicability and are governed by the values of three stress intensity factors. In other words, the applied loading, the crack length and the geometrical configuration of the cracked bodies influence the strength of these fields only through the stress intensity factors. We can have two cracked bodies with different geometries, crack lengths and applied loads with the same mode. The stress and deformation fields near the crack tip will be the same if the stress intensity factors are equal.

The singular stress fields represent the asymptotic fields as the distance from the crack tip tends to zero, and their realm of applicability is confined to a very small region around the crack tip. Let the singular solution dominate inside a circle of radius D surrounding the crack tip (Figure 3.1). Consider also that the region of inelastic deformation attending the crack tip is represented by R. When R is sufficiently small compared to D and any other characteristic geometric dimension such as notch radius, plate thickness, crack ligament, etc., the singular stress field governed by the stress intensity factors forms a useful approximation to the elastic field in the ring enclosed by radii R and D. This situation has been termed "small-scale yielding".

In the present chapter we present an elementary analysis of the elastic-plastic stress field in cracked bodies. The chapter includes an approximate calculation of the plastic zone for "small" applied loads; a description of the actual plastic enclaves revealed by experiments in plates of finite thickness; and the models of Irwin and Dugdale for the determination of the extent of plastic zone directly ahead of the crack.

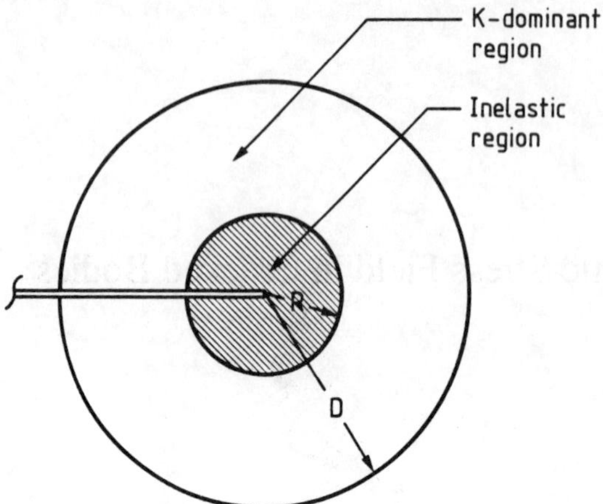

Fig. 3.1. Inelastic and K-dominant regions around a crack tip.

3.2. Approximate determination of the crack-tip plastic zone

A first estimate of the extent of the plastic zone attending the crack tip can be obtained by determining the locus of points where the elastic stress field satisfies the yield criterion. This calculation is very approximate, since yielding leads to stress redistribution and modifies the size and shape of the plastic zone. Strictly speaking, the plastic zone should be determined from an elastic-plastic analysis of the stress field around the crack tip. However, we can obtain some useful results regarding the shape of the plastic zone from the approximate calculation.

First consider opening-mode loading. Introducing the expressions for the singular principal stresses given by Equation (2.37) into the von Mises yield criterion expressed by Equation (1.3), we obtain the following expression for the radius of the plastic zone

$$r_p(\theta) = \frac{1}{4\pi}\left(\frac{K_{\mathrm{I}}}{\sigma_Y}\right)^2 \left(\frac{3}{2}\sin^2\theta + 1 + \cos\theta\right) \tag{3.1}$$

for plane stress, and

$$r_p(\theta) = \frac{1}{4\pi}\left(\frac{K_{\mathrm{I}}}{\sigma_Y}\right)^2 \left[\frac{3}{2}\sin^2\theta + (1-2\nu)^2(1+\cos\theta)\right] \tag{3.2}$$

for plane strain, where σ_Y is the yield stress.

The extent of the plastic zone along the crack axis ($\theta = 0$) is given by

$$r_p(0) = \frac{1}{2\pi}\left(\frac{K_{\mathrm{I}}}{\sigma_Y}\right)^2 \tag{3.3}$$

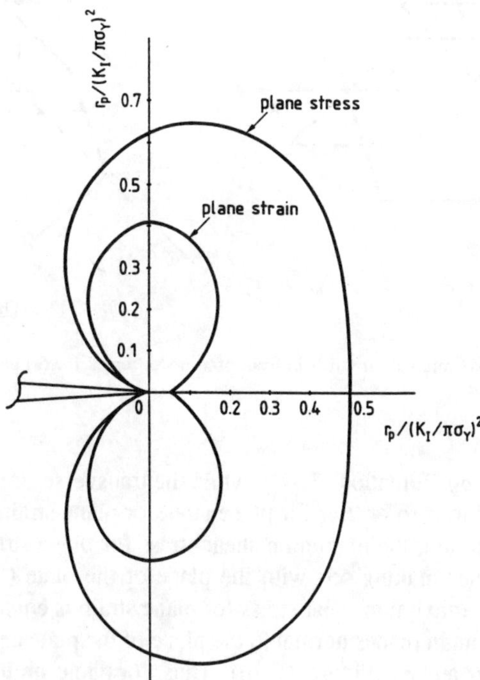

Fig. 3.2. Approximate estimation of the crack-tip plastic zones for mode-I loading under plane stress and plane strain. $\nu = 1/3$.

for plane stress, and

$$r_p(0) = \frac{1}{18\pi}\left(\frac{K_I}{\sigma_Y}\right)^2 \tag{3.4}$$

for plane strain, with $\nu = 1/3$.

Figure 3.2 shows the shapes of the plastic zones for plane stress and plane strain with $\nu = 1/3$. Observe that the plane stress zone is much larger than the plane strain zone because of the higher constraint for plane strain. Equations (3.3) and (3.4) show that the extent of the plastic zone along the crack axis for plane strain is 1/9 that of plane stress. We now present a few remarks concerning the qualitative nature of the plastic zones in plates of finite thickness.

Consider first a material element in front of the crack ($\theta = 0$) and assume conditions of plane stress or plane strain. The in-plane principal stresses σ_1 and σ_2 are

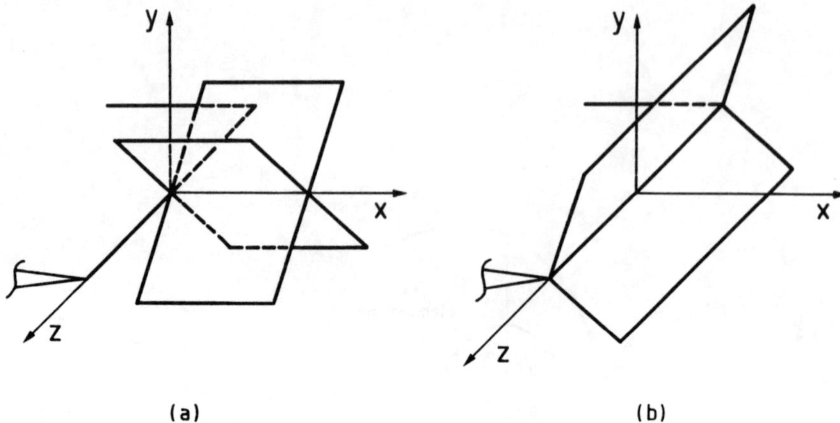

Fig. 3.3. Planes of maximum shear stress in front of a mode-I crack for (a) plane stress and (b) plane strain.

equal ($\sigma_1 = \sigma_2 = \sigma$) (Equation (2.37)), while the transverse stress σ_z is a principal stress and is equal to zero or $2\nu\sigma$ for plane stress or plane strain, respectively. For a Tresca yield criterion, the maximum shear stress for plane stress is equal to $\sigma/2$ and occurs in planes making 45° with the plane of the plate (Figure 3.3(a)). On the other hand, the maximum shear stress for plane strain is equal to $\sigma/6$ (assuming $\nu = 1/3$) and occurs in planes normal to the plane of the plate and making 45° with the directions of σ_1 and σ_2 (Figure 3.3(b)). Thus, for plane strain, not only is much more stress required to yield a material element than in plane stress, but also the planes of yielding are different. Analogous conditions hold for different θ angles.

Conditions of plane stress dominate in very thin plates where it can be assumed that the transverse stress σ_z is zero through the plate thickness. On the other hand, for thick plates, the state of stress is primarily one of plane strain. The type of plastic deformation associated with these two cases is shown in Figure 3.4. Under plane stress, slip takes place on planes at 45° to the plate surface, producing a rather large strain through the thickness (Figure 3.4(a)), while, in plane strain, slip occurs on planes perpendicular to the plate surface, giving a hinge-type deformation pattern (Figure 3.4(b)).

In cracked plates, conditions of plane stress dominate at the traction-free surfaces, while plane strain prevails in the interior. This results in a variation of the plastic zone through the plate thickness, which decreases from the surface to the interior of the plate (Figure 3.5). Although the state of stress is always a combination of plane stress and plane strain, some guidelines for determining the predominant type can be established. This is achieved by comparing the size of the plastic zone to the thickness, B, of the plate. When the length of the plastic zone c in front of the crack is of the order of B, then plane stress dominates. On the other hand, if c is much less than B the greatest part of the thickness is under plane strain. According to the

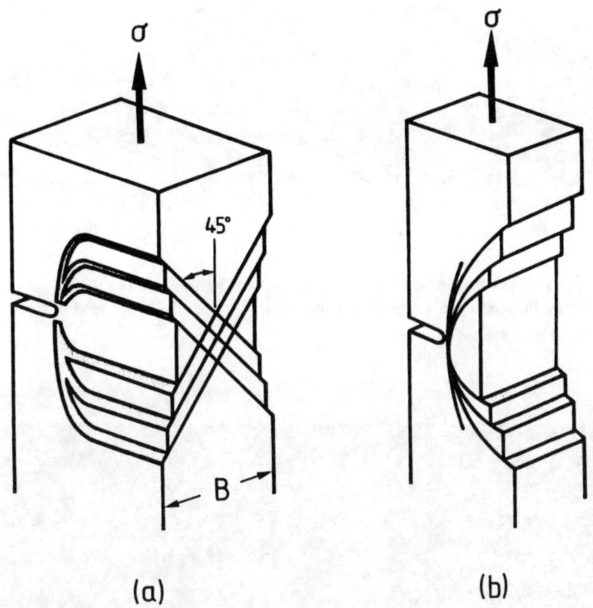

Fig. 3.4. Slip-planes around a mode-I crack for (a) plane stress and (b) plane strain.

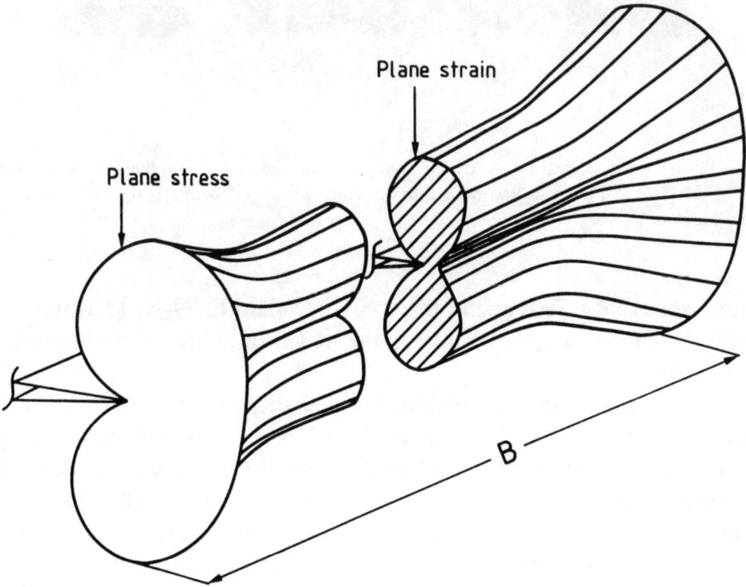

Fig. 3.5. Schematic representation of the three-dimensional nature of the crack-tip zone around a crack tip in a finite thickness plate.

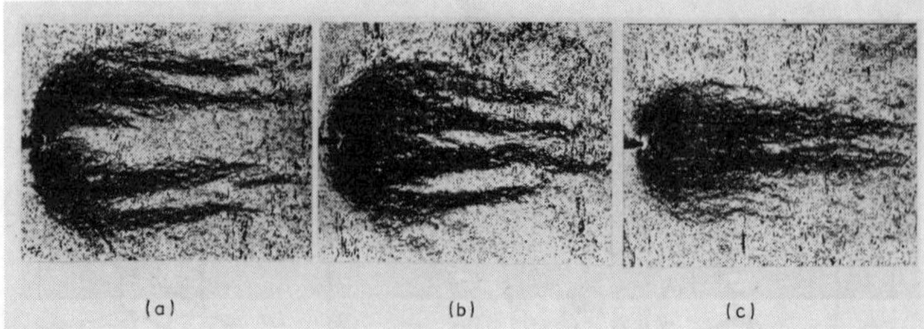

Fig. 3.6. Plastic zones, appearing as dark regions, in a cracked plate at (a) the surface of the specimen, (b) the section halfway between the surface and the midsection and (c) the midsection. (Photograph by P.N. Mincer, Battelle Memorial Institute.)

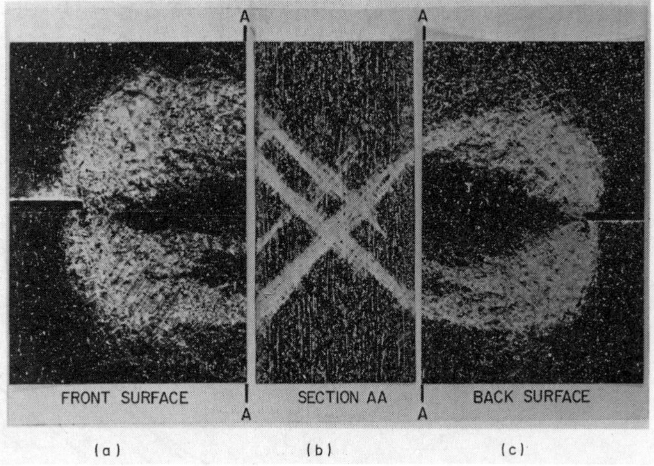

Fig. 3.7. Plastic zones, appearing as light regions, in a cracked plate at (a) the front and (c) the back surfaces of the plate and (b) a section normal to the crack plane. (Photograph by P.N. Mincer, Battelle Memorial Institute.)

American Society for Testing and Materials, Standard E399–81 [3.1], plane strain dominates when $c < B/25$, where c is the length of the plane strain plastic zone along the crack axis.

These qualitative predictions on the characteristics of the crack tip plastic zones were experimentally verified by Hahn *et al.* [3.2, 3.3]. They performed experiments on steel cracked specimens, and by etching their polished surfaces, revealed the three-dimensional character of the plastic zone in front of the crack. The plastic zones in the interior were obtained by sectioning, repolishing and etching. Figure 3.6 shows the plastic zones (appearing as dark regions) on (a) the surface of the specimen, (b) a section halfway between the surface and the midsection and (c) the midsection. The applied load produced a net section stress equal to 0.9 of the yield stress. The specimen has a thickness of 0.232 in and plane stress dominates. Observe that the

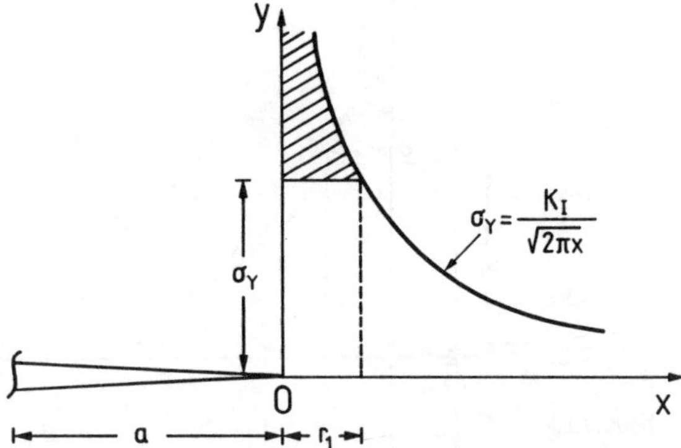

Fig. 3.8. Elasticity σ_y stress distribution ahead of a crack.

two yielded regions on the surface of the specimen merge into a single region on the midsection. Figure 3.7 shows the plastic zones (appearing as light regions) on (a) the front surface, (b) a section normal to the crack plane and (c) the back surface, of a 0.197 in thickness specimen at the same stress level as in Figure 3.6. The figure shows the slip bands on planes subtending 45° with the crack plane, which is indicative of plane stress. For lower stress levels, for which the plastic zones are much smaller than the plate thickness, it was found that the yield regions through the thickness remain almost the same; this is consistent with plane strain deformation.

3.3. Irwin's model

Irwin [3.4] presented a simplified model for the determination of the plastic zone attending the crack tip under small-scale yielding. He focused attention only on the extent along the crack axis and not on the shape of the plastic zone, for an elastic-perfectly plastic material.

To begin with, let us consider the elastic distribution of the $\sigma_y(= \sigma_x)$ stress along the crack axis in Figure 3.8 and assume that the plate is under plane stress. To estimate the extent of the plastic zone in front of the crack we use the approximate solution of Section 3.2 and find the distance r_1 from the crack tip to the point at which the yield stress σ_Y is exceeded. The value of r_1, determined from the condition $\sigma_y = \sigma_Y$, is given by Equation (3.3). The σ_y stress distribution along the x-axis is represented by the horizontal line $\sigma_y = \sigma_Y$ up to the point $x = r_1$ followed by the elastic singular σ_y-curve.

It is apparent in this determination that the equilibrium condition along the y-direction is violated, since the actual elastic stress distribution inside the plastic zone is replaced by a constant stress equal to σ_Y. The stresses in the shaded area in Figure 3.8 should produce a stress redistribution along the x-axis and the actual plastic zone length must be larger than r_1. Thus, as a result of the crack-tip plasticity,

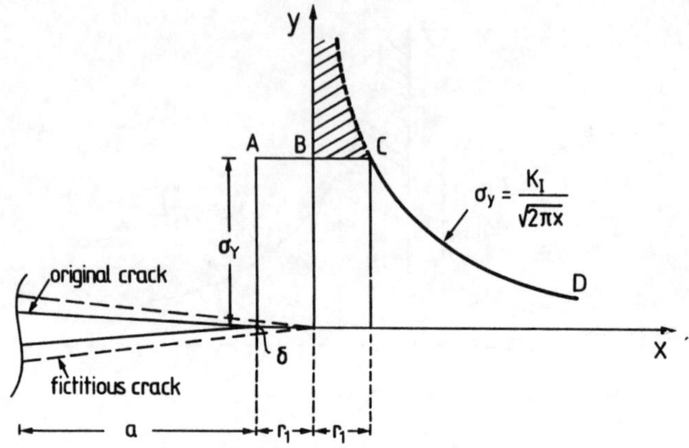

Fig. 3.9. Elastoplastic σ_y stress distribution ahead of a crack according to the Irwin model.

the displacements are larger and the stiffness of the plate is lower than in the elastic case.

These observations led Irwin to propose that the effect of plasticity makes the plate behave as if it had a crack longer than the actual crack size. The fictitious crack length is determined as follows: The area underneath the σ_y-curve up to the point $x = r_1$ is given by

$$\int_0^{r_1} \frac{K_\mathrm{I}\,\mathrm{d}x}{\sqrt{2\pi x}} = 2\sigma_Y r_1 \tag{3.5}$$

where the value of r_1 was introduced from Equation (3.3). Therefore, the shaded area in Figure 3.8 is $\sigma_Y r_1$. This result suggests that in order to satisfy equilibrium along the y-direction the original crack should be extended by a length r_1, as in Figure 3.9. In this case the σ_y stress distribution is represented by the curve ABCD, so that the area underneath this curve is equal to the area underneath the σ_y-curve in Figure 3.8 and equilibrium is maintained. Thus, the length of the plastic zone c in front of the crack is equal to $2r_1$, and is given by

$$c = \frac{1}{\pi}\left(\frac{K_\mathrm{I}}{\sigma_Y}\right)^2 \tag{3.6}$$

for plane stress. Equation (3.6) shows that the length of the plastic zone c, according to the Irwin model, is twice that determined from the approximate solution of Section 3.2. For plane strain, Irwin [3.5] suggested that due to existing constraint the stress required to produce yielding increases by a factor of $\sqrt{3}$. This results in a plastic zone length c in front of the crack, given by

$$c = \frac{1}{3\pi}\left(\frac{K_\mathrm{I}}{\sigma_Y}\right)^2. \tag{3.7}$$

The length of the plastic zone in front of the crack according to the Irwin model has been used to characterize the state of stress in a cracked plate as being either plane stress or plane strain. According to the ASTM Standard E399 [3.1] referred to in Section 3.2, the stress condition is characterized as plane stress when $c = B$ and as plane strain when $c < B/25$, where B is the thickness of the plate. Using Equation (3.7), we obtain for plane strain that

$$B > \frac{25}{3\pi}\left(\frac{K_\mathrm{I}}{\sigma_Y}\right)^2 \simeq 2.5\left(\frac{K_\mathrm{I}}{\sigma_Y}\right)^2 . \tag{3.8}$$

The distance δ of the faces of the fictitious crack at the tip of the initial crack of length a is given by the use of Equation (2.39)

$$\delta = 2v = \frac{\kappa+1}{2\mu}\,\sigma\,\sqrt{(a+c/2)^2 - a^2} \tag{3.9}$$

which gives

$$\delta = \frac{4}{\pi E}\,\frac{K_\mathrm{I}^2}{\sigma_Y} \tag{3.10}$$

for plane stress, and

$$\delta = \frac{4(1-\nu^2)}{3\pi E}\,\frac{K_\mathrm{I}^2}{\sigma_Y} \tag{3.11}$$

for plane strain.

The quantity δ given by Equation (3.11) has played an important role in characterizing the propensity of a crack to extend, and will be described in detail in Chapter 6.

3.4. Dugdale's model

A simplified model for plane stress yielding which avoids the complexities of a true elastic-plastic solution was introduced by Dugdale [3.6]. The model applies to very thin plates in which plane stress conditions dominate, and to materials with elastic-perfectly plastic behavior which obey the Tresca yield criterion. To analyze the model, we consider a crack of length $2a$ in an infinite plate subjected to uniaxial uniform stress σ at infinity perpendicular to the crack plane (Figure 3.10). We make the following hypotheses:

(i) All plastic deformation concentrates in a line in front of the crack.

(ii) The crack has an effective length which exceeds that of the physical crack by the length of the plastic zone.

The first hypothesis is justified from the fact that for plane stress, following the considerations of Section 3.2, yielding takes place on planes that subtend 45° with

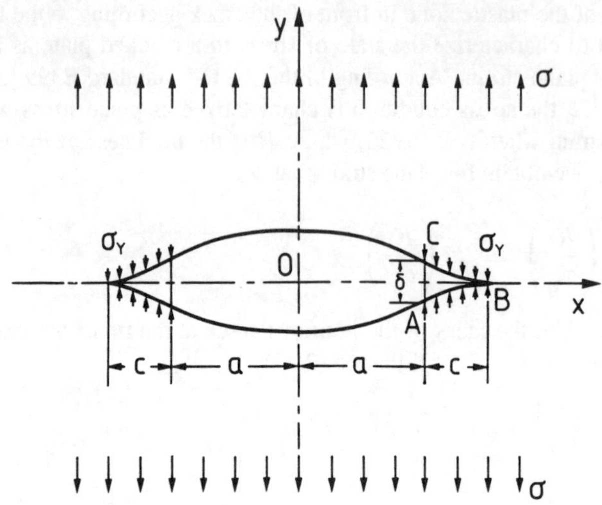

Fig. 3.10. Dugdale's model for a mode-I crack of length $2a$.

the plate surface and the height of the plastic zone is equal to the plate thickness. Thus, for very thin plates, the plastic zones approach line segments. With the Tresca yield criterion, there should be stresses equal to the yield stress σ_Y along the plastic zone. The length of the plastic zone c is determined from the condition that the σ_y stress at the tip of the effective crack should remain bounded and equal to the yield stress σ_Y.

Based on the above arguments, the solution of the elastic-plastic problem of Figure 3.10 is, according to the Dugdale model, reduced to an elastic problem. The Westergaard function Z of the problem is obtained by adding these functions for the following two problems:

(i) A crack of length $2(a+c)$ in an infinite plate subjected to a uniform stress σ at infinity. The Westergaard function is given by (Equation (2.19))

$$Z_{\mathrm{I}} = \frac{\sigma z}{\sqrt{z^2 - (a+c)^2}} . \tag{3.12}$$

(ii) A crack of length $2(a+c)$ in an infinite plate subjected to a uniform stress distribution equal to σ_Y along the plastic zone ($a < |x| < a+c$). The Westergaard function for this problem is given by (Problem 2.8)

$$Z_2 = \frac{2\sigma_Y}{\pi} \left[\frac{z}{\sqrt{z^2 - (a+c)^2}} \arccos\left(\frac{a}{a+c}\right) - \operatorname{arccot}\left(\frac{a}{z}\sqrt{\frac{z^2 - (a+c)^2}{(a+c)^2 - a^2}}\right) \right] . \tag{3.13}$$

The Westergaard function of the problem of Figure 3.10 is

$$Z = Z_1 - Z_2 = \frac{2\sigma_Y}{\pi} \operatorname{arccot}\left(\frac{a}{z}\sqrt{\frac{z^2 - (a+c)^2}{(a+c)^2 - a^2}}\right) \tag{3.14}$$

by zeroing the singular term of Z. For the length of the plastic zone, this condition gives

$$\frac{a}{a+c} = \cos\left(\frac{\pi}{2}\frac{\sigma}{\sigma_Y}\right) . \tag{3.15}$$

Equation (3.15) for small values of σ/σ_Y gives

$$c = \frac{\pi}{8}\left(\frac{K_\mathrm{I}}{\sigma_Y}\right)^2 . \tag{3.16}$$

By comparing Equations (3.16) and (3.6) we can deduce that the Irwin model underestimates the length of plastic zone by about 20%, compared to the Dugdale model.

The displacement of the crack faces, obtained by introducing the value of Z from Equation (3.14) into Equation (2.38), is given by

$$v = \frac{(a+c)\,\sigma_Y}{\pi E}\left[\frac{x}{a+c}\ \ln\frac{\sin^2(\theta_2 - \theta)}{\sin^2(\theta_2 + \theta)} + \cos\theta_2\ \ln\frac{(\sin\theta_2 + \sin\theta)^2}{(\sin\theta_2 - \sin\theta)^2}\right] \tag{3.17}$$

where we put

$$\theta = \arccos\frac{x}{a+c}\,, \quad \theta_2 = \frac{\pi}{2}\frac{\sigma}{\sigma_Y} . \tag{3.18}$$

The opening of the effective crack at the tip of the physical crack is given by

$$\delta = 2\lim_{x \to \pm a} v = \frac{8\sigma_Y a}{\pi E}\ln(\sec\theta_2) . \tag{3.19}$$

By expanding Equation (3.19) and retaining the first term for small values of σ/σ_Y we obtain

$$\delta = \frac{K_\mathrm{I}^2}{E\sigma_Y} . \tag{3.20}$$

By comparing Equations (3.20) and (3.10) we can deduce that the Irwin model overestimates δ by 27 per cent compared to the Dugdale model. The variation of the dimensionless quantities c/a and $(\pi E\delta)/(8\sigma_Y a)$ versus σ/σ_Y is shown in Figure 3.11.

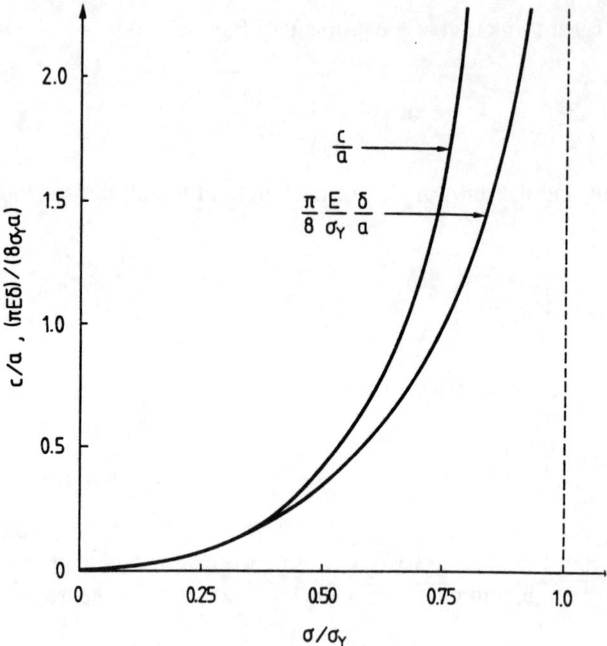

Fig. 3.11. Normalized plastic zone length and crack-tip opening displacement versus normalized applied stress for the Dugdale model.

Examples

Example 3.1.

Determine the radius of the plastic zone accompanying the crack tip for mixed-mode (opening-mode and sliding-mode) loading under plane strain conditions. Plot the resulting elastic-plastic boundary for a crack of length $2a$ in an infinite plate subtending an angle $\beta = 30°$ with the direction of applied uniaxial stress at infinity. $\nu = 0.3$.

Solution: By superimposing the stresses for opening-mode and sliding-mode loading and omitting the constant term we obtain, after some algebra, for plane strain conditions ($\sigma_z = \nu(\sigma_x + \sigma_y)$, see Equations (2) of Example 2.3) for the radius r of the plastic zone

$$r = \frac{1}{2\pi\sigma_Y^2}\left[K_{\mathrm{I}}^2 \cos^2\frac{\theta}{2}\left[(1-2\nu)^2 + 3\sin^2\frac{\theta}{2}\right] + K_{\mathrm{I}}K_{\mathrm{II}}\sin\theta[3\cos\theta - (1-2\nu)^2] + K_{\mathrm{II}}^2\left[3 + \sin^2\frac{\theta}{2}\left\{(1-2\nu)^2 - 9\cos^2\frac{\theta}{2}\right\}\right]\right]. \tag{1}$$

Equation (1) for opening-mode ($K_{\mathrm{II}} = 0$) coincides with Equation (3.2).

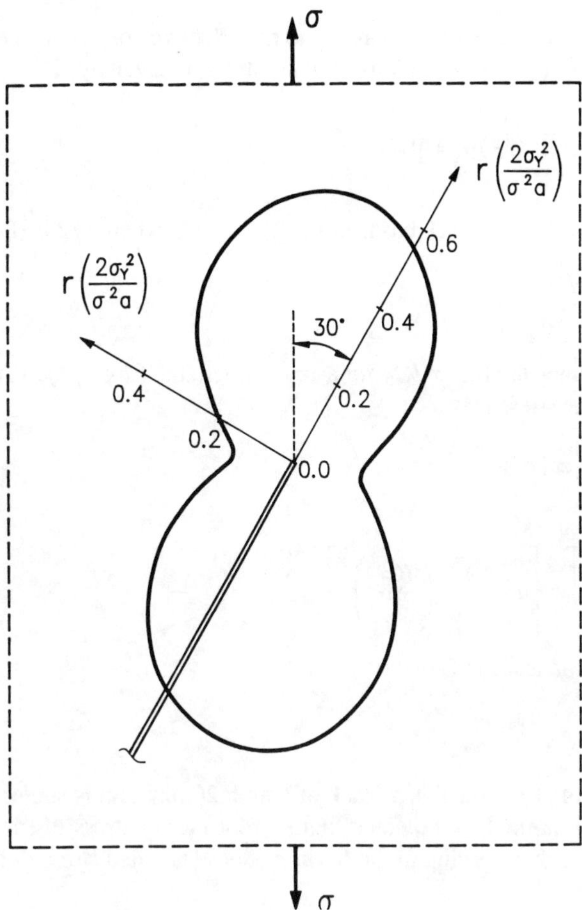

Fig. 3.12. Elastic-plastic boundary surrounding the tip of an inclined crack in an infinite plate.

For a crack of length $2a$ in an infinite plate subtending an angle $\beta = 30°$ with the direction of applied uniaxial stress at infinity the stress intensity factors K_{I}, K_{II} are given by (Problem 2.22)

$$K_{\mathrm{I}} = \sigma \sqrt{\pi a} \sin^2 \beta \,, \quad K_{\mathrm{II}} = \sigma \sqrt{\pi a} \sin \beta \cos \beta \,. \tag{2}$$

Introducing these values into Equation (1) we obtain the radius of the plastic zone. For $\beta = 30°$ and $\nu = 0.3$ it is shown in Figure 3.12.

Example 3.2.

Consider a central crack of length $2a$ in an infinite plate subjected to uniaxial stress σ at infinity perpendicular to the crack plane. According to the Irwin model, the effective crack is larger than the actual crack by the length of plastic zone. Show that the stress intensity factor corresponding to the effective crack, called effective stress intensity factor K_{eff}, for conditions of plane stress, is given by

$$K_{\text{eff}} = \frac{\sigma(\pi a)^{1/2}}{[1 - 0.5(\sigma/\sigma_Y)^2]^{1/2}} \,. \tag{1}$$

Solution: The effective crack has a length $2(a + c/2)$ where $c/2$ is (Equation (3.6))

$$\frac{c}{2} = \frac{1}{2\pi}\left(\frac{K_{\text{eff}}}{\sigma_Y}\right)^2 . \tag{2}$$

The stress intensity factor K_{eff} for a crack of length $2(a + c/2)$ in an infinite plate subjected to the stress σ is

$$K_{\text{eff}} = \sigma\left[\pi\left(a + \frac{c}{2}\right)\right]^{1/2} \tag{3}$$

or

$$K_{\text{eff}} = \sigma\left[\pi\left[a + \frac{1}{2\pi}\left(\frac{K_{\text{eff}}}{\sigma_Y}\right)^2\right]\right]^{1/2} . \tag{4}$$

This Equation leads to Equation (1).

Example 3.3.

A large plate of steel contains a crack of length 20 mm and is subjected to a stress $\sigma = 500$ MPa normal to the crack plane. Plot the σ_y stress distribution directly ahead of the crack according to the Irwin model. The yield stress of the material is 2000 MPa.

Solution: Since the plate is large the effective stress intensity factor K_{eff} is computed from Equation (1) of Example 3.2. We have

$$K_{\text{eff}} = \frac{\sigma[\pi(0.01)]^{1/2}}{\left[1 - 0.5\left(\dfrac{500}{2000}\right)^2\right]^{1/2}} = 90 \text{ MPa}\sqrt{m} \,.$$

The length of the plastic zone c is

$$c = \frac{1}{\pi}\left(\frac{90}{2000}\right)^2 = 0.64 \text{ mm} \,.$$

The σ_y stress is constant along the length of plastic zone, while in the elastic region it varies according to

$$\sigma_y = \frac{K_{\text{eff}}}{(2\pi x)^{1/2}}$$

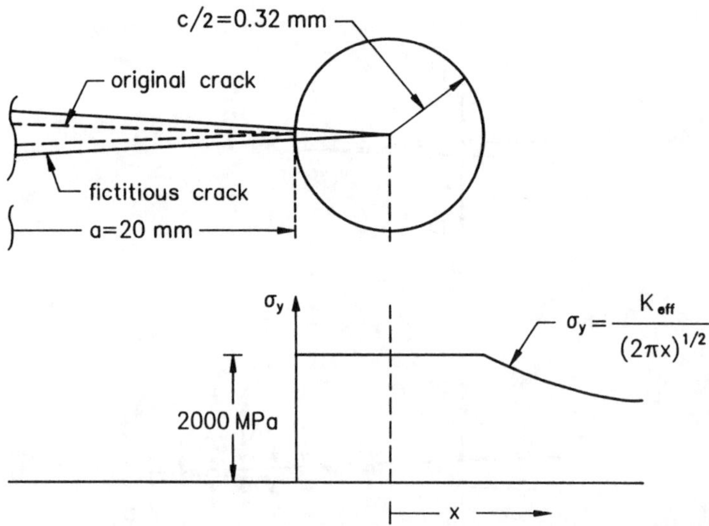

Fig. 3.13. Original and fictitious crack and σ_y stress distribution according to the Irwin model.

where x is measured from the tip of the effective crack ($x > 0.32$ mm).

The σ_y stress distribution is shown in Figure 3.13.

Example 3.4.

The stress intensity factor for an infinite plate with a semi-infinite crack subjected to concentrated loads P at distance L from the crack tip (Figure 3.14a) is given by

$$K_{\mathrm{I}} = \frac{2P}{(2\pi L)^{1/2}} \, . \tag{1}$$

For this situation determine the length of the plastic zone according to the Dugdale model.

Solution: According to the Dugdale model there is a fictitious crack equal to the real crack plus the length of plastic zone (Figure 3.14b). This crack is loaded by the applied loads P and an additional uniform compressive stress equal to the yield stress σ_Y along the plastic zone.

The length of plastic zone c is determined from the condition that the stresses should remain bounded at the tip of the fictitious crack. This condition is expressed by zeroing the stress intensity factor.

The stress intensity factor $K_{\mathrm{I}}^{(P)}$ due to applied loads P is

$$K_{\mathrm{I}}^{(P)} = \frac{2P}{[2\pi(c+L)]^{1/2}} \, . \tag{2}$$

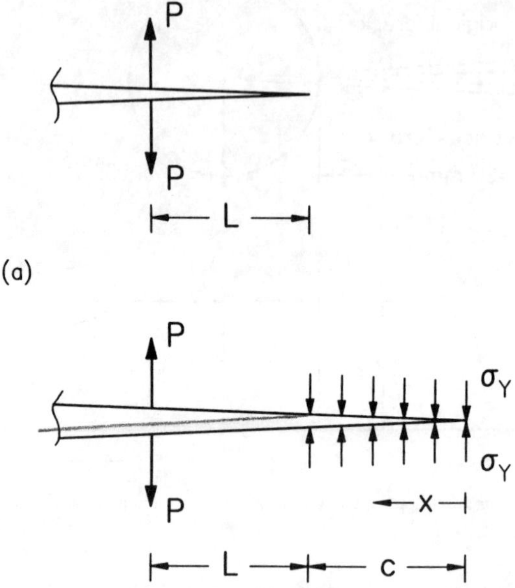

Fig. 3.14. (a) A semi-infinite crack subjected to concentrated loads P and (b) calculation of the length of plastic zone according to the Dugdale model.

The stress intensity factor $K_{\mathrm{I}}^{(\sigma_Y)}$ due to the uniform stress σ_Y along the length of plastic zone is determined as

$$K_{\mathrm{I}}^{(\sigma_Y)} = -\int_0^c \frac{2\sigma_Y}{(2\pi x)^{1/2}}\,\mathrm{d}x \tag{3}$$

or

$$K_{\mathrm{I}}^{(\sigma_Y)} = -\frac{4\sigma_Y c^{1/2}}{(2\pi)^{1/2}}\,. \tag{4}$$

The condition that the stress intensity factor be zero at the tip of the fictitious crack is expressed as

$$K_{\mathrm{I}}^{(P)} + K_{\mathrm{I}}^{(\sigma_Y)} = 0 \tag{5}$$

which leads to

$$c = \frac{L}{2}\left[\left(1 + \left(\frac{P}{L\sigma_Y}\right)^2\right)^{1/2} - 1\right]. \tag{6}$$

Problems

3.1. Show that the radius of the plastic zone surrounding the tip of a mode-II crack is given by

$$r_p(\theta) = \frac{1}{8\pi}\left(\frac{K_{\text{II}}}{\sigma_Y}\right)^2 (14 - 2\cos\theta - 9\sin^2\theta)$$

for plane stress, and by

$$r_p(\theta) = \frac{1}{8\pi}\left(\frac{K_{\text{II}}}{\sigma_Y}\right)^2 [12 + 2(1-\nu)(1-\cos\theta) - 9\sin^2\theta]$$

for plane strain. Plot the resulting curves for $\nu = 1/3$.

3.2. Show that the radius of the plastic zone surrounding the tip of a mode-III crack is a circle centered at the crack tip with radius

$$r_p = \frac{3}{2\pi}\left(\frac{K_{\text{III}}}{\sigma_Y}\right)^2 .$$

3.3. Find the equation of the plastic zone ahead of a crack for opening-mode loading under conditions of plane stress and plane strain for a material obeying the Tresca yield criterion expressed by Equation (1.2). Compare the resulting equation with Equations (3.1) and (3.2).

3.4. As in Problem 3.3 for sliding-mode loading.

3.5. Determine the crack tip plastic zone for opening-mode loading for a pressure modified von Mises yield criterion expressed by

$$(\sigma_1^2 + \sigma_2^2 + \sigma_3^2 - \sigma_1\sigma_2 - \sigma_2\sigma_3 - \sigma_3\sigma_1)^{1/2} + \frac{R-1}{R+1}(\sigma_1 + \sigma_2 + \sigma_3) = \frac{2R}{R+1}\sigma_T$$

where $R = \sigma_c/\sigma_T$ and σ_T and σ_c are the yield stress of the material in tension and compression, respectively. Plot the resulting elastic-plastic boundaries for plane stress and plane strain conditions when $R = 1.2$ and 1.5. Compare the results with those obtained by the von Mises criterion.

3.6. As in Problem 3.5 for a pressure modified von Mises yield criterion expressed by

$$(\sigma_1^2 + \sigma_2^2 + \sigma_3^2 - \sigma_1\sigma_2 - \sigma_2\sigma_3 - \sigma_3\sigma_1) + (\sigma_c - \sigma_T)(\sigma_1 + \sigma_2 + \sigma_3) = \sigma_c\sigma_T .$$

3.7. Determine the minimum plate thickness required for plane strain conditions to prevail at the crack tip according to the ASTM specifications for the following steels:
a. 4340, with $K_{IC} = 100$ MPa $\sqrt{m}$ and $\sigma_Y = 860$ MPa.
b. A533, with $K_{IC} = 180$ MPa $\sqrt{m}$ and $\sigma_Y = 350$ MPa.
Discuss the results.

3.8. Consider a crack in a finite width plate subjected to opening-mode loading. Establish an iterative process for determining the effective stress intensity factor K_{eff} according to the Irwin model.

3.9. A thin steel plate of width $2b = 40$ mm contains a central crack of length $2a = 20$ mm and is subjected to a stress $\sigma = 500$ MPa normal to the crack plane. Plot the σ_y stress distribution directly ahead of the crack according to the Irwin model. The yield stress of the material is 2000 MPa. Consult Appendix 2.1.

3.10. A thin steel plate of width $b = 100$ mm contains an edge crack of length $a = 20$ mm and is subjected to a stress $\sigma = 400$ MPa normal to the crack plane. Compute the length of the plastic zone and plot the σ_y stress distribution directly ahead of the crack according to the Irwin model. The yield stress of the material is 2000 MPa. Consult Appendix 2.1.

3.11. Show that the length of plastic zone c according to Dugdale model for a crack of length $2a$ in an infinite plate subjected to uniform stress σ at infinity normal to the crack for small applied stresses σ is given by

$$\frac{a}{c} = 0.8106\left(\frac{\sigma_Y}{\sigma}\right)^2 - 1\,.$$

3.12. Use Equation (3.14) to show that, at the tip of the fictitious crack of the Dugdale model, the stresses σ_x and σ_y are equal to the yield stress σ_Y.

3.13. Establish Equation (3.17) for the displacement of the crack faces according to the Dugdale model.

3.14. Show that according to the Dugdale model the stress σ_y outside the plastic zone along the crack axis is given by

$$\sigma_y = \sigma_Y\left[1 - \frac{1}{\theta_2}\arctan\left(\frac{\sin 2\theta_2}{\cos 2\theta_2 - e^{2\theta}}\right)\right]$$

where

$$\theta_2 = \frac{\pi}{2}\frac{\sigma}{\sigma_Y}\,, \quad \theta = \operatorname{arccos}\frac{x}{a+c}\,.$$

Plot the σ_y distribution for $\sigma/\sigma_Y = 0.5$.

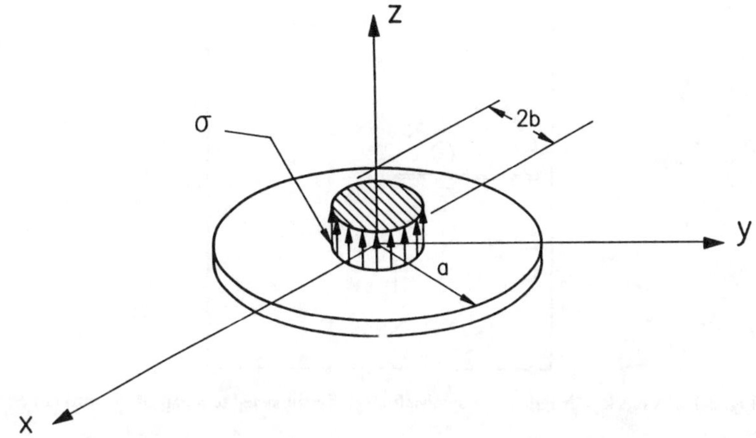

Fig. 3.15. A penny-shaped crack in an infinite plate subjected to a uniform stress σ over a concentric circular area of radius b.

3.15. For Example 3.3 plot the σ_y stress distribution directly ahead of the crack according to the Dugdale model. Compare the results with those obtained by the Irwin model.

3.16. Show that according to the Dugdale model the curve $v = v(x)$ (Equation (3.17)) which gives the vertical displacements of the crack faces has a vertical slope at the tips of the actual crack. Draw a sketch of this curve.

3.17. The stress intensity factor for a penny-shaped crack of radius a in an infinite solid subjected to a uniform stress σ over a concentric circular area of radius b ($b < a$) (Figure 3.15) is given by

$$K_{\mathrm{I}} = \frac{2\sigma}{\sqrt{\pi a}} \left[a - (a^2 - b^2)^{1/2}\right].$$

Determine the length of the plastic zone according to the Dugdale model.

3.18. The stress intensity factor for an edge crack of length a in a semi-infinite solid subjected to a pair of concentrated shear forces S applied to the crack at a distance b from the solid edge (Figure 3.16) is

$$K_{\mathrm{III}} = \frac{2S(\pi a)^{1/2}}{\pi(a^2 - b^2)^{1/2}}.$$

Determine the length of plastic zone according to the Dugdale model, and plot the variation of c/a versus $S/a\tau_Y$ for different values of b/a, where τ_Y is the yield stress in shear.

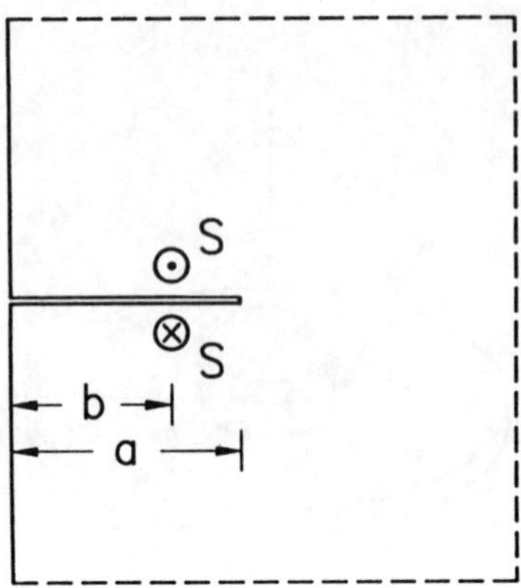

Fig. 3.16. A crack of length a in a semi-infinite solid subjected to a pair of shear forces S.

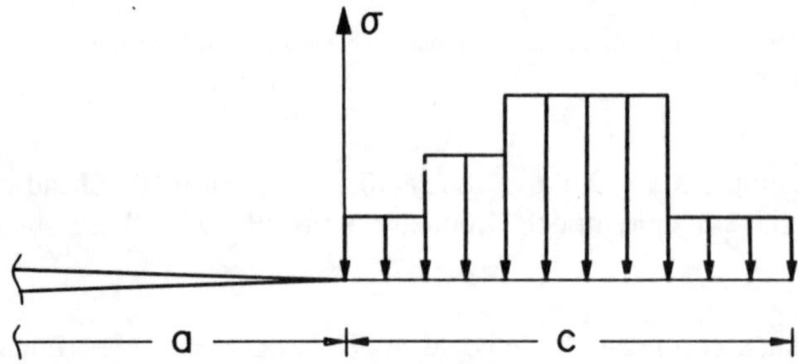

Fig. 3.17. A stepwise stress distribution inside the plastic zone for a crack in an infinite plate for the determination of fracture quantities according to the Dugdale model.

3.19. The Dugdale model was modified by taking a variable stress distribution inside the plastic zone to take into account the strain-hardening of the material. Consider a stepwise stress distribution inside the plastic zone (Figure 3.17) for a crack of length $2a$ in an infinite plate loaded by uniform stress σ at infinity perpendicular to the crack plane. Determine:

1. The Westergaard function.
2. The length of the plastic zone.
3. The crack shape.

References

3.1. Standard test method for plane-strain fracture toughness of metallic materials, *Annual Book of ASTM Standards*, Part 10, E 399–81, American Society for Testing and Materials, Philadelphia, pp. 592–621 (1981).

3.2. Hahn, G.T. and Rosenfield, A.R. (1965) 'Local yielding and extension of a crack under plane stress', *Acta Metallurgica* **13**, 293–306.

3.3. Rosenfield, A.R., Dai, P.K., and Hahn, G.T. (1966) 'Crack extension and propagation under plane stress', *Proceedings of the First International Conference on Fracture* (eds. T. Yokobori, T. Kawasaki and J.L. Swedlow), Sendai, Japan, Vol. 1, pp. 223–258.

3.4. Irwin, G.R. (1960) 'Plastic zone near a crack tip and fracture toughness', *Proceedings of the Seventh Sagamore Ordnance Material Conference*, pp. IV63–IV78.

3.5. Irwin, G.R. (1968) 'Linear fracture mechanics, fracture transition, and fracture control', *Engineering Fracture Mechanics* **1**, 241–257.

3.6. Dugdale, D.S. (1960) 'Yielding of steel sheets containing slits', *Journal of the Mechanics and Physics of Solids* **8**, 100–104.

Chapter 4

Crack Growth Based on Energy Balance

4.1. Introduction

When a solid is fractured new surfaces are created in the medium in a thermodynamically irreversible manner. Material separation is caused by the rupture of atomic bonds due to high local stresses. The phenomenon of fracture may be approached from different points of view, depending on the scale of observation. At one extreme is the atomic approach where the phenomena of interest take place in the material within distances of the order of 10^{-7} cm. At the other extreme is the continuum approach, which considers material behavior at distances greater than 10^{-2} cm. In the atomic approach, the problem is studied using the concepts of quantum mechanics; the continuum approach uses the theories of continuum mechanics and classical thermodynamics. The complex nature of fracture prohibits a unified treatment of the problem, and the existing theories deal with the subject either from the microscopic or from the macroscopic point of view. A major objective of fracture mechanics is to bridge the gap between these two approaches.

The continuum mechanics approach to fracture assumes the existence of defects large compared to the characteristic dimensions of the microstructure and considers the material as a homogeneous continuum. To study the problem of growth of an existing crack, void or other defect we need stress analysis coupled with a postulate predicting when fracture will occur. A number of such failure criteria have been advanced over the years. Each criterion involves a quantity that has to be related to the loss of continuity, and has a critical value that serves as a measure of the resistance of the material to separation.

In the present chapter we will develop the theory of crack growth based on the global energy balance of the entire system. This approach was proposed by Griffith [1.2, 1.3] more than six decades ago, and is the earliest attempt to formulate a linear elastic theory of crack propagation. The chapter starts with the global energy balance in a continuum during crack growth, from which the Griffith criterion is deduced. In an attempt to extend the principles of linear elastic analysis to situations of highly localized yielding at the crack front, the various irreversibilities associated with fracture are lumped together to define the fracture toughness of the material. This approach allows Griffith's theory to be applied to metals and other engineering materials. We present a graphical representation of the various terms appearing in the energy balance equation, and establish the equivalence of the energy approach and that based on the intensity of the local stress field. The chapter concludes with a study of crack stability.

4.2. Energy balance during crack growth

Consider a crack with area A in a deformable continuum subjected to arbitrary loading. According to the law of conservation of energy we have

$$\dot{W} = \dot{E} + \dot{K} + \dot{\Gamma} \tag{4.1}$$

where $\dot{W}$ is the work performed per unit time by the applied loads, $\dot{E}$ and $\dot{K}$ are the rates of change of the internal energy and kinetic energy of the body, and $\dot{\Gamma}$ is the energy per unit time spent in increasing the crack area. A dot over a letter denotes differentiation with respect to time.

The internal energy E can be put in the form

$$E = U^e + U^p \tag{4.2}$$

where U^e represents the elastic strain energy and U^p the plastic work.

If the applied loads are time independent and the crack grows slowly the kinetic term K is negligible and can be omitted from the energy balance Equation (4.1). Since all changes with respect to time are caused by changes in crack size, we can state

$$\frac{\partial}{\partial t} = \frac{\partial A}{\partial t}\frac{\partial}{\partial A} = \dot{A}\frac{\partial}{\partial A}, \quad \dot{A} \geq 0 \tag{4.3}$$

and Equation (4.1) becomes

$$\frac{\partial W}{\partial A} = \left(\frac{\partial U^e}{\partial A} + \frac{\partial U^p}{\partial A}\right) + \frac{\partial \Gamma}{\partial A}. \tag{4.4}$$

Equation (4.4) represents the energy balance during crack growth. It indicates that the work rate supplied to the continuum by the applied loads is equal to the rate of the elastic strain energy and plastic strain work plus the energy dissipated in crack propagation. Equation (4.4) may be put in the form

$$-\frac{\partial \Pi}{\partial A} = \frac{\partial U^p}{\partial A} + \frac{\partial \Gamma}{\partial A} \tag{4.5}$$

where

$$\Pi = U^e - W \tag{4.6}$$

is the potential energy of the system. Equation (4.5) shows that the rate of potential energy decrease during crack growth is equal to the rate of energy dissipated in plastic deformation and crack growth. Both energy balance Equations (4.4) and (4.5) will be used in the sequel.

4.3. Griffith theory

For an ideally brittle material, the energy dissipated in plastic deformation is negligible and can be omitted from Equation (4.4). If γ represents the energy required to form a unit of new material surface, then Equation (4.4) takes the form

$$G = \frac{\partial W}{\partial A} - \frac{\partial U^e}{\partial A} = \frac{\partial \Gamma}{\partial A} = 2\gamma \tag{4.7}$$

where the factor 2 appearing on the right-hand side of the equation refers to the two new material surfaces formed during crack growth.

The left hand side of the equation represents the energy available for crack growth, and is given the symbol G in honor of Griffith. Because G is derived from a potential function, just like a conservative force, it is often referred to as the crack driving force. The right-hand side of Equation (4.7) represents the resistance of the material that must be overcome for crack growth, and is a material constant.

Equation (4.7) represents the fracture criterion for crack growth. Two limiting cases, the "fixed-grips" and "dead-load" loading, are usually encountered in practice. In the fixed-grips loading the surface of the continuum on which the loads are applied is assumed to remain stationary during crack growth. If the work of the body forces is ignored, the work performed by the applied loads vanishes and Equation (4.7) takes the form

$$G = -\frac{\partial U^e}{\partial A} = 2\gamma\,. \tag{4.8}$$

Equation (4.8) indicates that the energy rate for crack growth is supplied by the existing elastic strain energy of the solid. Because of this property, the symbol G is usually referred to as the "elastic strain energy release rate".

In the dead-load situation the applied loads on the surface of the solid are kept constant during crack growth. Clapeyron's theorem of linear elastostatics states that the work performed by the constant applied loads is twice the increase of elastic strain energy ($\partial W/\partial A = 2\partial U^e/\partial A$). Thus Equation (4.7) takes the form

$$G = \frac{\partial U^e}{\partial A} = 2\gamma\,. \tag{4.9}$$

Contrary to the previous case of "fixed-grips" the energy required for crack growth is not supplied by the existing elastic strain energy of the solid, but by the work performed by the external loads; the elastic strain energy of the solid is increased. Thus, the term "strain energy release rate" for G in this case is physically inappropriate.

Equations (4.8) and (4.9) show that the magnitude of the elastic strain energy release rate necessary for crack growth is the same for either "fixed-grips" or "dead-load" loading. However, the elastic strain energy of the system decreases for "fixed-grips" and increases for "dead-load" conditions. Equations (4.8) and (4.9) can be put in the form

$$G = -\frac{\partial \Pi}{\partial A} = 2\gamma \,. \tag{4.10}$$

where the potential energy Π is defined from Equation (4.6). Equation (4.10) may be written as

$$\frac{\partial(\Pi + \Gamma)}{\partial A} = 0 \tag{4.11}$$

which, in Griffith's terminology, states that the "total potential energy" of the system $(\Pi + \Gamma)$ is stationary.

Consider a line crack of length $2a$ in an infinite plate subjected to a uniform stress σ perpendicular to the crack. The change in elastic strain energy due to the presence of a crack, is given by [1.2, 1.3]

$$U^e = \frac{\pi a^2 \sigma^2}{8\mu} (\kappa + 1) \,. \tag{4.12}$$

where $\kappa = 3 - 4\nu$ for plane strain and $\kappa = (3 - \nu)/(1 + \nu)$ for generalized plane stress.

For $A = 2a \times 1$, Equation (4.9) gives the critical stress required for unstable crack growth as

$$\sigma_c = \sqrt{\frac{2E\gamma}{\pi a(1 - \nu^2)}} \tag{4.13}$$

for plane strain, and

$$\sigma_c = \sqrt{\frac{2E\gamma}{\pi a}} \tag{4.14}$$

for generalized plane stress.

Observe that the stress σ_c is inversely proportional to the square root of the half crack length. This result was verified experimentally by Griffith on glass for a wide range of crack lengths.

4.4. Graphical representation of the energy balance equation

The graphical representation of the various terms appearing in the energy balance equation is useful as it provides a better insight into the variation of the relevant quantities during crack growth, and helps the interpretation of experimental results. The load-displacement response of the body during crack growth, as obtained from

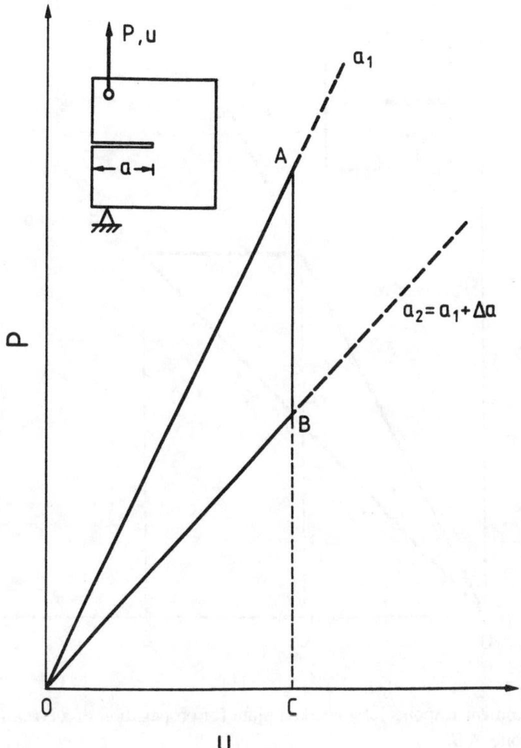

Fig. 4.1. Load-displacement response of a cracked plate for propagation of a crack from length a_1 to a_2 under "fixed grips" conditions along AB.

a testing machine, is examined separately for the cases of "fixed-grips", "dead-load" and the general case of changing both the load and displacement during crack propagation. Finally, the graphical representation in $G - a$ coordinates is introduced. In the analysis it is assumed that the load-displacement response of the body is linear elastic.

(a) "Fixed-grips" loading

The load-displacement response of a body of unit thickness with an initial crack of length a_1 is represented in Figure 4.1 by the straight line OA. During loading up to the point A elastic strain energy represented by the area (OAC) is stored in the body. This energy is released when the body is unloaded. Let us assume that at point A the crack starts to propagate under constant displacement to a new length $a_2 = a_1 + \Delta a$. The straight line OB represents the load-displacement response of the body with a longer crack of length a_2. During crack propagation the load drops from point A to point B. Line OB should lie below line OA since the stiffness of the body decreases

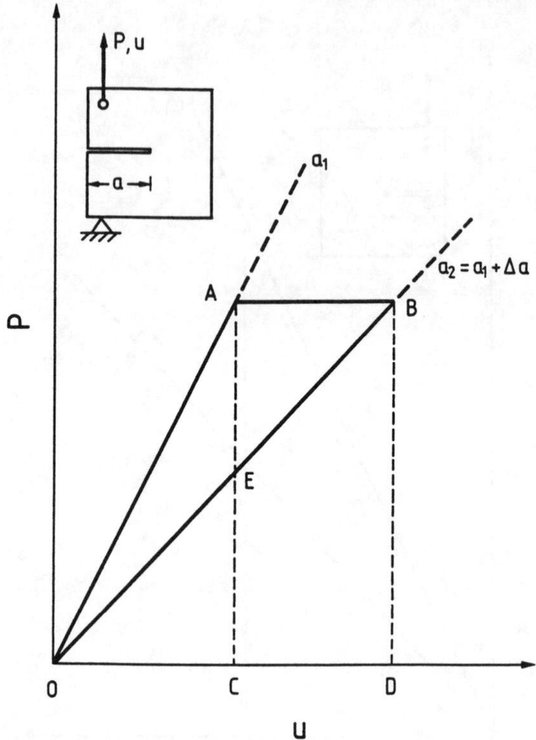

Fig. 4.2. Load-displacement response of a cracked plate for propagation of a crack from length a_1 to a_2 under constant load along AB.

with increase of the crack length. The elastic strain energy stored in the body at point B is represented by the area (OBC). If the applied load is removed at point B the unloading path will follow the line BO. Since the point of application of the load remains fixed during crack growth, no extra work is supplied to the body. The reduction in strain energy during crack growth is represented by the area (OAB). It is that obtained for the elastic energy release rate from Equation (4.8), and is balanced by the material resistance to crack growth 2γ

$$G = \frac{(OAB)}{\Delta a} = 2\gamma\,. \tag{4.15}$$

(b) "Dead-load" loading

The graphical representation of the load-displacement response of a cracked body during crack growth under constant loading is represented in Figure 4.2. The displacement increases from A to B as the crack length increases from a_1 to $a_2 = a_1 + \Delta a$. The energy at the beginning of crack growth is represented by the area (OAC) and at the end by the area (OBD). During crack growth the load P performs work

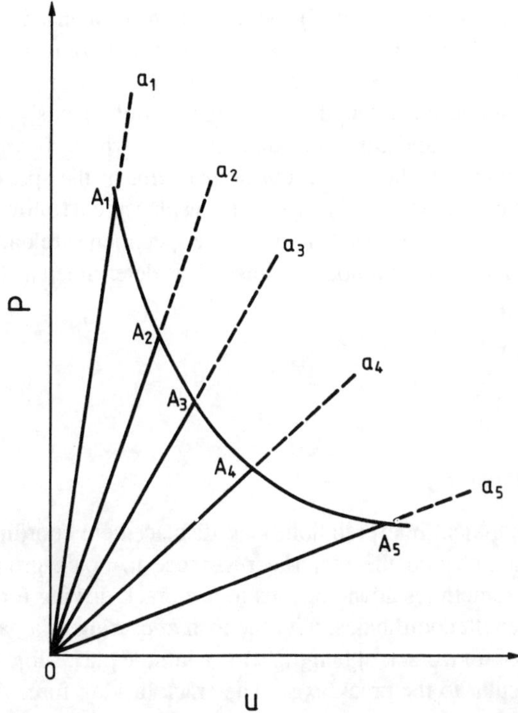

Fig. 4.3. Load-displacement response of a cracked plate for propagation of a crack from an initial length a_1 to a final length a_5 under general load-displacement conditions along $A_1A_2A_3A_4A_5$.

represented by the area $(ABDC)$. The energy supplied to the body for fracture is equal to $(OAC) + (ABDC) - (OBD) = (OAB)$. Equation (4.9) takes the form

$$G = \frac{(OAB)}{\Delta a} = 2\gamma\,. \tag{4.16}$$

Note that the work supplied for crack growth under "dead-load" loading differs from that necessary for crack growth under "fixed-grips" loading by the amount (ABE) which disappears as the crack growth increment Δa tends to zero.

(c) General load-displacement relation

Usually, both load and displacement change during crack growth. The load-displacement response depends mainly on the form of the specimen and the type of testing machine. In this case there is no mathematical relation between the crack driving force and the change in elastic strain energy. The load-displacement response during quasi-static growth of a crack of initial length a_1 to a final length a_5 is presented by the curve $A_1A_2A_3A_4A_5$ in Figure 4.3. The equation which expresses the crack driving force in terms of segmental areas of the load-displacement curve is Equation (4.15) or

(4.16). This still holds for the case of a general relation between load and displacement during crack growth. This equation can be used for the experimental determination of the resistance of the material to crack growth. During stable crack growth, the load, the displacement and the crack length are recorded simultaneously. This allows us to construct the $P-u$ curve and draw the radial lines OA_i which correspond to different crack lengths. A check of the overall elastic behavior of the specimen is made by removing the applied load and verifying that the displacement follows unloading lines A_iO which should revert to the origin. When the specimen is reloaded the reloading lines should coincide with the unloading lines. G is determined as (Figure 4.3)

$$G = \frac{(OA_iA_j)}{(a_j - a_i)} \tag{4.17}$$

with $i, j = 1, 2, 3, 4, 5$.

(d) $G - a$ representation

In the previous graphical interpretation, load-displacement coordinates were used, while the crack length and the material resistance to crack growth appeared as parameters. It is sometimes advantageous to use crack driving force/crack growth resistance-crack length coordinates, with the load appearing as a parameter. This is shown in Figure 4.4 for a crack of length $2a$ in an infinite plate subjected to a uniform stress σ perpendicular to the crack axis. The crack driving force G obtained from Equation (4.12) is

$$G = \frac{\kappa + 1}{8\mu} \pi a \sigma^2 . \tag{4.18}$$

The $G - a$ relation is represented in Figure 4.4 by straight lines for the three different values of the applied stress σ. The intersection of these lines with the constant line $G = 2\gamma$ gives the critical crack length for crack growth. Or, inversely, for a given crack length a_3 the applied stress should be increased to σ_3 for crack growth. For a larger crack length a_2 a lower stress σ_2 is required for crack growth.

4.5. Equivalence between strain energy release rate and stress intensity factor

The connection between the strain energy release rate, which is a global quantity, and the stress intensity factor, which expresses the strength of the local elastic stress field in the neighborhood of the crack tip, is very important. Consider the case of an opening-mode where the crack extends along its own direction in a self-similar manner. Due to symmetry, only normal stresses will be present in elements along the crack direction (Figure 4.5). Assume the crack extends by a length δ; the energy released during crack extension is the work performed by the stresses $\sigma_y(\delta - \beta^*, 0)$ acting through the displacements $u_y(\beta, \pi)$. For $\delta \to 0$ the conditions $u_y \to u_y^*$ and $\beta \to \beta^*$ are satisfied, and the work performed is

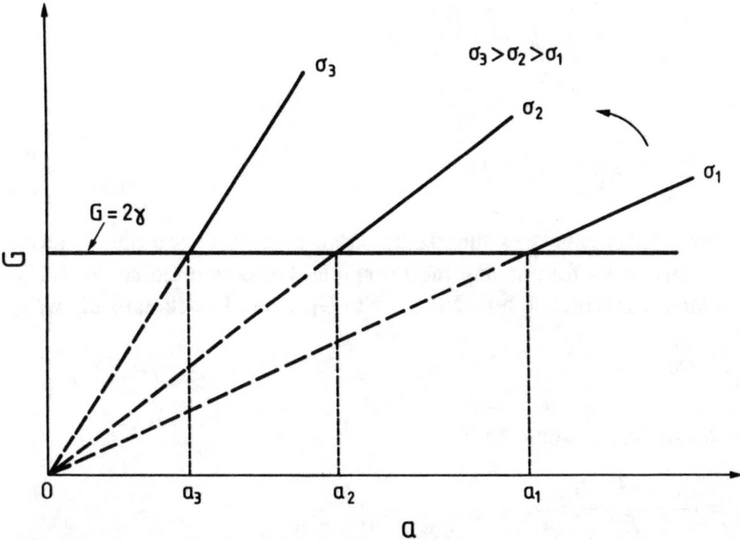

Fig. 4.4. Crack driving force G versus crack length a curves for a crack of length $2a$ in an infinite plate subjected to a uniform stress σ perpendicular to the crack axis.

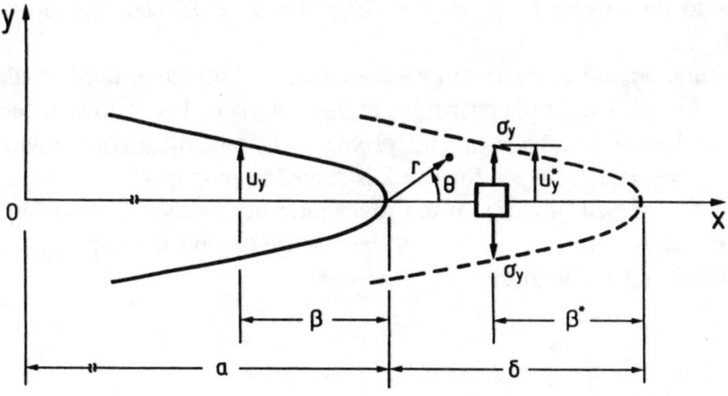

Fig. 4.5. Self-similar crack growth.

$$G_{\mathrm{I}} = 2 \lim_{\delta \to 0} \frac{1}{\delta} \int_0^{\delta} \frac{1}{2} \sigma_y(\delta - \beta, 0)\, u_y(\beta, \pi)\, \mathrm{d}\beta \tag{4.19}$$

where the subscript I was inserted to denote mode-I loading.

Introducing the expressions of σ_y and u_y from Equations (2.28) and (2.30), we obtain for G_{I}

$$G_{\mathrm{I}} = \frac{\kappa + 1}{4\mu} \frac{K_{\mathrm{I}}^2}{\pi} \frac{1}{\delta} \int_0^{\delta} \sqrt{\frac{\beta}{\delta - \beta}} \, \mathrm{d}\beta \tag{4.20}$$

or

$$G_{\mathrm{I}} = \frac{\kappa + 1}{8\mu} K_{\mathrm{I}}^2 . \tag{4.21}$$

Equation (4.21) expresses the elastic strain energy release rate in terms of the stress intensity factor for opening-mode loading. Observe in Equation (4.19) that the nonsingular stress terms do not contribute to G_{I}. From Equation (4.21) we get

$$G_{\mathrm{I}} = \frac{K_{\mathrm{I}}^2}{E} \tag{4.22}$$

for generalized plane stress, and

$$G_{\mathrm{I}} = \frac{(1 - \nu^2) \, K_{\mathrm{I}}^2}{E} \tag{4.23}$$

for plane strain.

Equation (4.21) allows us to find G_{I} when the stress intensity factor K_{I} is known. Thus, for a crack in an infinite plate subjected to a uniform uniaxial stress σ perpendicular to the crack, $K_{\mathrm{I}} = \sigma \sqrt{\pi a}$ (Equation (2.26)), and Equation (4.12) is recovered.

The calculation of the strain energy release rate G_{II} for sliding-mode loading is not easy, since the crack does not propagate in its own plane, but follows a curved path which is not known in advance. This prohibits the analytical computation of G_{II}. Only for the special case when the crack is forced to propagate along its own plane can G_{II} be determined in terms of the stress intensity factor. For this hypothetical situation the shear stresses τ_{xy} have to be released along the segment δ of crack growth, and G_{II} takes the form

$$G_{\mathrm{II}} = 2 \lim_{\delta \to 0} \frac{1}{\delta} \int_0^{\delta} \frac{1}{2} \tau_{xy}(\delta - \beta, 0) \, u_x(\beta, \pi) \, \mathrm{d}\beta . \tag{4.24}$$

Introducing Equations (2.43) and (2.44) into Equation (4.24) we find

$$G_{\mathrm{II}} = \frac{\kappa + 1}{8\mu} K_{\mathrm{II}}^2 . \tag{4.25}$$

For out-of-plane shear the direction of crack growth is predetermined and G_{III} is computed as previously by

$$G_{\mathrm{III}} = 2 \lim_{\delta \to 0} \frac{1}{\delta} \int_0^{\delta} \frac{1}{2} \tau_{yz}(\delta - \beta, 0) \, w(\beta, \pi) \, \mathrm{d}\beta \tag{4.26}$$

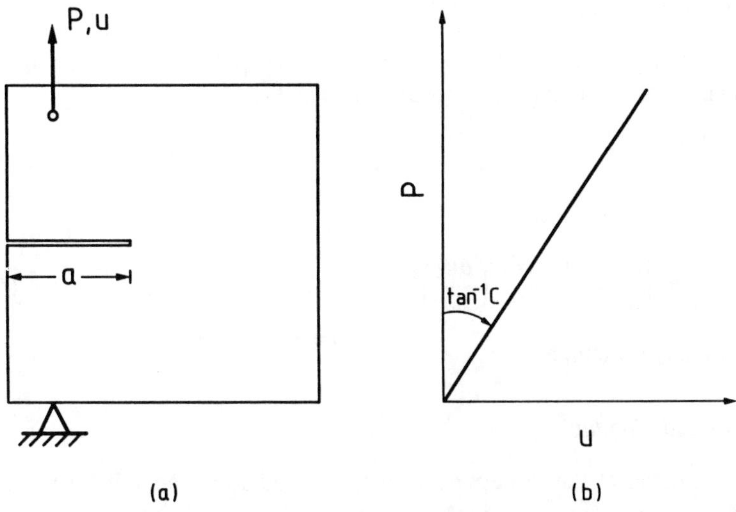

Fig. 4.6. Load-displacement response of a cracked plate.

which, together with Equation (2.56), gives

$$G_{\text{III}} = \frac{K_{\text{III}}^2}{2\mu} \, . \tag{4.27}$$

Equations (4.21), (4.25) and (4.27) establish the equivalence of the strain energy release rate and the stress intensity factor approach in fracture mechanics and form the basis for the critical stress intensity factor fracture criterion.

4.6. Compliance

Let us now consider the load-displacement response of a cracked plate of thickness B subjected to a concentrated force P (Figure 4.6 (a)). As long as there is no crack growth, and the elastic behavior is linear, the load-displacement relation is

$$u = CP \tag{4.28}$$

where C is the compliance (reciprocal of stiffness) of the plate. In Figure 4.6(b) the compliance is represented by the tangent of the angle between the load-displacement curve and the P-axis. We will seek analytical expressions for the strain energy release rate on the left-hand side of Equation (4.7) in terms of the compliance. The case of "fixed-grips" and "dead-load" conditions during crack growth are considered separately.

(a) "Fixed-grips" loading

For a constant displacement u during crack growth the applied load does not perform work, and the elastic strain energy stored in the plate is

$$U^e = \frac{Pu}{2}\,. \tag{4.29}$$

Equation (4.7) takes the form

$$G = -\frac{1}{2}\,u\,\frac{\mathrm{d}P}{\mathrm{d}A} = \frac{1}{2B}\,\frac{u^2}{C^2}\left(\frac{\mathrm{d}C}{\mathrm{d}a}\right)_u \tag{4.30}$$

where a is the crack length.

(b) "Dead-load" loading

For a constant load P, the work performed by the load during an infinitesimal crack growth is

$$\mathrm{d}W = P\,\mathrm{d}u \tag{4.31}$$

while the change in elastic strain energy is

$$\mathrm{d}U^e = \mathrm{d}\left(\frac{Pu}{2}\right) = \frac{P\,\mathrm{d}u + u\,\mathrm{d}P}{2}\,. \tag{4.32}$$

Equation (4.7) becomes

$$G = \frac{1}{2}\left(P\,\frac{\mathrm{d}u}{\mathrm{d}A} - u\,\frac{\mathrm{d}P}{\mathrm{d}A}\right) = \frac{1}{2B}\,P^2\left(\frac{\mathrm{d}C}{\mathrm{d}a}\right)_P\,. \tag{4.33}$$

Equations (4.30) and (4.33) express the strain energy release rate in terms of the derivative of the compliance of the cracked plate with respect to the crack length for "fixed-grips" or "dead-load" loading. By combining Equations (4.30) and (4.33) with Equations (4.22) and (4.23) we obtain an opening-mode stress intensity factor

$$K_{\mathrm{I}}^2 = \frac{EP^2}{2B}\left(\frac{\mathrm{d}C}{\mathrm{d}a}\right)_P = \frac{Eu^2}{2BC^2}\left(\frac{\mathrm{d}C}{\mathrm{d}a}\right)_u \tag{4.34}$$

for generalized plane stress, and

$$K_{\mathrm{I}}^2 = \frac{EP^2}{2(1-\nu^2)\,B}\left(\frac{\mathrm{d}C}{\mathrm{d}a}\right)_P = \frac{Eu^2}{2(1-\nu^2)\,BC^2}\left(\frac{\mathrm{d}C}{\mathrm{d}a}\right)_u \tag{4.35}$$

for plane strain.

Equations (4.34) and (4.35) can be used for the analytical or experimental determination of the K_{I} stress intensity factor.

For the experimental determination of the stress intensity factor from Equation (4.34) or (4.35), a series of specimens with different crack lengths are used

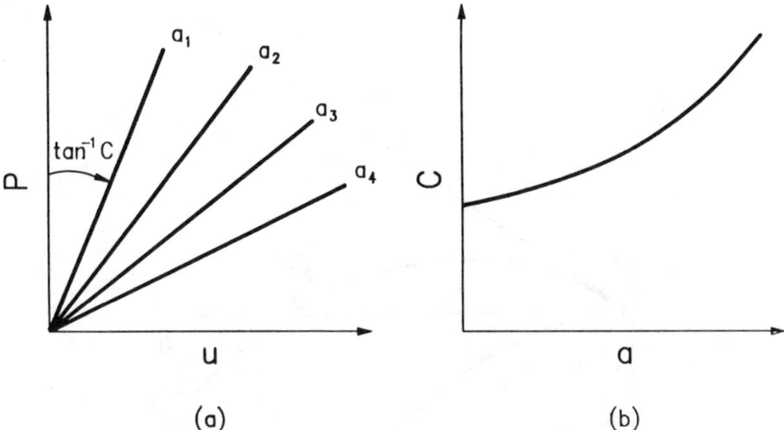

Fig. 4.7. (a) Load-displacement response for different crack lengths and (b) compliance versus crack length.

to calculate the derivative of the compliance with respect to the crack length. This is shown in Figure 4.7.

The accuracy of the experimental determination of K_{I} depends on the changes in the displacement between loading points remote from the crack as crack extends. This experimental technique provides a quick way of determining K_{I} when the crack geometry is complicated and the mathematical solution is difficult.

4.7. Crack stability

In the Griffith energy balance approach to crack growth the critical load is determined from Equation (4.7); this arose from the conservation of energy in the entire body. Crack growth is considered unstable when the system energy at equilibrium is maximum and stable when it is minimum. A sufficient condition for crack stability is

$$\frac{\partial^2(\Pi+\Gamma)}{\partial A^2}\begin{cases} <0: & \text{unstable fracture} \\ >0: & \text{stable fracture} \\ =0: & \text{neutral equilibrium} \end{cases} \tag{4.36}$$

where the potential energy Π of the system is defined by Equation (4.6).

Two example problems will now be considered with respect to crack stability. The first concerns a line crack in an infinite plate subjected to a uniform stress perpendicular to the crack axis. The potential energy of the system $\Pi = -U^e$, where U^e is given by Equation (4.12) and $\Gamma = 4\gamma a$. The terms Π, Γ and $(\Pi+\Gamma)$ are plotted in Figure 4.8 against half crack length a. Observe that the total potential energy of the system $(\Pi+\Gamma)$ at the critical crack length presents a maximum, which

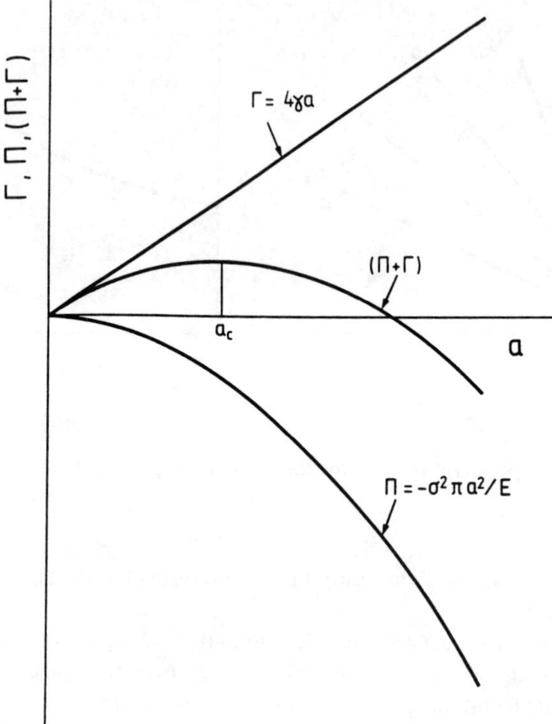

Fig. 4.8. Potential energy, Π, surface energy, Γ, and the sum of potential and surface energy, (Π + Γ), versus crack length a for a line crack in an infinite medium subjected to a uniform stress perpendicular to the crack axis.

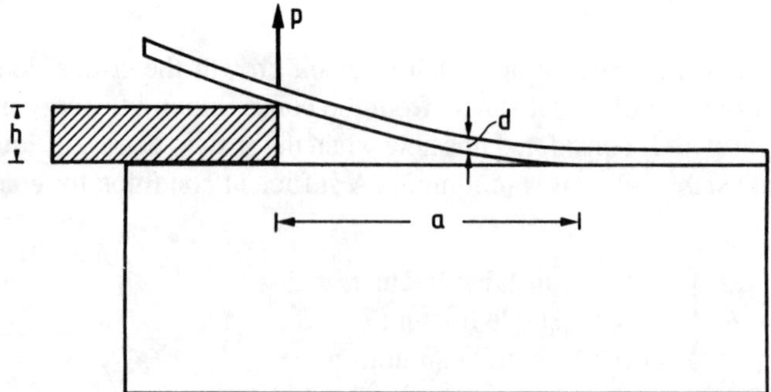

Fig. 4.9. Wedge inserted to peel off mica, according to Obreimoff's experiment.

corresponds to unstable equilibrium. This result is also verified by Equation (4.36).

The second problem concerns the experiment carried out by Obreimoff [4.1] on the cleavage of mica (Figure 4.9). A wedge of thickness h is inserted underneath a flake of mica which is detached from a mica block along a length a. The energy of the

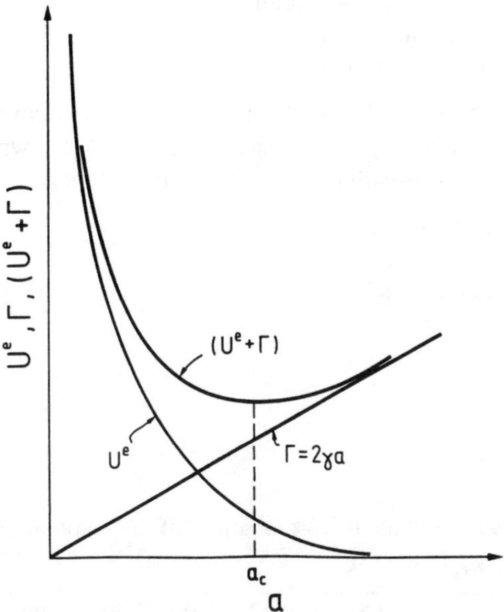

Fig. 4.10. Potential energy, Π $(= U^e)$, surface energy, Γ, and the sum of potential and surface energy, $(U^e + \Gamma)$, versus crack length a for Obreimoff's experiment.

system is calculated by considering the mica flake as a cantilever beam with height d built-in at distance a from the point of application of the wedge. During propagation of the crack the force P does not perform work. According to the elementary theory of beam bending, the elastic energy stored in the cantilever beam is

$$U^e = \frac{Ed^3h^2}{8a^3}\,. \tag{4.37}$$

The surface energy Γ is given by

$$\Gamma = 2\gamma a \tag{4.38}$$

and from Equation (4.7) the equilibrium crack length a_c is obtained as

$$a_c = \left(\frac{3Ed^3h^2}{16\gamma}\right)^{1/4}\,. \tag{4.39}$$

The quantities $\Pi = U^e$, Γ and $(U^e + \Gamma)$ are plotted in Figure 4.10 versus the crack length a. The total potential energy of the system $(U^e + \Gamma)$ at critical crack length a_c is a minimum, which corresponds to stable equilibrium. This result is also verified by Equation (4.36).

The stability condition (4.36) can be expressed in terms of the crack driving force G, given by Equation (4.10)

$$\frac{\partial(G-R)}{\partial A}\begin{cases} >0: & \text{unstable fracture}\\ <0: & \text{stable fracture}\\ =0: & \text{neutral equilibrium}\end{cases} \tag{4.40}$$

where $R = \mathrm{d}\Gamma/\mathrm{d}A$. For an ideally brittle material $R = 2\gamma = \text{const}$ and the R-term disappears from Equation (4.40). Referring to Equation (4.21), we may express the stability condition of crack growth in terms of the stress intensity factor as follows:

$$\frac{\partial K}{\partial A}\begin{cases} >0: & \text{unstable fracture}\\ <0: & \text{stable fracture}\\ =0: & \text{neutral equilibrium.}\end{cases} \tag{4.41}$$

Examples

Example 4.1.

The following data were obtained from a series of tests conducted on precracked specimens of thickness 1 mm.

Crack length a(mm)	Critical load P(kN)	Critical displacement u(mm)
30.0	4.00	0.40
40.0	3.50	0.50
50.5	3.12	0.63
61.6	2.80	0.78
71.7	2.62	0.94
79.0	2.56	1.09

where P and u are the critical load and displacement at crack growth. The load-displacement record for all crack lengths is linearly elastic up to the critical point.

Determine the critical value of the strain energy release rate $G_c = R$ from: (a) the load-displacement records, and (b) the compliance-crack length curve.

Solution: The load-displacement ($P - u$) records up to the point of crack growth for the crack lengths of the problem are shown in Figure 4.11. Note that the $P - u$ curves are linear and revert to the origin when the load is removed. The critical value of the strain energy release rate $G_c = R$ is calculated from Equation (4.17) for the various segmental areas. When the points (P_i, u_i) and (P_j, u_j) are joined by straight lines we have (Figure 4.12)

$$(OA_iA_j) = (OA_iA_i') + (A_iA_i'A_j'A_j) - (OA_jA_j')$$

or

$$(OA_iA_j) = \frac{1}{2}\,P_iu_i + \frac{1}{2}\,(P_i + P_j)\,(u_j - u_i) - \frac{1}{2}\,P_ju_j$$

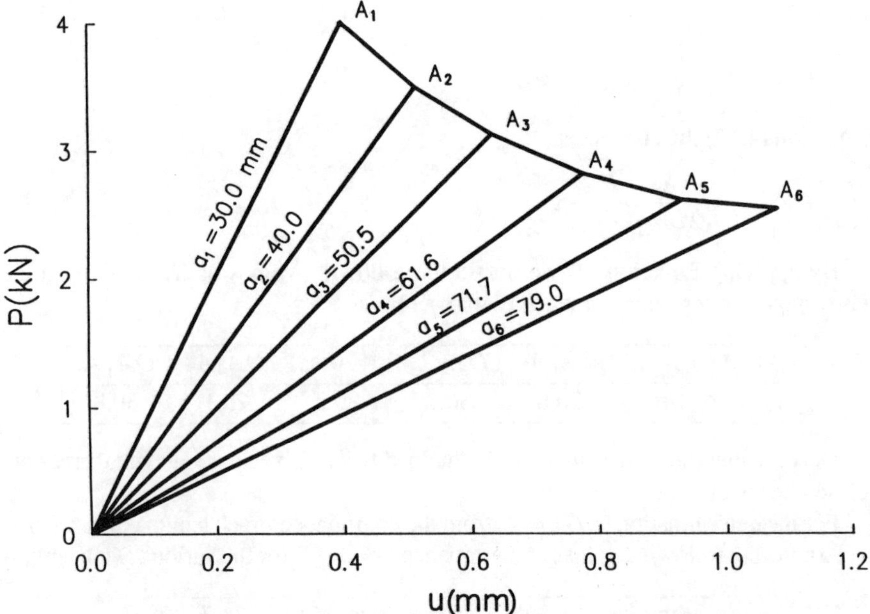

Fig. 4.11. Load-displacement $(P - u)$ records up to the point of crack growth for different crack lengths.

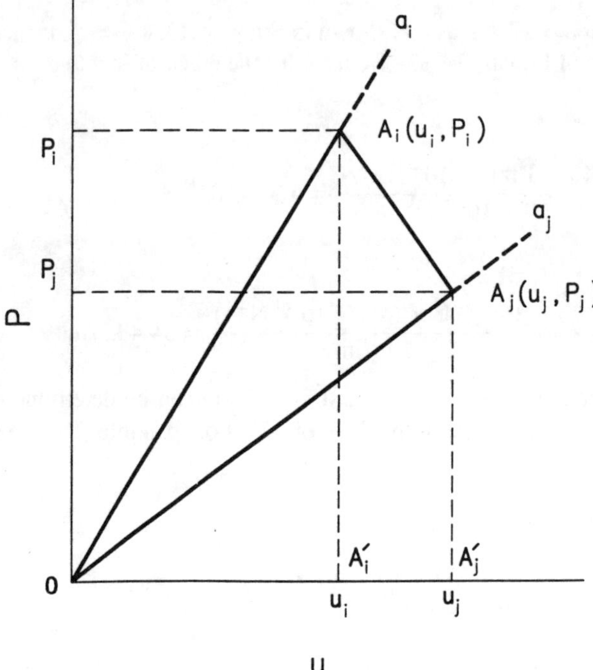

Fig. 4.12. Calculation of the critical strain energy release rate $G_c = R$ from the segmental area OA_iA_j.

or

$$(OA_iA_j) = \frac{1}{2}(P_iu_j - P_ju_i)\,. \qquad (1)$$

Equation (4.17) thus becomes

$$G_c = R = \frac{P_iu_j - P_ju_i}{2B(a_j - a_i)}\,. \qquad (2)$$

By applying Equation (2) we obtain the following values of $G_c = R$ from the corresponding segmental areas of Figure 4.11

Area	OA_1A_2	OA_2A_3	OA_3A_4	OA_4A_5	OA_5A_6
$G_c = R$(kJ/m^2)	30.0	30.7	30.2	29.1	30.8

Observe that the values of $G_c = R$ obtained from the various segmental areas are close to each other.

For the determination of $G_c = R$ from the compliance-crack length curve we first determine the following values of compliance $C = u/P$ for the various crack lengths

a(mm)	30.0	40.0	50.5	61.6	71.7	79.0
$C(\times 10^{-7}$ m/N)	1.00	1.43	2.02	2.79	3.59	4.26

The variation of C versus a is shown in Figure 4.13. $G_c = R$ is then determined by application of Equation 4.33. We have for the crack of length $a_1 = 30$ mm

$$P = 4\text{ kN}$$

$$\frac{\mathrm{d}C}{\mathrm{d}a} = \frac{(1.43 - 1.00) \times 10^{-7}\text{ m/N}}{10 \times 10^{-3}\text{ m}} = 4.3 \times 10^{-6}\text{ N}^{-1}$$

and

$$G_c = R = \frac{(4 \times 10^3)^2\text{ N}^2 \times (4.3 \times 10^{-6}\text{ N}^{-1})}{2 \times 10^{-3}\text{ m}} = 34.4\text{ kJ/m}^2\,.$$

For the crack lengths a_2, a_3, a_4 and a_5, $\mathrm{d}C/\mathrm{d}a$ can be determined as the mean value of the left and right derivatives of C. For example, for the crack length $a = 50.5$ mm, we have

$$P = 3.12\text{ kN}$$

$$\left(\frac{\mathrm{d}C}{\mathrm{d}a}\right)_l = \frac{(2.02 - 1.43) \times 10^{-7}\text{ m/N}}{10.5 \times 10^{-3}\text{ m}} = 5.6 \times 10^{-6}\text{ N}^{-1}$$

$$\left(\frac{\mathrm{d}C}{\mathrm{d}a}\right)_r = \frac{(2.77 - 2.02) \times 10^{-7}\text{ m/N}}{11.1 \times 10^{-3}\text{ m}} = 6.8 \times 10^{-6}\text{ N}^{-1}$$

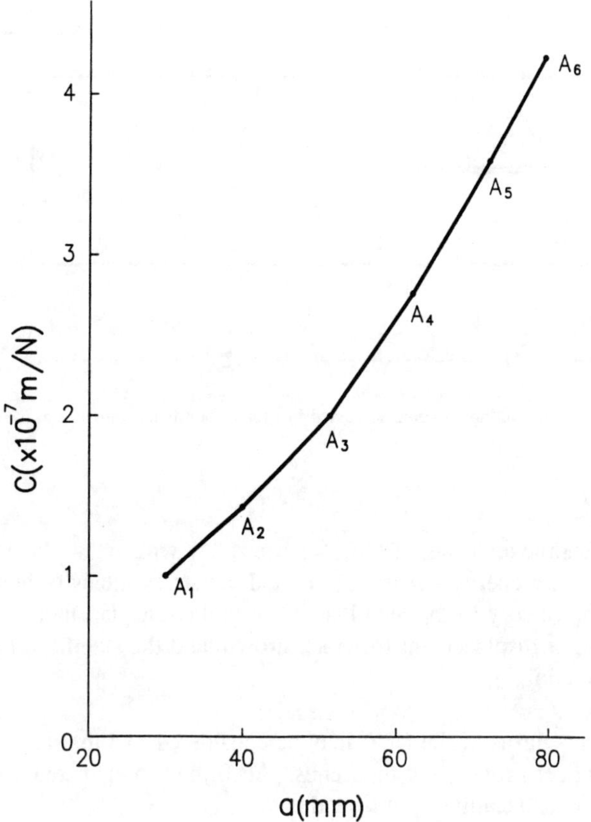

Fig. 4.13. Variation of the compliance C versus crack length a.

$$\left(\frac{\mathrm{d}C}{\mathrm{d}a}\right) = \frac{(5.6 + 6.8) \times 10^{-6}}{2} = 6.2 \times 10^{-6}\ \mathrm{N}^{-1}$$

and

$$G_c = R = \frac{(3.12 \times 10^3)^2\ \mathrm{N}^2 \times (6.2 \times 10^{-6})\ \mathrm{N}^{-1}}{2 \times 10^{-3}} = 30.2\,\mathrm{kJ/m}^2\ .$$

Using this procedure we obtain the following values of $G_c = R$ at various crack lengths

a(mm)	30.0	40.0	50.5	61.6	71.7	79.0
$G_c = R$(kJ/m^2)	34.4	30.4	31.2	29.2	29.7	30.0

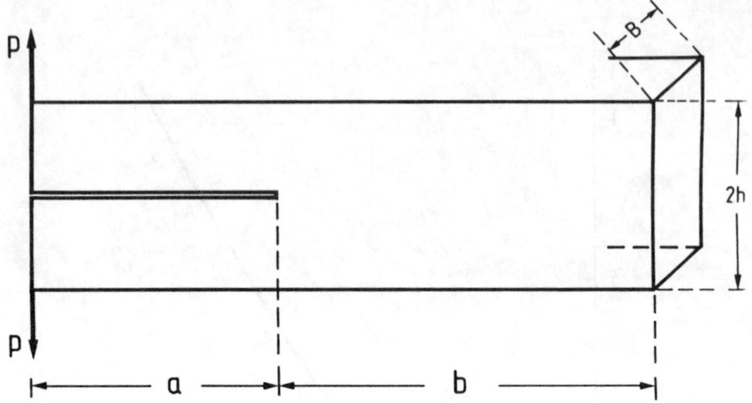

Fig. 4.14. A double cantilever beam subjected to a load P or displacement u at the end point.

Example 4.2.

For a double cantilever beam (DCB) (Figure 4.14) with $a \gg 2h$ and $b \gg 2h$, determine the strain energy release rate G and the stress intensity factor K_{I} using elementary beam theory for applied load P or applied displacement u. Determine the critical load or displacement for crack growth and the stability of cracking for both cases of loading.

Solution: The two arms of the DCB may be considered to a first approximation as cantilevers with zero rotation at their ends. According to elementary beam theory the deflection of each cantilever at its end is

$$u' = \eta \frac{Pa^3}{3EI}, \quad I = \frac{Bh^3}{12} \tag{1}$$

where $\eta = 1$ or $\eta = 1 - \nu^2$ for generalized plane stress or plane strain, respectively. The relative displacement u of the points of application of the loads P is

$$u = 2u' = \frac{2\eta Pa^3}{3EI}. \tag{2}$$

The compliance of the DCB is

$$C = \frac{u}{P} = \frac{2\eta a^3}{3EI}. \tag{3}$$

For applied displacement u, G is determined from Equation (4.30) as

$$G = \frac{3u^2Eh^3}{16\eta a^4} \tag{4}$$

and for applied load P, G is determined from Equation (4.33) as

$$G = \frac{12\eta P^2 a^2}{EB^2 h^3} \,. \tag{5}$$

$K_{\rm I}$ is determined from Equation (4.22) or (4.23) for generalized plane stress or plane strain, respectively. For applied displacement u we obtain

$$K_{\rm I} = \frac{\sqrt{3h^3}\, Eu}{4\eta a^2} \tag{6}$$

and for constant load P we obtain

$$K_{\rm I} = \sqrt{\frac{12}{h^3}}\, \frac{Pa}{B} \,. \tag{7}$$

The critical displacement or load for crack growth is determined from Equation (4.9). The critical applied displacement u_c is determined as

$$u_c = a^2 \sqrt{\frac{32\eta\gamma}{3Eh^3}} \tag{8}$$

and the critical applied load P_c as

$$P_c = \frac{B}{a} \sqrt{\frac{\gamma E h^3}{6}} \,. \tag{9}$$

From Equations (4) and (5) or (6) and (7) we obtain that for applied displacement or applied load we have $\partial G/\partial A < 0$, $\partial K_{\rm I}/\partial A < 0$ or $\partial G/\partial A > 0$, $\partial K_{\rm I}/\partial A > 0$, respectively. According to Equation (4.40) and (4.41) this implies that crack growth is stable for controlled displacement and unstable for controlled load.

Example 4.3.

Design a contoured double cantilever beam (CDCB) so that $K_{\rm I}$ is independent of crack length (the crack grows under neutral equilibrium).

Solution: Putting

$$h = h_0\, a^n \tag{1}$$

where h is the height of the CDCB at position $x = a$ (Figure 4.15), the compliance of the CDCB for conditions of generalized plane stress according to elementary beam theory is

$$C = \frac{8a^{3(1-n)}}{(1-n)\, EBh_0^3} \,. \tag{2}$$

From Equation (4.34) it is deduced that $K_{\rm I}$ is constant (independent of crack length) if $\mathrm{d}C/\mathrm{d}a$ is constant. We have

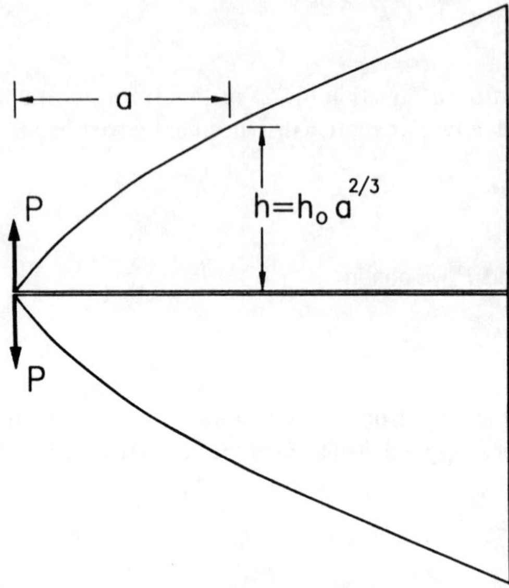

Fig. 4.15. A contoured double cantilever beam designed in such a way that K_{I} is independent of crack length.

$$\frac{\mathrm{d}C}{\mathrm{d}a} = \frac{24a^{2-3n}}{EBh_0^3} \tag{4}$$

implying that $\mathrm{d}C/\mathrm{d}a$ is constant for $n = 2/3$.

Thus, K_{I} for a CDCB is independent of crack length when

$$h = h_0 a^{2/3} \,. \tag{5}$$

A CDCB designed according to Equation (5) is shown in Figure 4.15.

Example 4.4.

Consider a long strip of height $2h$ and thickness B with a crack of length $2a$ subjected to a uniform stress σ along its upper and lower faces (Figure 4.16a). For $a \gg h$ use elementary analysis to determine the stress intensity factor at the crack tip.

Solution: Since $a \gg h$, the parts of the strip above and below the crack may be considered as two beams AB of length $2a$ built-in at their ends (Figure 4.16b). According to elementary beam theory the bending moment at the built-in end of the beam is

$$M_A = -\frac{qBa^2}{3} \tag{1}$$

and the bending moment M for half length of the beam is

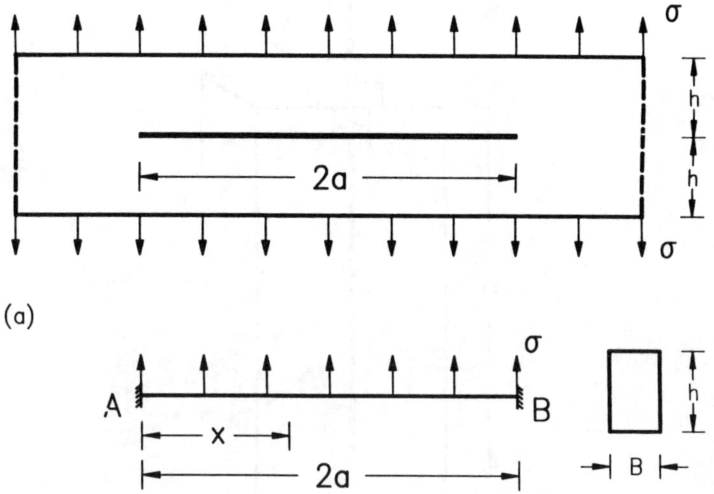

Fig. 4.16. (a) A long strip with a crack and (b) its idealization as two beams AB built-in at their ends.

$$M = -\frac{qB}{6}(2a^2 - 6ax + 3x^2)\,, \quad 0 \le x \le a\,. \tag{2}$$

The elastic strain energy U contained in the two beams is

$$U = 2\int_0^a \frac{M^2\,\mathrm{d}x}{2EI} = 4\int_0^{a/2} \frac{M^2}{2EI}\,\mathrm{d}x \tag{3}$$

with

$$I = \frac{Bh^3}{12}\,. \tag{4}$$

Substituting the value of M from Equation (2) into Equation (3) and performing the integration we obtain

$$U = \frac{24q^2a^5B}{45Eh^3}\,. \tag{5}$$

G_{I} is computed from Equation (4.9) as

$$G_{\mathrm{I}} = \frac{\partial U}{\partial A} = \frac{1}{2B}\frac{\partial U}{\partial a} \tag{6}$$

or

$$G_{\mathrm{I}} = \frac{4q^2a^4}{3Eh^3}\,. \tag{7}$$

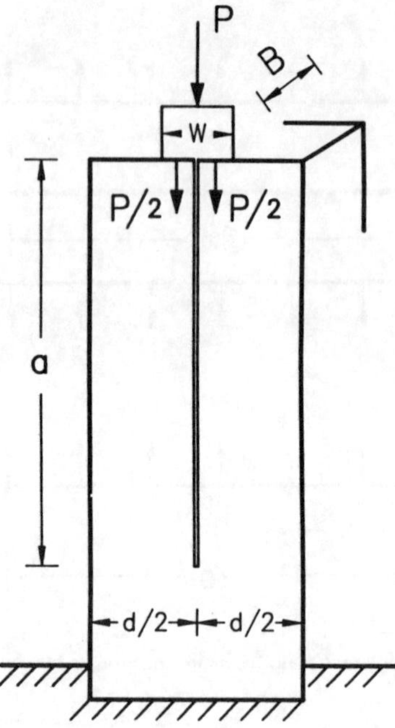

Fig. 4.17. A long block with a crack of length a split by a force P.

For a thin plate (conditions of generalized plane stress) K_{I} is computed from Equation 4.22 as

$$K_{\mathrm{I}} = (EG_{\mathrm{I}})^{1/2} = \frac{2qa^2}{\sqrt{3}\,h^{3/2}} \,. \tag{8}$$

Example 4.5.

Determine the force P required for growth of the crack of Figure 4.17.

Solution: Each part of the body to the left and to the right of the crack may be considered as a cantilever beam of length a and cross-sectional area $B\mathrm{d}/2$ built-in at its end. The bending moment M applied to each cantilever beam is

$$M = \frac{P(d-w)}{8} \,.$$

In calculating the elastic strain energy contained in each beam we ignore the energy due to axial forces and take only the energy due to bending moments. For the total strain energy contained in both beams we have

$$U = 2\int_0^a \frac{M^2}{2EI}\,\mathrm{d}x \tag{1}$$

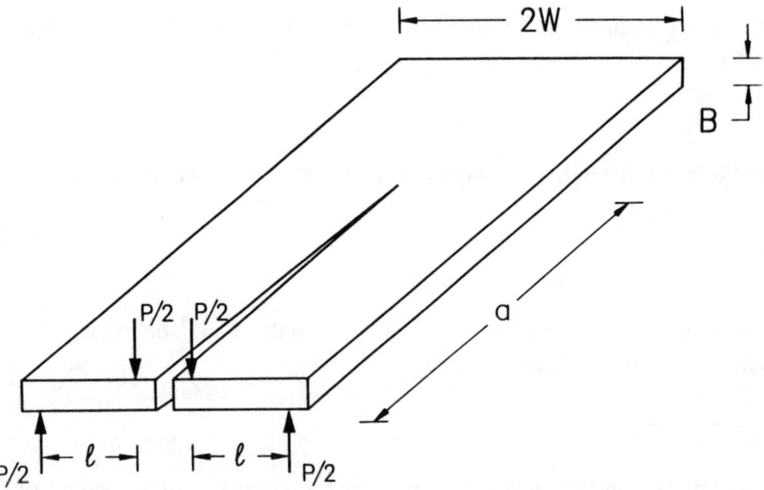

Fig. 4.18. A double torsion specimen.

or

$$U = \frac{P^2(d-w)^2\,a}{64EI}, \quad \mathrm{I} = \frac{Bd^3}{96}. \tag{2}$$

The strain energy release rate G is computed from Equation (4.9) as

$$G = \frac{\partial U}{\partial A} = \frac{3P^2(d-w)^2}{2EB^2\,\mathrm{d}^3}. \tag{3}$$

Note that G is independent of crack length, that is, crack grows under neutral equilibrium.

The crack growth condition is

$$G = R \tag{4}$$

where R is a material parameter.

From Equations (3) and (4) we obtain the critical load P_c required for crack growth to be

$$P_c = \frac{B}{1-w/d}\left(\frac{2ERd}{3}\right)^{1/2}. \tag{5}$$

Example 4.6.

Determine the stress intensity factor for the double torsion specimen shown in Figure 4.18 using strength of materials analysis.

Solution: Each arm of the specimen may be considered as a bar built-in at its end of length a, and rectangular cross-section of dimensions W and B. It is subjected to a torsion moment M_t of magnitude

$$M_t = \frac{Pl}{2} . \tag{1}$$

The elastic strain energy U contained in both arms of the specimen is

$$U = 2 \int_0^a \frac{M_t \theta_t}{2} \, \mathrm{d}x \tag{2}$$

where θ_t is the angle of rotation of each arm due to the torsion moment.

From strength of materials we have

$$\theta_t = \frac{M_t}{k\mu B^3 W} \tag{3}$$

where k depends on the ratio W/B and μ is the shear modulus. Values of k for different ratios W/B are given below.

Introducing Equation (3) into Equation (2) we obtain

$$U = \frac{P^2 l^2 a}{4k\mu B^3 W} . \tag{4}$$

It is evident that the crack is under tearing mode of deformation. G_{III} is computed from Equation (4.9) as

$$G_{\mathrm{III}} = \frac{\partial U}{\partial A} = \frac{P^2 l^2}{4k\mu B^4 W} , \quad A = aB . \tag{5}$$

K_{III} is computed from Equation (4.27) as

$$K_{\mathrm{III}} = (2\mu G_{\mathrm{III}})^{1/2} = \frac{Pl}{(2kW)^{1/2} B^2} . \tag{6}$$

For $W \gg B$ $k = 1/3$ and Equation (6) becomes

$$K_{\mathrm{III}} = \left(\frac{3}{2W}\right)^{1/2} \frac{Pl}{B^2} . \tag{7}$$

Note that K_{III} is independent of crack length, indicating that crack grows under neutral equilibrium.

W/B	∞	10	5	2.5	1
k	0.330	0.312	0.291	0.249	0.141

Example 4.7.

For nonlinear elastic behavior the load-displacement response of a cracked plate subjected to a concentrated load P or to a load-point displacement u is

$$C_n = \frac{u^n}{P} \,. \tag{1}$$

Determine the strain energy release rate G_{I} for "fixed-grips" and "dead-load" loadings.

Solution: The strain energy release rate G_{I} is equal to the work performed by the applied loads minus the elastic strain energy stored in the system during crack growth. It is computed as

$$G_{\mathrm{I}} = P\,\frac{\mathrm{d}u}{\mathrm{d}A} - \frac{\mathrm{d}U}{\mathrm{d}A} \,. \tag{2}$$

U is given by

$$U = \int_0^u P\,\mathrm{d}u = \frac{Pu}{1+n} \,. \tag{3}$$

From Equations (1) to (3) we obtain

$$G_{\mathrm{I}} = \frac{1}{1+n}\left(nP\,\frac{\mathrm{d}u}{\mathrm{d}A} - u\,\frac{\mathrm{d}P}{\mathrm{d}A}\right) . \tag{4}$$

For "fixed-grips" loading (constant displacement) Equation (4) becomes

$$G_{\mathrm{I}} = \frac{1}{1+n}\,\frac{u^{1+n}}{C_n^2}\,\frac{\mathrm{d}C_n}{\mathrm{d}A} \tag{5}$$

while for "dead-load" loading (constant load) Equation (4) becomes

$$G_{\mathrm{I}} = \frac{1}{1+n}\,P^{(1+n)/n}\,C_n^{(1+n)/n}\,\frac{\mathrm{d}C_n}{\mathrm{d}A} \,. \tag{6}$$

Equations (5) and (6) for linear behavior ($n = 1$) coincide with Equations (4.30) and (4.33).

Example 4.8.

Determine the compliance of a thin plate of thickness B, height $2h$, width $2b$ with a center crack of length $2a$ subjected to uniform tension σ normal to the crack (Case 1 of Appendix 2.1). Take the approximate expression for the stress intensity factor K_{I} as (Problem 2.21)

$$K_{\mathrm{I}} = \sigma(\pi a)^{1/2}\left(\frac{2b}{\pi a}\,\tan\frac{\pi a}{2b}\right)^{1/2} . \tag{1}$$

Solution: For generalized plane stress (thin plate) we obtain from Equation (4.34)

$$\frac{\mathrm{d}C}{\mathrm{d}A} = \frac{2K_{\mathrm{I}}^2}{EP^2} \tag{2}$$

with

$$A = 2aB\,, \quad P = 2b\,B\sigma\,. \tag{3}$$

Equation (2) becomes

$$\frac{\partial C}{\partial a} = \frac{2}{EBb}\tan\left(\frac{\pi a}{2b}\right) \tag{4}$$

and integrating with respect to a we have

$$C - C_0 = \frac{2}{EBb}\int_0^a \tan\left(\frac{\pi a}{2b}\right)\mathrm{d}a \tag{5}$$

where C_0 is the compliance of the plate without a crack. C_0 is given by

$$C_0 = \frac{h}{EBb}\,. \tag{6}$$

Equation (5) gives for the compliance of the plate

$$C = -\frac{4}{\pi EB}\log\cos\left(\frac{\pi a}{2b}\right) + \frac{h}{EBb}\,. \tag{7}$$

Problems

4.1. A plate of thickness 50 mm with a crack of length 100 mm is subjected to a progressively increasing load. The crack starts to grow at a load of 1 kN and continues to propagate under constant load, until the crack length becomes 150 mm, when it stops. At the beginning of crack growth the displacement of the load was measured to be 5 mm, and at crack arrest, 10 mm. Calculate the work done during crack growth, and find the critical value of the strain energy release rate for crack growth.

4.2. A plate of thickness 20 mm with a crack of length 50 mm is subjected to displacement controlled loading. The crack starts to grow at a displacement u = 10 mm and continues to propagate under constant displacement until the crack length is 100 mm, when it stops. At the beginning of crack growth the load was measured to be 2 kN and at crack arrest 1.5 kN. Calculate the elastic strain energy released during crack growth.

4.3. A series of identical specimens of thickness 10 mm with different crack lengths were used to determine the compliance versus crack length curve. It was found that the slope of this curve at a crack length of 5 mm was 10^{-6} mN^{-1}. The specimen with this crack length was then loaded up to crack growth. The critical load for crack growth was measured to be 2 kN. Determine the critical value of strain energy release rate for crack growth.

4.4. The following data were obtained from a series of tests conducted on precracked specimens of thickness 10 mm.

Crack length a(mm)	Compliance $C \times 10^{-7}$(m/N)	Critical load P(kN)
50.0	1.00	10.00
66.7	1.43	8.75
84.2	2.02	7.80
102.7	2.79	7.00
119.5	3.59	6.55

where P is the critical load at crack growth. The load-displacement record for all crack lengths is linearly elastic up to the critical point.

Determine the critical value of the strain energy release rate for crack growth.

4.5. The following data were obtained from a series of tests conducted on precracked specimens of thickness 20 mm.

Crack length a(mm)	Critical load P(kN)	Critical displacement u(mm)
15	8.0	1.00
20	7.0	1.25
40	5.0	6.28

where P and u are the critical load and displacement at crack growth. The load-displacement record for all three cracks is linearly elastic up to the critical point.

Determine the critical value of the strain energy release rate for crack growth.

4.6. For the double cantilever beam (DCB) shown in Figure 4.19 with thickness B and $a \gg 2h$ and $b \gg 2h$ determine the strain energy release rate and the stress intensity factor at the crack tip using elementary beam theory; comment on the stability of cracking. Consider plane strain and generalized plane stress.

4.7. As in Problem 4.6 for the DCB of Figure 4.20.

4.8. As in Problem 4.6 for the DCB of Figure 4.21 subjected to uniform displacement u along its upper side.

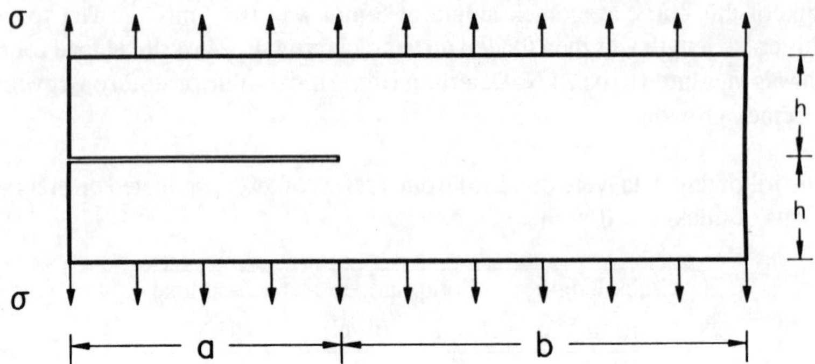

Fig. 4.19. A double cantilever beam subjected to a uniform stress distribution along its upper and lower faces.

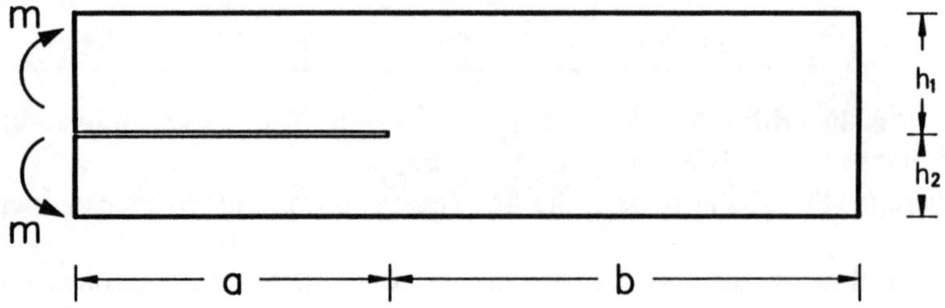

Fig. 4.20. A double cantilever beam with arms of different heights subjected to end moments.

4.9. As in Problem 4.6 for the DCB of Figure 4.22.

4.10. As in Problem 4.6 for the strip of Figure 4.23.

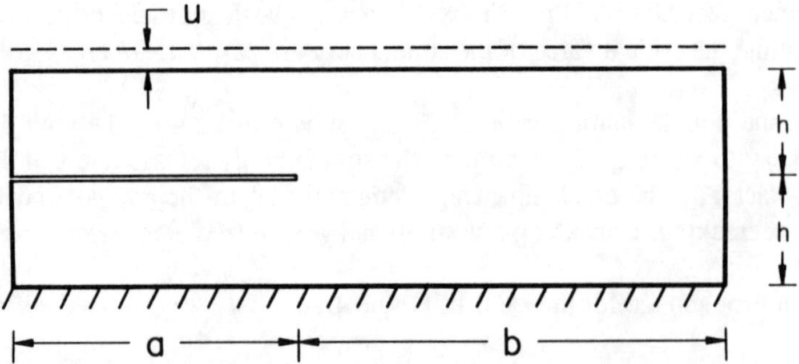

Fig. 4.21. A double cantilever beam subjected to uniform displacement u along its upper side.

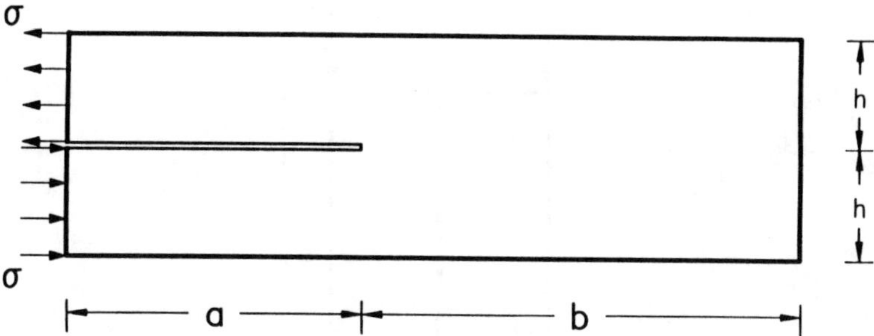

Fig. 4.22. A double cantilever beam subjected to uniform splitting stresses σ along its left side.

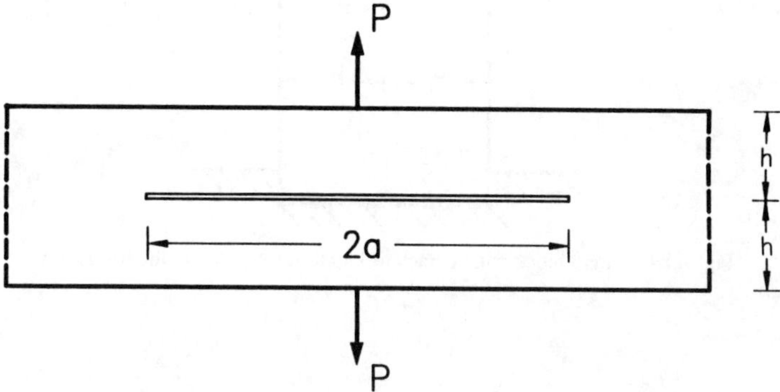

Fig. 4.23. A long strip with a crack subjected to the forces P.

4.11. As in Problem 4.6 for the DCB of Figure 4.24.

4.12. As in Problem 4.6 for the configuration of Figure 4.25.

4.13. Consider the double cantilever beam of Figure 4.14. Show that when the effect of shear force is taken into consideration the stress intensity factor is given by

$$K_{\mathrm{I}} = \frac{\sqrt{12}\,P}{Bh^{1/2}}\left(\frac{a^2}{h^2} + \frac{1}{3}\right)^{1/2}.$$

4.14. In an improved model for studying the double cantilever beam (DCB) (Figure 4.26a) we suppose that the crack ligament is a beam on an elastic foundation (Figure 4.26b). Using such conditions, show that the compliance of the DCB is

$$C = \frac{2}{EB\lambda^3 h^3}\left[2\lambda^3 a^3 + 6\lambda^2 a^2\left(\frac{\sinh \lambda b \cosh \lambda b + \sinh \lambda c \cosh \lambda c}{\sinh^2 \lambda c - \sin^2 \lambda c}\right) + \right.$$

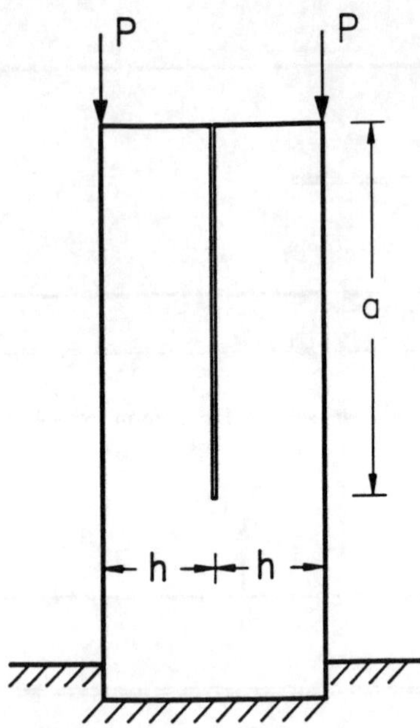

Fig. 4.24. A long block with a crack of length a subjected to the forces P.

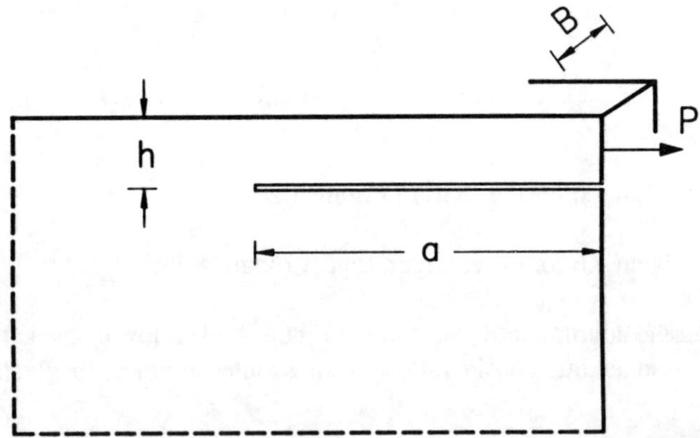

Fig. 4.25. A half-plane with a crack of length a at a small distance h from its boundary subjected to the force P.

$$+6\lambda a \left(\frac{\sinh^2 \lambda b + \sin^2 \lambda b}{\sinh^2 \lambda b - \sin^2 \lambda b} \right) + 3 \left(\frac{\sinh \lambda b \cosh \lambda b - \sin \lambda b \cos \lambda b}{\sinh^2 \lambda b - \sin^2 \lambda b} \right) \Bigg]$$

where

$$\lambda^4 = \frac{3k}{EBh^3}$$

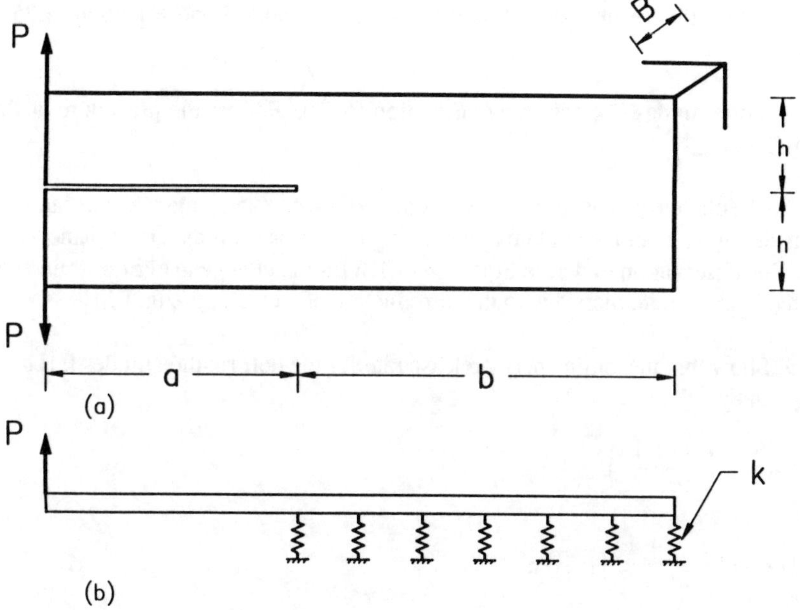

Fig. 4.26. (a) A double cantilever beam and (b) modelling the crack ligament as a beam on elastic foundation.

k is the constant of the spring ($\sigma = ky$, σ is the stress and y is the deflection).

4.15. Consider a double cantilever beam (Figure 4.14) made of a nonlinear elastic material obeying the equation

$$\sigma = E_0 \, \varepsilon^n \, .$$

Show that the strain energy release rate G_{I} for constant applied load P or constant applied displacement u is given by

$$G_{\mathrm{I}} = \frac{4n}{1+n} \, \frac{P^{1/(1+n)}}{B} \left[\frac{2(2+n)\,a}{Bh^2 E_0} \right]^{1/2} \frac{a}{h}$$

or

$$G = \frac{3n}{2(2+n)\,(1+n)} \, u^{1+n} \left[\frac{(1+2n)\,h}{4na^3} \right]^n \frac{hE_0}{a} \, .$$

4.16. Consider a circle of radius r centered at the tip of a mode-I or mode-II crack and show that the strain energy release rate is given by

$$G = \frac{1}{2} \int_{-\pi}^{\pi} \left[\left(\sigma^r \, \frac{\mathrm{d}u_r}{\mathrm{d}a} - u_r \, \frac{\mathrm{d}\sigma_r}{\mathrm{d}a} \right) + \left(\tau_{r\theta} \, \frac{\mathrm{d}u_\theta}{\mathrm{d}a} - u_\theta \, \frac{\mathrm{d}\tau_{r\theta}}{\mathrm{d}a} \right) \right] r \, \mathrm{d}\theta \, .$$

Based on this result derive Equation (4.21) for mode-I and Equation (4.25) for mode-II.

4.17. For a mode-III crack derive Equation (4.27). Follow the procedure of Problem 4.16.

4.18. The change in elastic strain energy stored in a cracked plate may be obtained by considering an uncracked plate and relaxing the stresses on the crack plane to zero. Use this observation to derive Equation (4.12) for the change in elastic strain energy stored in an infinite plate due to the introduction of a crack of length $2a$.

4.19. Show that the strain energy release rate for the deformation modes I, II and III is given by

$$G_{\mathrm{I}} = 2\left(\frac{\kappa+1}{2\kappa-1}\right) r_0 W_{\mathrm{I}}$$

$$G_{\mathrm{II}} = 2\left(\frac{\kappa+1}{2\kappa+3}\right) r_0 W_{\mathrm{II}}$$

$$G_{\mathrm{III}} = r_0 W_{\mathrm{III}} \,.$$

Here W_{I}, W_{II}, W_{III} are the strain energies contained in a small circle of radius r_0 surrounding the crack tip.

4.20. For a certain experiment, Gurney and Ngan (Proc. Roy Soc. Lond., **A325**, 207, 1971) expressed the load-displacement-crack area relation in the form

$$P = \sum_m \frac{C_m u^m}{A^{(m+1)/2}}$$

where C_m $(m = 1, 2, 3, \ldots)$ are constants.

Use this expression to calculate U and then G. Show that

$$G = \frac{Pu}{2A} \,.$$

4.21. For a specific test the load-displacement-crack area relation of the previous problem takes the form

$$P = 350\frac{u}{A} - 1890\frac{u^2}{A^{3/2}} + 5250\frac{u^3}{A^2}$$

where P is in Kgf (1 Kgf = 9.807 N), u is in cm and A in cm^2. In the test a crack of area $A = 50$ cm^2 starts to grow when $u = 0.5$ cm. Determine $R = G_c$ without resorting to the previous problem. Then compare the value of R with that obtained by the previous problem.

4.22. The stress intensity factor for a center cracked specimen of width $2b$ and crack length $2a$ is given by

$$K_{\mathrm{I}} = \sigma\sqrt{\pi a}\left[1 - 0.1\left(\frac{a}{b}\right) + \left(\frac{a}{b}\right)^2\right].$$

Determine the compliance of the specimen. Compare results with Example 4.8.

4.23. According to the British Standard BS 5447:1977 the stress intensity factor of a compact tension specimen of width W, thickness B and crack length a is given by

$$K_{\mathrm{I}} = \frac{P}{BW^{1/2}}\left[29.6\left(\frac{a}{W}\right)^{1/2} - 185.5\left(\frac{a}{W}\right)^{3/2} + 655.7\left(\frac{a}{W}\right)^{5/2} - \right.$$

$$\left. - 1017.0\left(\frac{a}{W}\right)^{7/2} + 638.9\left(\frac{a}{W}\right)^{9/2}\right].$$

Determine the compliance of the specimen.

4.24. According to the British Standard BS 5447:1977 the stress intensity factor for a single-edge-cracked three-point bend specimen of width W, span $L = 2W$, thickness B and crack length a is given by

$$K_{\mathrm{I}} = \frac{3PL}{BW^{3/2}}\left[1.93\left(\frac{a}{W}\right)^{1/2} - 3.07\left(\frac{a}{W}\right)^{3/2} + 14.53\left(\frac{a}{W}\right)^{5/2} - \right.$$

$$\left. - 25.11\left(\frac{a}{W}\right)^{7/2} + 25.80\left(\frac{a}{W}\right)^{9/2}\right].$$

Determine the compliance of the specimen.

4.25. The stress intensity factor for the tapered semi-infinite double cantilever beam of Figure 4.27 is given by

$$K_{\mathrm{I}} = \frac{P\beta}{Bh^{1/2}}\left(\frac{a}{h} + 0.7\right)$$

where B is the thickness of the beam and β is a parameter depending on the angle a.

Determine the compliance of the beam.

4.26. An elastic rod of diameter d and large length is embedded in a semi-infinite elastic medium (Figure 4.28). The rod separates from the medium over a length L.

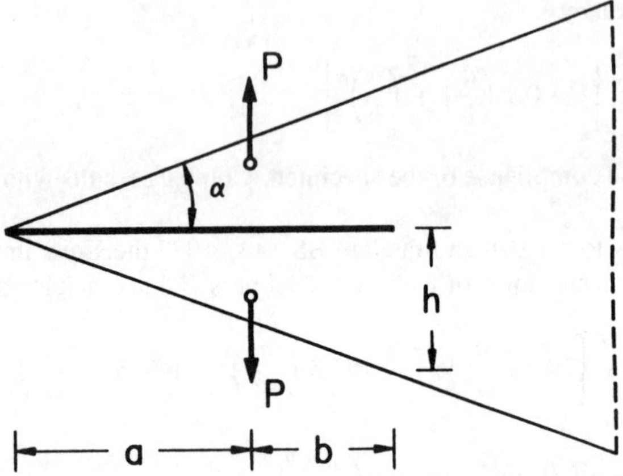

Fig. 4.27. A tapered semi-infinite double cantilever beam.

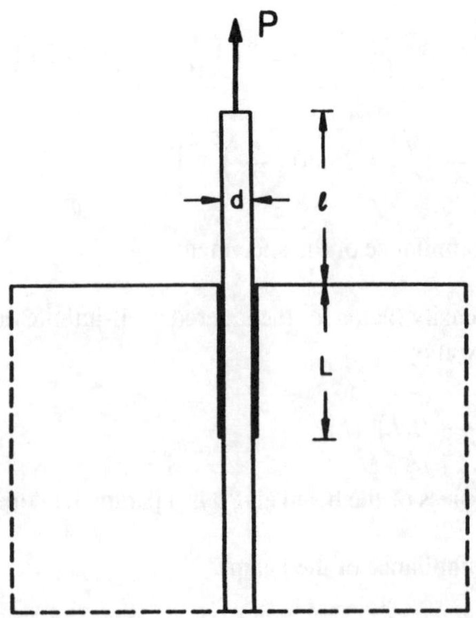

Fig. 4.28. A long rod embedded in a semi-infinite elastic medium except from the length *L*.

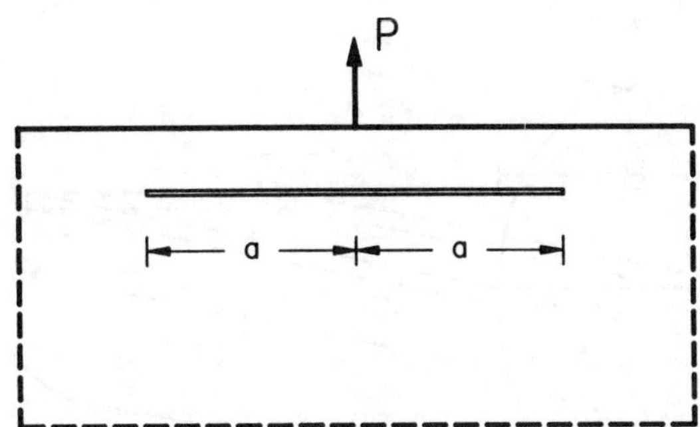

Fig. 4.29. A crack parallel to the edge of a semi-infinite plate subjected to a concentrated load P.

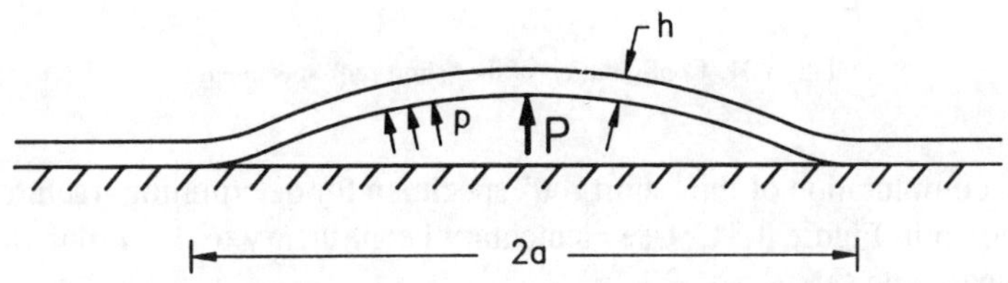

Fig. 4.30. Configuration of the "blister test" specimen.

Show that the load P required for increasing the length L of separation between the rod and the medium is

$$P = \left(\frac{\pi^2 d^3 RE}{2}\right)^{1/2}.$$

4.27. A crack of length $2a$ is parallel to the edge of a thin semi-infinite plate subjected to a concentrated load P (Figure 4.29). Use elementary beam theory to determine the stress intensity factor at the crack tip and comment on the stability of crack growth.

4.28. The configuration of the so-called "blister test" used in adhesion testing is shown in Figure 4.30. The adhered layer of thickness h may be considered as a circular plate of radius a built-in at its periphery. Use elementary plate theory to show that the strain energy release rate G for a constant pressure p or a central load P is given by

$$G = \frac{3}{32}\,\frac{(1-\nu^2)\,p^2 a^4}{Eh^3}\,, \quad G = \frac{3}{8\pi^2}\,\frac{(1-\nu^2)\,P^2}{Eh^3}\,.$$

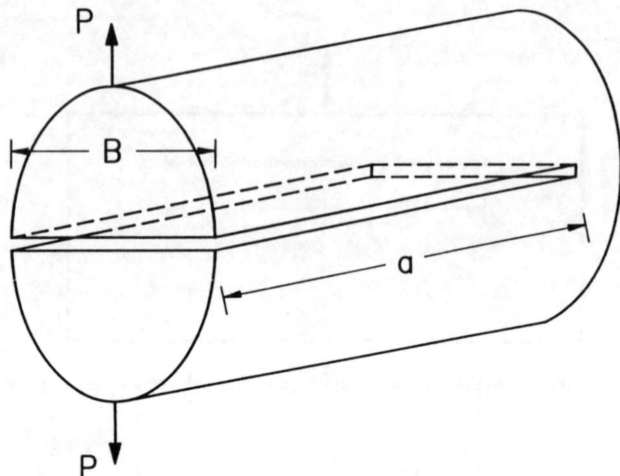

Fig. 4.31. Configuration of the "short rod" specimen.

4.29. The configuration of the "short rod" specimen for determining fracture toughness is shown in Figure 4.31. Use elementary beam theory to determine the strain energy release rate G.

Reference

1.1. Obreimoff, J.W. (1930) 'The splitting strength of mica', *Proceedings of the Royal Society of London* **A127**, 290–297.

Chapter 5

Critical Stress Intensity Factor Fracture Criterion

5.1. Introduction

When a solid is fractured, work is performed to create new material surfaces in a thermodynamically irreversible manner. In Griffith's theory of ideally brittle materials, the work of fracture is spent in the rupture of cohesive bonds. The fracture surface energy γ, which represents the energy required to form a unit of new material surface, corresponds to a normal separation of atomic planes. For the fracture of polycrystals, however, the work required for the creation of new surfaces should also include: dissipation associated with nonhomogeneous slip within and between the grains; plastic and viscous deformation; and possible phase changes at the crack surfaces. The energy required for the rupture of atomic bonds is only a small portion of the dissipated energy in the fracture process. There are situations where the irreversible work associated with fracture is confined to a small process zone adjacent to the crack surfaces, while the remaining material is deformed elastically. In such a case the various work terms associated with fracture may be lumped together in a macroscopic term R (resistance to fracture) which represents the work required for the creation of a unit of new material surface. R may be considered as a material parameter. The plastic zone accompanying the crack tip is very small and the state of affairs around the crack tip can be described by the stress intensity factor.

The present chapter presents a fracture criterion based on the critical value of the strain energy release rate. Due to the equivalence between the strain energy release rate and the stress intensity factor, this criterion is referred to as the critical stress intensity factor fracture criterion. We discuss the variation of the critical stress intensity factor with the specimen thickness for unstable crack growth and present the experimental procedure for determining the plane strain critical stress intensity factor. The chapter concludes with a description of the crack growth resistance curve method for the study of fracture in situations where small, slow, stable crack growth, usually accompanied by inelastic deformation, is observed before global instability.

5.2. Fracture criterion

When the zones of plastic deformation around the crack tip are very small, the plastic strain term appearing in the energy balance equation (Equation (4.4)) can be omitted, and the work rate supplied to the body for crack growth is represented by Equation (4.7). In such circumstances, fracture is assumed to occur when the strain energy release rate G, which represents the energy pumped into the fracture zone from the elastic bulk of the solid, becomes equal to the energy required for the creation of a unit area of new material R. The fracture condition is

$$G_{\mathrm{I}} = G_c = R \,. \tag{5.1}$$

Equation (5.1) is usually expressed in terms of the opening-mode stress intensity factor K_{I}. By introducing a new material parameter K_c from the equation

$$K_c = \sqrt{\frac{ER}{\beta}}\,, \tag{5.2}$$

where $\beta = 1$ for plane stress and $\beta = 1 - \nu^2$ for plane strain, and by substituting G_{I} in terms of K_{I} from Equation (4.22) or (4.23), we can write Equation (5.2) as

$$K_{\mathrm{I}} = K_c \,. \tag{5.3}$$

Equation (5.3) expresses the critical stress intensity factor fracture criterion. The left-hand side of the equation depends on the applied load, the crack length and the geometrical configuration of the cracked plate. The right-hand side is a material parameter and can be determined experimentally. Note that Equation (5.3) was derived from the global energy balance of the continuum; it expresses the law of conservation of energy.

5.3. Variation of K_c with thickness

Laboratory experiments indicate that K_c varies with the thickness B of the specimen tested. The form of variation of K_c with B is shown in Figure 5.1. Three distinct regions, corresponding to "very thin", "very thick" and "intermediate range thickness" specimens can be distinguished. Study of the load-displacement response and the appearance of the fracture surfaces of the specimen are helpful in understanding the mechanisms of fracture in each of these three regions. The fractures are classified as square or slant according to whether the fracture surface is normal to, or forms a 45° inclination angle with, the direction of the applied tensile load. We now analyze the state of affairs in the three regions of Figure 5.1.

In region I, corresponding to thin specimens, the critical fracture toughness G_c (which is proportional to K_c^2) increases almost linearly with B up to a maximum value at a critical thickness B_m. The load-displacement response is linear and the fracture surface is completely slant (Figure 5.2(a)). In this case, as explained in

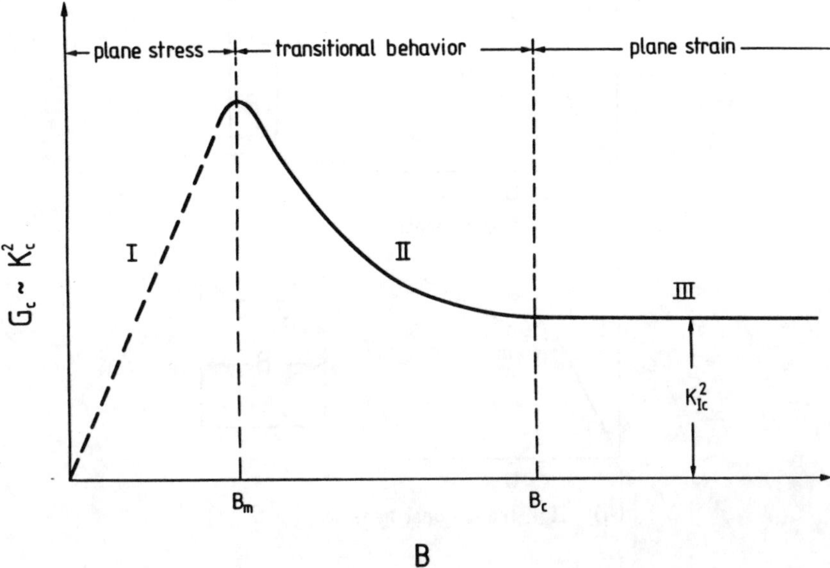

Fig. 5.1. Critical fracture toughness G_c (or K_c^2) versus plate thickness B.

Section 3.2, plane stress predominates in the specimen and yielding occurs on planes through thickness at 45° to the specimen flat surfaces. In such circumstances the crack extends in an antiplane shear mode.

For very thick specimens (region III) the load-displacement response is linear, and the state of stress is predominantly plane strain, except for a thin layer at the free surfaces where plane stress dominates (see Figure 5.2(c)). The fracture surface is almost completely square with very small slant parts at the free surfaces. A triaxial state of stress is produced in most parts of the specimen, which reduces the ductility of the material, and fracture takes place at the lowest value of the critical strain energy release rate G_c. For increasing thickness beyond a critical minimum value, B_c, plane strain conditions dominate and the fracture toughness remains the same. The critical value of stress intensity factor in region III for plane strain conditions is denoted by K_{Ic} in Figure 5.1 and is independent of the specimen thickness. K_{Ic} is the so-called fracture toughness and represents an important material property. The larger the value of K_{Ic}, the larger the resistance of the material to crack propagation. Experimental determination of K_{Ic} takes place according to the ASTM specifications described in the next section.

For intermediate values of specimen thickness (region II) the fracture behavior is neither predominantly plane stress nor predominantly plane strain. The thickness is such that the central and edge region, under plane strain and plane stress conditions respectively, are of comparable size. The fracture toughness in this region changes between the minimum plane strain toughness and the maximum plane stress tough-

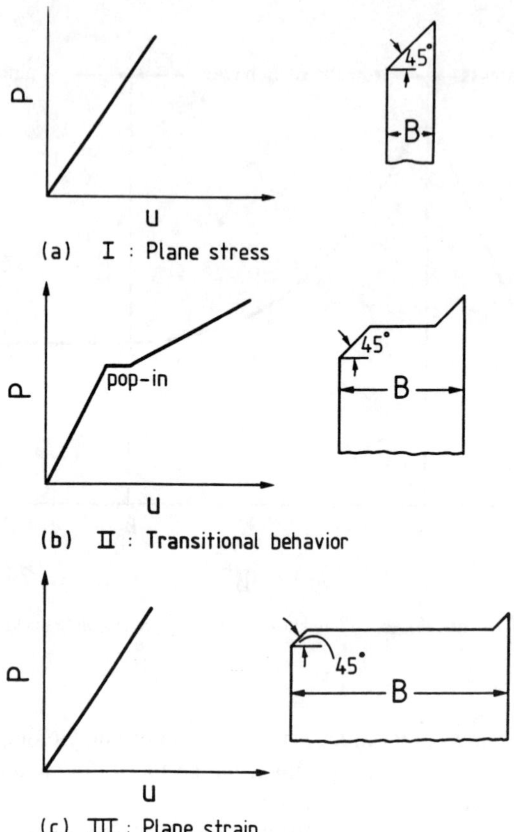

Fig. 5.2. Load-displacement response for (a) plane stress, (b) transitional behavior and (c) plane strain.

ness. In the load-displacement curve (Figure 5.2(b)) at some value of the applied load the crack extends mainly from the center of the thickness of the specimen, while the edge regions are plastically deformed. The crack grows in a "thumbnail" shape (Figure 5.3) under constant or decreasing load while the overall displacement is increased. This behavior is known as "pop-in" (Figure 5.2). After crack growth at pop-in, the stiffness of the load-displacement curve decreases, since it corresponds to a longer crack.

A simplified model was proposed by Krafft *et al.* [5.1] to explain the decrease of fracture toughness with increase of depth of square fracture. Figure 5.4 shows that the square fracture occupies the part $(1 - S)B$ of the specimen thickness and that the slant fracture surface is at 45°. The work for plastic deformation $(\mathrm{d}W_p/\mathrm{d}V)$ is assumed constant. If $\mathrm{d}W_f/\mathrm{d}A$ is the work consumed to produce a unit area of flat fracture, the work done for an advance of crack length by $\mathrm{d}a$ is

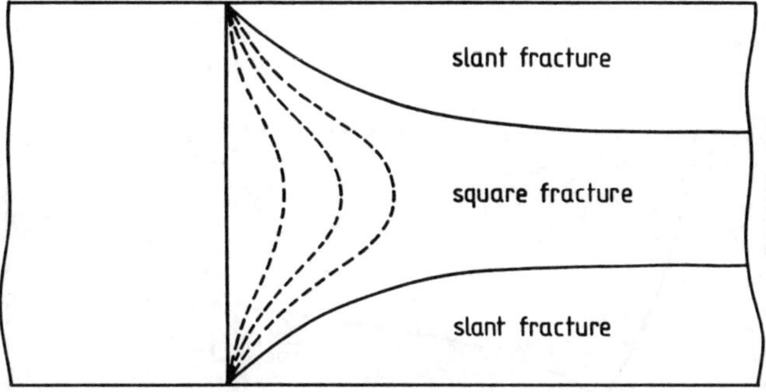

Fig. 5.3. Thumbnail crack growth with square and slant fracture.

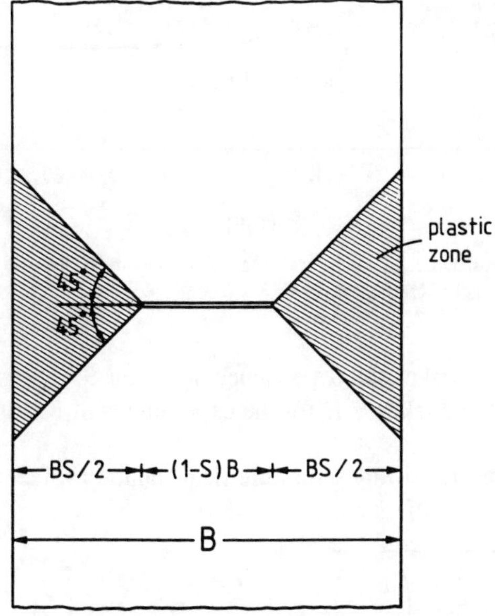

Fig. 5.4. Calculation of crack growth resistance according to the model of Krafft *et al.* [5.1].

$$\mathrm{d}W = \left(\frac{\mathrm{d}W_f}{\mathrm{d}A}\right)(1-S)\,B\,\mathrm{d}a + \left(\frac{\mathrm{d}W_p}{\mathrm{d}V}\right)\frac{B^2S^2}{2}\,\mathrm{d}a\,. \tag{5.4}$$

From this equation the strain energy release rate is

$$G = \frac{\mathrm{d}W}{\mathrm{d}A} = \left(\frac{\mathrm{d}W_f}{\mathrm{d}A}\right)(1-S) + \left(\frac{\mathrm{d}W_p}{\mathrm{d}V}\right)\frac{B^2S^2}{2}\,. \tag{5.5}$$

By fitting the experimental data to this equation and assuming that the slant fracture has a thickness of 2 mm, Krafft *et al.* obtained the expression

$$G_c = 20(1-S) + 200S^2\,. \tag{5.6}$$

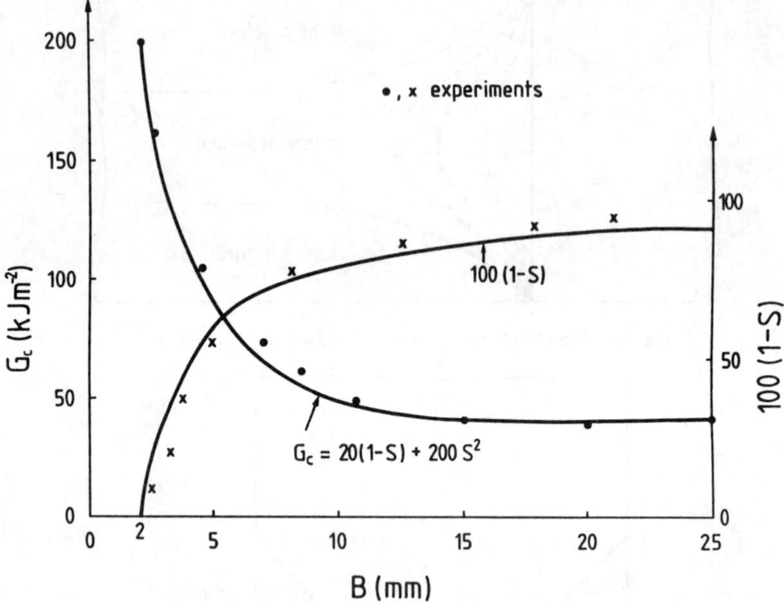

Fig. 5.5. Crack growth resistance G_c and percentage square fracture $100(1-S)$ versus plate thickness B according to experiments by Krafft *et al.* [5.1].

Equation (5.6) establishes the dependence of G_c on S. The variation of G_c and $100(1-S)$ versus the thickness B for the experiments of Krafft *et al.* is shown in Figure 5.5.

Irwin suggested the following semi-empirical equation which relates K_c with the plane strain fracture toughness K_{Ic}:

$$K_c = K_{\mathrm{Ic}} \sqrt{1 + \frac{1.4}{B^2}\left(\frac{K_{\mathrm{Ic}}}{\sigma_Y}\right)^4} \tag{5.7}$$

where σ_Y is the yield stress.

5.4. Experimental determination of K_{Ic}

For the experimental determination of the plane strain fracture toughness special requirements must be fulfilled to obtain reproducible values of K_{Ic} under conditions of maximum constraint around the crack tip. Furthermore, the size of the plastic zone accompanying the crack tip must be very small relative to the specimen thickness and the K_{I}-dominant region. The procedure for measuring K_{Ic} has been standardized by the American Society for Testing and Materials (ASTM) [3.1] to meet these requirements in small specimens that can easily be tested in the laboratory. In this

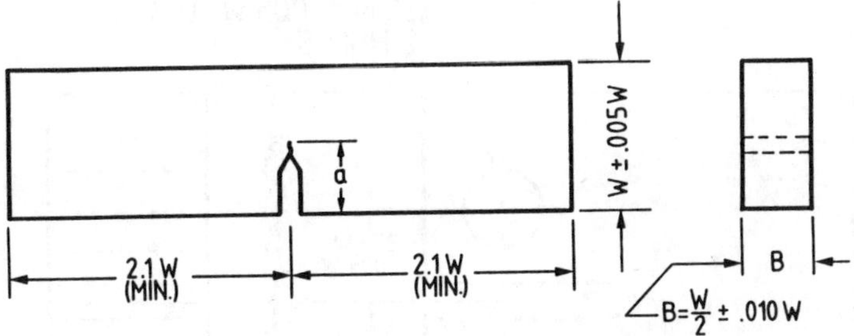

Fig. 5.6. Three-point bend specimen according to ASTM standards.

section we present the salient points of the ASTM standard test method so that the reader may understand the meaning of K_{Ic} and become familiar with the experimental procedure.

(a) Test specimens

The specimens used to measure K_{Ic} must be designed to ensure that the size of the plastic zone accompanying the crack tip be very small relative to the specimen thickness, and plane strain conditions dominate around the crack tip. According to the ASTM standard, the minimum characteristic specimen dimensions, including the specimen thickness B, the crack length a and the specimen width W, must be fifty times greater than the radius of the plane strain plastic zone at fracture. When the plastic zone size is determined according to the Irwin model this condition implies (see Equation (3.8)) that

$$B \geq 2.5 \left(\frac{K_{Ic}}{\sigma_Y} \right)^2 \tag{5.8a}$$

$$a \geq 2.5 \left(\frac{K_{Ic}}{\sigma_Y} \right)^2 \tag{5.8b}$$

$$W \geq 2.5 \left(\frac{K_{Ic}}{\sigma_Y} \right)^2 \tag{5.8c}$$

Many precracked test specimens are described in the ASTM Specification E399–81. These include the three-point bend specimen; the compact tension specimen; the arc-shaped specimen; and the disk-shaped compact specimen. The geometrical configurations of the most widely used three-point bend specimen and compact tension specimen are shown in Figures 5.6 and 5.7. Several formulas have been proposed for the calculation of stress intensity factor for the standard specimens. According to ASTM standards the following expressions for the computation of K_I are used:

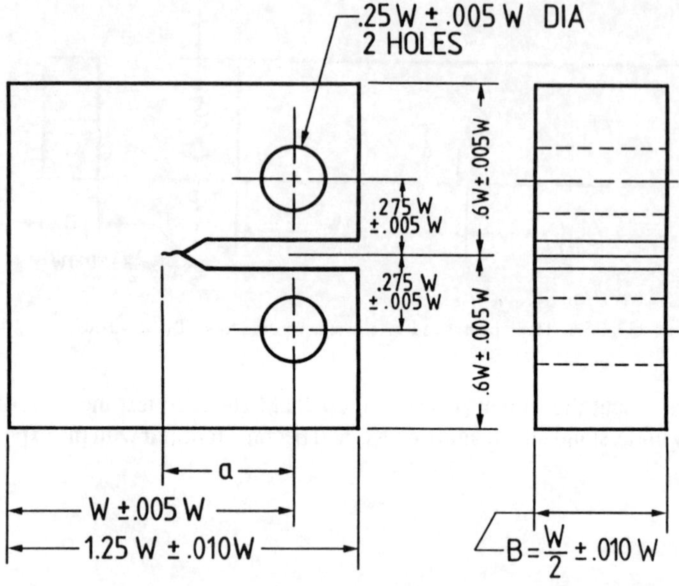

Fig. 5.7. Compact tension specimen according to ASTM standards.

$$K_I = \frac{PS}{BW^{3/2}} \frac{3\left(\frac{a}{W}\right)^{1/2}\left[1.99 - \frac{a}{W}\left(1 - \frac{a}{W}\right)\left(2.15 - 3.93\frac{a}{W} + 2.7\frac{a^2}{W^2}\right)\right]}{2\left(1 + 2\frac{a}{W}\right)\left(1 - \frac{a}{W}\right)^{3/2}} \tag{5.9}$$

for the bend specimen, and

$$K_I = \frac{P}{BW^{1/2}} \frac{\left(2 + \frac{a}{W}\right)\left[0.886 + 4.64\frac{a}{W} - 13.32\left(\frac{a}{W}\right)^2 + 14.72\left(\frac{a}{W}\right)^3 - 5.6\left(\frac{a}{W}\right)^4\right]}{\left(1 - \frac{a}{W}\right)^{3/2}} \tag{5.10}$$

for the compact tension specimen. The quantities a, W and B are shown in Figures 5.6 and 5.7, and S is the distance between the points of support of the beam in Figure 5.6.

Equation (5.9) is accurate to within 0.5 per cent, over the entire range of a/W ($a/W < 1$), while Equation (5.10) is accurate to within 0.5 per cent for $0.2 < a/W < 1$.

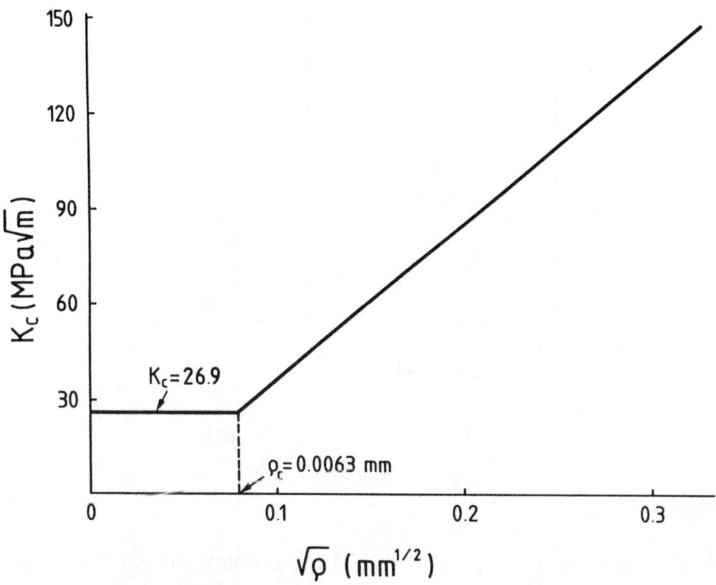

Fig. 5.8. Effect of notch radius ρ on the critical stress intensity factor K_c [1.23].

(b) Precrack

The precrack introduced in the specimen must simulate the ideal plane crack with zero root radius, as was assumed in the stress intensity factor analysis. The effect of the notch radius ρ on the critical value of the stress intensity factor K_c is shown in Figure 5.8. K_c decreases with decreasing ρ until a limiting radius ρ_c is obtained. Below ρ_c, K_c is approximately constant, which shows that a notch with radius smaller than ρ_c can simulate the theoretical crack. The crack front must be normal to the specimen free surfaces, and the material around the crack should experience little damage. To meet these requirements, a special technique is used for the construction of the precrack in the specimen.

A chevron starter notch (Figure 5.9) of length $0.45W$ is first machined in the specimen. The notch is then extended by fatigue at a length $0.05W$ beyond the notch root. The advantage of the chevron notch is that it forces crack initiation in the center, so that a straight machined crack front is obtained. If the initial machined notch front were straight, it would be difficult to produce a final straight crack front. The crack length a used in the calculations is the average of the crack lengths measured at the center of the crack front, and midway between the center and the end of the crack front, on each surface ($a = (a_1 + a_2 + a_3)/3$). The surface crack length should not differ from the average length by more than 10 per cent.

In order to ensure that the material around the crack front does not experience large plastic deformation or damage, and that the fatigue crack is sharp, the fatigue

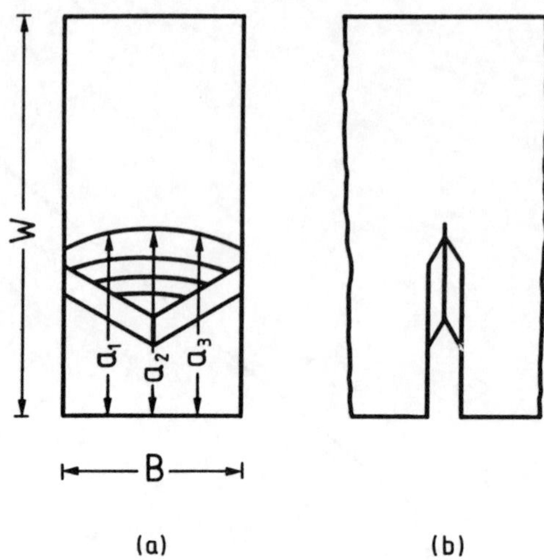

Fig. 5.9. Chevron notch.

loading should satisfy some requirements. The maximum stress intensity factor to which the specimen is subjected during fatigue must not exceed 60 per cent of K_{Ic} and the last 2.5 per cent of the crack length should be loaded at a maximum K_I such that $K_I/E < 0.002.\sqrt{\text{in}}(0.32 \times 10^{-3}\sqrt{\text{m}})$.

(c) Experimental procedure

The precracked standard specimen is loaded by special fixtures recommended by ASTM. The load, and the relative displacement of two points located symmetrically on opposite sides of the crack plane, are recorded simultaneously during the experiment. The specimen is loaded at a rate such that the rate of increase of stress intensity, K_I, is within the range $0.55 - 2.75$ MPa $\text{m}^{1/2}$/s. A test record consisting of an autographic plot of the output of the load-sensing transducers versus the output of the displacement gage is obtained. A combination of load-sensing transducer and autographic recorder is selected so that the maximum load can be determined from the test record with an accuracy of 1 per cent. The specimen is tested until it can sustain no further increase of load.

(d) Interpretation of test record and calculation of K_{Ic}

For perfectly elastic behavior until fracture, the load-displacement curve should be a straight line. Most structural materials, however, present elastoplastic behavior which, combined with some stable crack growth before catastrophic fracture, leads to nonlinear load-displacement diagrams. The principal types of load-displacement curve observed in experiments are shown in Figure 5.10. Type I corresponds to nonlinear behavior, type III to purely linear response and type II reflects the phe-

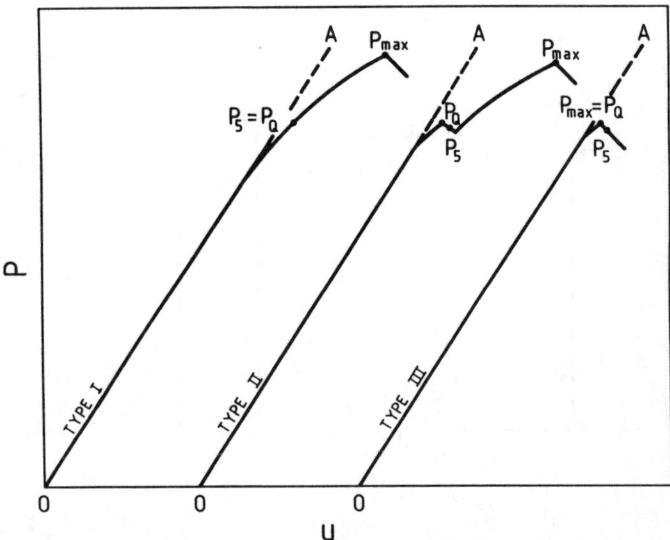

Fig. 5.10. Determination of P_Q for three types of load-displacement response according to ASTM standards.

nomenon of pop-in. For the determination of a valid K_{Ic}, a conditional value K_Q is obtained first. This involves a geometrical construction on the test record, consisting of drawing a secant line OP through the origin with slope equal to 0.95 of the slope of the tangent to the initial linear part of the record. The load P_5 corresponds to the intersection of the secant with the test record. The load P_Q is then determined as follows: if the load at every point on the record which precedes P_5 is lower than P_Q then $P_Q = P_5$ (type I); if, however, there is a maximum load preceding P_5 which is larger than P_5 then P_Q is equal to this load (types II and III). The test is not valid if $P_{\max}/P_Q$ is greater than 1.10, where $P_{\max}$ is the maximum load the specimen was able to sustain. In the geometrical construction, the 5 per cent secant offset line represents the change in compliance due to crack growth equal to 2 per cent of the initial length (Problems 5.7 and 5.8).

After determining P_Q, we calculate K_Q using Equation (5.9) or (5.10) for the bend specimen or the compact tension specimen. When K_Q satisfies the inequalities (5.8), then K_Q is equal to K_{Ic} and the test is a valid K_{Ic} test. When these inequalities are not satisfied, it is necessary to use a larger specimen to determine K_{Ic}. The dimensions of the larger specimen can be estimated on the basis of K_Q.

Values of the critical stress intensity factor K_{Ic} together with the ultimate stress σ_u for some common metals and alloys are given in Appendix 5.1.

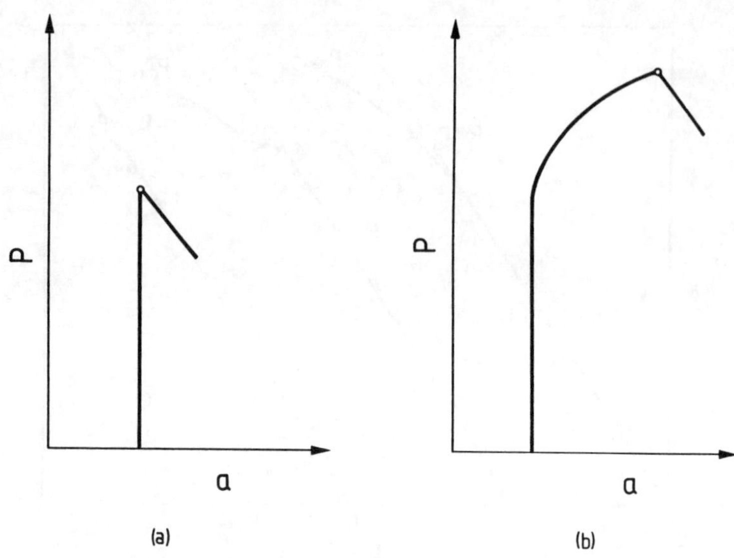

Fig. 5.11. Typical load-crack size curves for (a) plane strain and (b) plane stress.

5.5. Crack growth resistance curve (R-curve) method

The crack growth resistance curve, or R-curve, method is a one-parameter method for the study of fracture in situations where small, slow, stable crack growth – usually accompanied by inelastic deformation – is observed before global instability. Under such circumstances, fracture resistance of thin specimens is represented by a resistance curve, rather than a single resistance parameter. A brief description of the method follows.

(a) General remarks

As noted in Section 5.3, the fracture resistance of a material under plane strain conditions with small-scale crack-tip plasticity is described by the critical stress intensity factor K_{Ic}. Under such conditions, fracture of the material is sudden, and there is either no, or very little, crack growth before final instability. On the other hand, in thin specimens, insufficient material exists to support a triaxial constraint near the crack tip, and plane stress dominates. The crack tip plastic enclaves are no longer negligible, and final instability is preceded by some slow stable crack growth. As shown in Section 5.3, the fracture resistance depends upon thickness. In such circumstances, it has been observed experimentally that the fracture resistance increases with increasing crack growth. Typical curves representing the variation of crack size with load, under plane strain and plane stress conditions, are shown in Figure 5.11.

(b) R-curve

The theoretical basis for the R-curve can be provided by the energy balance equation (Equation (4.4)) which applies during stable crack growth. For situations in which the energy dissipated to plastic deformation U^p is not negligible, Equation (4.4) takes the form

$$G = R \tag{5.11a}$$

with

$$G = \frac{\partial W}{\partial A} - \frac{\partial U^e}{\partial A} \tag{5.11b}$$

and

$$R = \frac{\partial \Gamma}{\partial A} + \frac{\partial U^p}{\partial A}\,. \tag{5.11c}$$

R represents the rate of energy dissipation during stable crack growth. It is composed of two parts: the first corresponds to the energy consumed in the creation of new material surfaces; the second refers to the energy dissipated in plastic deformation. In situations where the crack-tip zones of plastic irreversibility are relatively small, the two dissipation terms in Equation (5.11c) may be lumped together to form a new material parameter associated with the resistance of the material to fracture. Following crack initiation, the plastic zone around the crack tip increases nonlinearly with crack size. Thus, the rate of the energy dissipated to plastic deformation, which constitutes the major part of the dissipation term in Equation (5.11c), increases nonlinearly with crack size. The graphical representation of the variation of R, or the critical stress intensity factor K_R plotted against crack extension, is called the crack growth resistance curve (R-curve). A typical form of the R-curve is shown in Figure 5.12. The R-curve is considered to be a characteristic of the material for a given thickness, temperature and strain rate, independent of the initial crack size and the geometry of the specimen.

(c) Determination of the critical load

During stable crack growth, Equation (5.11a) and inequality (4.40) should be satisfied. The strain energy release rate G, according to the R-curve method, is calculated from Equation (4.21) for "fixed-grips" or "dead-load" loading conditions. For example, G is given by Equation (4.18) for a line crack of length $2a$ in an infinite plate subjected to a stress σ perpendicular to the crack. For a crack of length $2a$ in a finite plate of width W, G is calculated from Equation (4.22) under plane stress as (see Problem (2.21))

$$G(a, \sigma) = \frac{\sigma^2 W}{E} \tan\left(\frac{\pi a}{W}\right)\,. \tag{5.12}$$

In graphical form both parts of Equation (5.11a) are represented in Figure 5.13 in $G - a$ coordinates. The R-curve is displayed at the initial crack length a_0 while

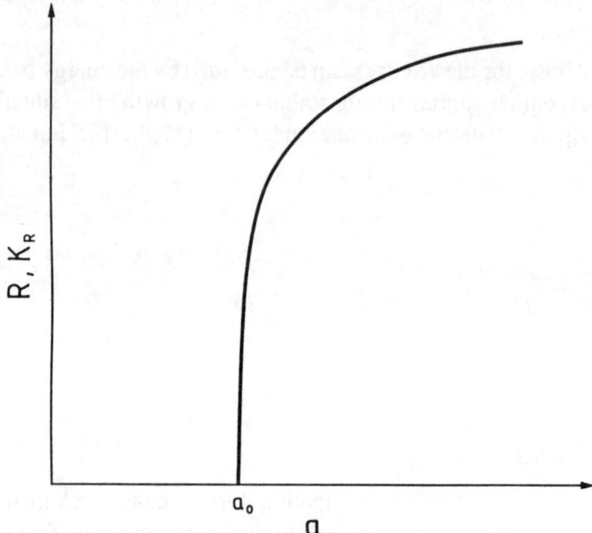

Fig. 5.12. R-curve.

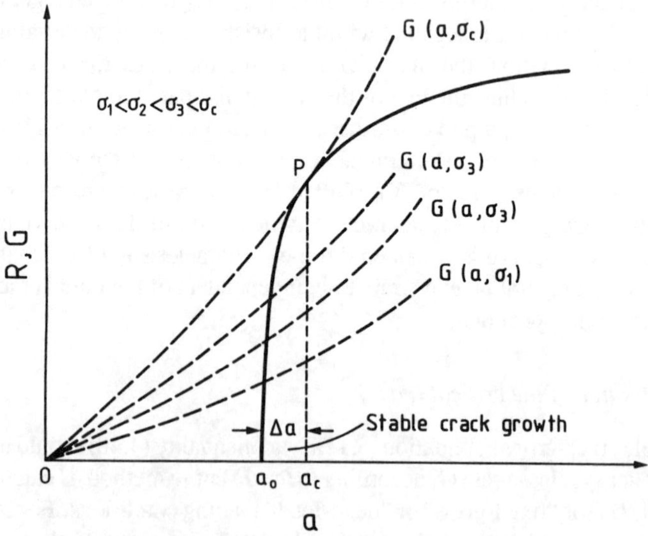

Fig. 5.13. R-curve and a family of rising G-curves.

the $G(a, \sigma_i)$-curves correspond to different values of the applied stress σ. The points of intersection of the G- and R-curves refer to stable crack growth, since Equation (5.11a) and inequality (4.40) are satisfied. Stable crack growth continues up to the point P at which the $G(a, \sigma_c)$-curve, that corresponds to the value σ_c of

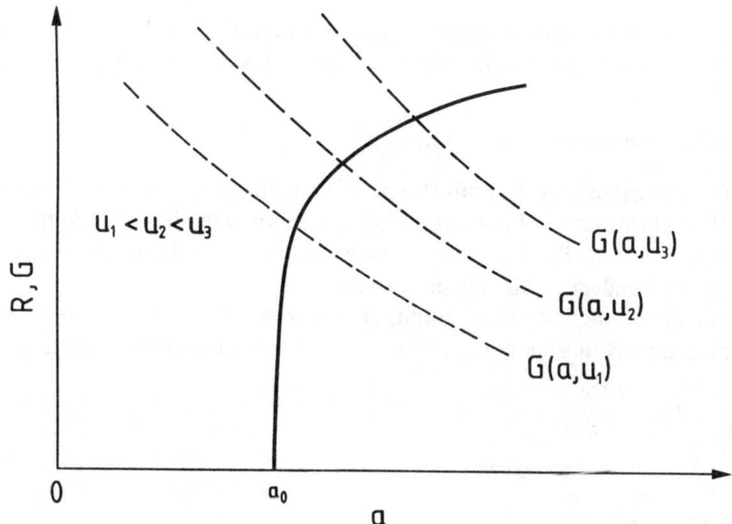

Fig. 5.14. *R*-curve and a family of decreasing *G*-curves.

the applied stress, is tangent to the R-curve. Beyond point P, crack propagation is unstable, according to inequality (4.40). Point P defines the critical stress σ_c, and the critical crack length a_c at instability. σ_c and a_c are determined from Equation (5.11a) and the third Equation (4.40).

For specimen configurations for which K, and therefore G, decrease with increasing crack length, the graphical representation of the quantities $G(a, u)$, where u is the applied displacement, and R is shown in Figure 5.14. Observe that if R increases monotonically, then crack growth is stable for all crack lengths up to a plateau level.

(d) Experimental determination of the R-curve

ASTM [5.2] issued a standard for the experimental determination of the R-curve of a material for given thickness, temperature and strain rate. The R-curve is assumed to be independent of the starting crack length and the specimen configuration. It is a function of crack extension only. Three types of fatigue precracked standard specimens are recommended by ASTM: the center cracked tension specimen, the compact specimen; the crack line wedge-loaded specimen. The specimen dimensions are chosen so that the ligament in the plane of the crack is predominantly elastic at all values of the applied load. The first two types of specimen are tested under load control, while the third is displacement controlled. The specimens are loaded incrementally to specially designed fixtures, which are described in detail in [5.2]. During the test, the load and the crack length are recorded simultaneously. The physical crack length is measured using optical microscopy or the electrical potential method. The effective crack length is obtained by adding the physical crack length and the Irwin plastic zone radius r_y $(= c/2)$ given by Equation (3.6). Calibration formulas are used to obtain the stress intensity factor during stable crack growth. The K–Δa or G–Δa relationship thus obtained constitutes the R-curve.

(e) Irwin-Orowan theory

In an effort to extend the Griffith theory to situations of semi-brittle fracture, Irwin [5.3] and Orowan [5.4] introduced independently a modification to the Griffith formula (4.13) or (4.14). The Irwin-Orowan theory can easily be described by the energy balance analysis of the R-curve method.

For a crack of length $2a$ in an infinite plate subjected to a stress σ perpendicular to the crack axis, G in Equation (5.11a) is given by Equation (4.18). Putting

$$\frac{\partial \Gamma}{\partial A} = 2\gamma\,, \quad \frac{\partial U^p}{\partial A} = 2\gamma_p \tag{5.13}$$

we find the crack growth resistance R as

$$R = 2(\gamma + \gamma_p) \tag{5.14}$$

Equation (5.11a) gives the critical stress

$$\sigma_c = \sqrt{\frac{2E(\gamma + \gamma_p)}{\pi a}} \tag{5.15}$$

for plane stress, and

$$\sigma_c = \sqrt{\frac{2E(\gamma + \gamma_p)}{(1 - \nu^2)\,\pi a}} \tag{5.16}$$

for plane strain.

In Equations (5.15) and (5.16) γ is much smaller (usually three orders of magnitude) than γ_p and can be omitted.

(f) Crack stability

Study of crack stability usually takes place when the specimen is loaded in hard (displacement-controlled) or soft (load-controlled) testing machines. From Equations (4.30) or (4.33) and (5.1) we derive

$$P^2 = \frac{2R}{\dfrac{\mathrm{d}}{\mathrm{d}A}\left(\dfrac{u}{P}\right)}\,. \tag{5.17}$$

For stability in soft testing machines ($\mathrm{d}P/P > 0$) Equation (5.17) gives,

$$\frac{1}{R}\frac{\mathrm{d}R}{\mathrm{d}A} \geq \frac{\dfrac{\mathrm{d}^2}{\mathrm{d}A^2}\left(\dfrac{u}{P}\right)}{\dfrac{\mathrm{d}}{\mathrm{d}A}\left(\dfrac{u}{P}\right)} \tag{5.18}$$

while for stability in hard testing machines ($\mathrm{d}u/u > 0$) Equation (5.17) gives

$$\frac{1}{R}\frac{\mathrm{d}R}{\mathrm{d}A} \geq \frac{\dfrac{\mathrm{d}^2}{\mathrm{d}A^2}\left(\dfrac{P}{u}\right)}{\dfrac{\mathrm{d}}{\mathrm{d}A}\left(\dfrac{P}{u}\right)}. \tag{5.19}$$

Equations (5.18) and (5.19) define the stability conditions in soft and hard testing machines. The right-hand side of Equations (5.18) and (5.19) depends on the geometry of the specimen and is called the geometry stability factor of the specimen.

5.6. Fracture mechanics design methodology

The objective of engineering design is the determination of the geometry and dimensions of the machine or structural elements, and the selection of the material, in such a way that the elements perform their operating function in an efficient, safe and economic manner. To achieve these objectives, we use an appropriate failure criterion which consists of comparing a critical quantity, depending on the geometry of the element, the loading and environmental conditions, with a material characteristic parameter. To select the appropriate failure criterion the designer should know the most probable mode of failure. Possible failure modes consist of (i) yielding or excessive plastic deformations; (ii) general instability (e.g. buckling); and (iii) fracture. In engineering design, however, all possible failure modes should be taken into consideration. For example, it was found that structures designed according to the first two failure modes failed in a sudden catastrophic manner, due to unstable crack propagation. This, as was explained in Chapter 1, gave impetus to the development of fracture mechanics.

Conventional design analysis of engineering components assumes a defect-free structural geometry and determines a relationship between the applied loading and the maximum stress that is developed in the component. For this reason a stress analysis is performed, based on the theory of elasticity and strength of materials. Safe design is achieved by making sure that the maximum stress is less than the ultimate stress of the material (divided by a factor of safety).

Fracture mechanics design methodology is based on the realistic assumption that all materials contain initial defects that can affect the load-carrying capacity of engineering structures. Defects are initiated in the material by manufacturing procedures or can be created during the service life, by fatigue, environment effects or creep. We analyze the component with a crack placed in the most probable, or dangerous, site and determine a characteristic quantity defining the propensity of the crack to extend. Usually we assume there is one dominant crack. The characteristic quantity depends on the particular failure criterion used, and in our case, it is the stress intensity factor, K_{I}. The failure criterion is expressed by Equation (5.3). Methods for determining K_{I}, the left-hand part of Equation (5.3), were developed in Chapter 2. The right-hand part of Equations (5.3), K_c, is a material parameter and is determined experimentally.

By this procedure we determine the maximum allowable applied loads for a specified crack size, or the maximum permissible crack size for specified applied loads.

Other fracture criteria based on quantities as, the J-integral, the crack opening displacement or the strain energy density factor will be developed in Chapters 6 and 7 of the book.

Examples

Example 5.1.

A three-point bend specimen was tested according to the ASTM E399 procedure. The 0.2 per cent offset yield stress of the material is $\sigma_Y = 1200$ MPa and the modulus of elasticity is $E = 210$ GPa. The specimen was tested at a loading rate of 100 kN/min and a type-I load-displacement record was obtained. A chevron starter notch was machined and the specimen was subjected to 30,000 cycles at $P_{\max} = 45$ kN and $P_{\min} = 0$. The final stage of fatigue crack growth was conducted for 50,000 cycles at $P_{\max} = 30$ kN and $P_{\min} = 0$. The specimen dimensions were measured as

$$S = 30 \text{ cm}, \; W = 8 \text{ cm}, \; B = 4 \text{ cm}$$
$$a_1 = 3.996 \text{ cm}$$
$$a_2 = 4.007 \text{ cm}$$
$$a_3 = 3.997 \text{ cm}$$
$$a \text{ (surface)} = 3.915 \text{ cm}$$
$$a \text{ (surface)} = 3.952 \text{ cm}.$$

The maximum load and the secant load of the test record were measured as $P_{\max} = 86$ kN and $P_Q = 80$ kN.

Determine K_{Ic}.

Solution: We first determine a conditional $K_{\mathrm{Ic}}(K_Q)$. We have:

$$a = \frac{a_1 + a_2 + a_3}{3} = \frac{3.996 + 4.007 + 3.997}{3} = 4.0 \text{ cm}$$

$$\frac{a - a \text{ (surface)}}{a} = \frac{4.0 - 3.915}{4.0} = 0.02 < 0.1$$

$$\frac{a - a \text{ (surface)}}{a} = \frac{4.0 - 3.952}{4.0} = 0.012 < 0.1$$

$$\frac{P_{\max}}{P_Q} = \frac{86 \text{ kN}}{80 \text{ kN}} = 1.075 < 1.10$$

K_Q is determined from Equation (5.9). This equation can be put in the form

$$K_Q = \frac{PS}{BW^{3/2}} f(a/W) \tag{1}$$

$$f(a/W) = \frac{3\left(\frac{a}{W}\right)^{1/2}\left[1.99 - \frac{a}{W}\left(1-\frac{a}{W}\right)\left(2.15 - 3.93\frac{a}{W} + 2.7\frac{a^2}{W^2}\right)\right]}{2\left(1+2\frac{a}{W}\right)\left(1-\frac{a}{W}\right)^{3/2}} . \quad (2)$$

To facilitate calculation of K_Q, values of $f(a/W)$ are tabulated for specific values of a/W, according to ASTM

a/W	$f(a/W)$	a/W	$f(a/W)$
0.450	2.29	0.500	2.66
0.455	2.32	0.505	2.70
0.460	2.35	0.510	2.75
0.465	2.39	0.515	2.79
0.470	2.43	0.520	2.84
0.475	2.46	0.525	2.89
0.480	2.50	0.530	2.94
0.485	2.54	0.535	2.99
0.490	2.58	0.540	3.04
0.495	2.62	0.545	3.09
–	–	0.550	3.14

For our case $(a/W) = 0.5$. Thus, $f(a/W) = 2.66$. The specimen is loaded at a K_{I} rate, $\Delta K_{\mathrm{I}}/\Delta t$, as

$$\frac{\Delta K_{\mathrm{I}}}{\Delta t} = \frac{(\Delta P)S}{BW^{3/2}}\, f(a/W)$$

or

$$\frac{\Delta K_{\mathrm{I}}}{\Delta t} = \frac{(100\ \mathrm{kN}/60\ \mathrm{s}) \times (0.3\ \mathrm{m})}{(0.04\ \mathrm{m}) \times (0.08\ \mathrm{m})^{3/2}} \times 2.66 = 1.47\ \mathrm{MPa}\sqrt{\mathrm{m}}/\mathrm{s} .$$

We have

$$0.55 < \frac{\Delta K_{\mathrm{I}}}{\Delta t} < 2.75\ \mathrm{MPa}\sqrt{\mathrm{m}}/\mathrm{s} .$$

K_Q is determined as

$$K_Q = \frac{(80\ \mathrm{kN}) \times (0.3\ \mathrm{m})}{(0.04\ \mathrm{m}) \times (0.08\ \mathrm{m})^{3/2}} \times 2.66 = 70.5\ \mathrm{MPa}\sqrt{\mathrm{m}} .$$

We have

$$B, a, W > 2.5\left(\frac{K_Q}{\sigma_Y}\right)^2 = 2.5\left(\frac{70.5\ \mathrm{MPa}\sqrt{\mathrm{m}}}{1200\ \mathrm{MPa}}\right)^2 = 0.86 \times 10^{-2}\ \mathrm{m} .$$

The maximum stress intensity factor during the first stage of fatigue growth of precrack is

$$K_{f(\max)} = \frac{(45\text{ kN}) \times (0.3\text{ m})}{(0.04\text{ m}) \times (0.08\text{ m})^{3/2}} \times 2.66 = 39.6\text{ MPa}\sqrt{\text{m}}$$

so that

$$K_{f(\max)} < 0.6\, K_Q\,.$$

During the final stage of fatigue growth of precrack K_{I} is

$$K_{\text{I}} = \frac{(30\text{ kN}) \times (0.3\text{ m})}{(0.04\text{ m}) \times (0.08\text{ m})^{3/2}} \times 2.66 = 26.4\text{ MPa}\sqrt{\text{m}}$$

and

$$\frac{K_{\text{I}}}{E} = \frac{26.4 \quad \text{MPa}\sqrt{\text{m}}}{210 \quad \text{GPa}} = 0.13 \times 10^{-3}\,\sqrt{\text{m}} < 0.32 \times 10^{-3}\,\sqrt{\text{m}}\,.$$

The above results indicate that all requirements of a valid K_{Ic} test according to ASTM E399 standard are satisfied and the conditional value K_Q is equal to K_{Ic}, that is

$$K_{\text{Ic}} = K_Q = 70.5\text{ MPa}\sqrt{\text{m}}\,.$$

Example 5.2.

Figures 5.15 shows the load-displacement record of a compact tension specimen tested according to ASTM E399 procedure to determine K_{Ic}. The 0.2 per cent offset yield stress of the material is 800 MPa. The specimen dimensions were measured as: $W = 12$ cm, $B = 5$ cm, $a = 6$ cm. Determine K_{Ic}.

Solution: A conditional $K_{\text{Ic}}(K_Q)$ is first determined from Equation (5.10). This equation can be put in the form

$$K_{\text{I}} = \frac{P_Q}{BW^{1/2}}\, f(a/W) \tag{1}$$

$$f(a/W) = \frac{\left(2 + \frac{a}{W}\right)\left[0.886 + 0.64\frac{a}{W} - 13.32\left(\frac{a}{W}\right)^2 + 14.72\left(\frac{a}{W}\right)^3 - 5.6\left(\frac{a}{W}\right)^4\right]}{\left(1 - \frac{a}{W}\right)^{3/2}} \tag{2}$$

To facilitate calculation of K_Q, values of $f(a/W)$ are tabulated for specific values of a/W, according to ASTM

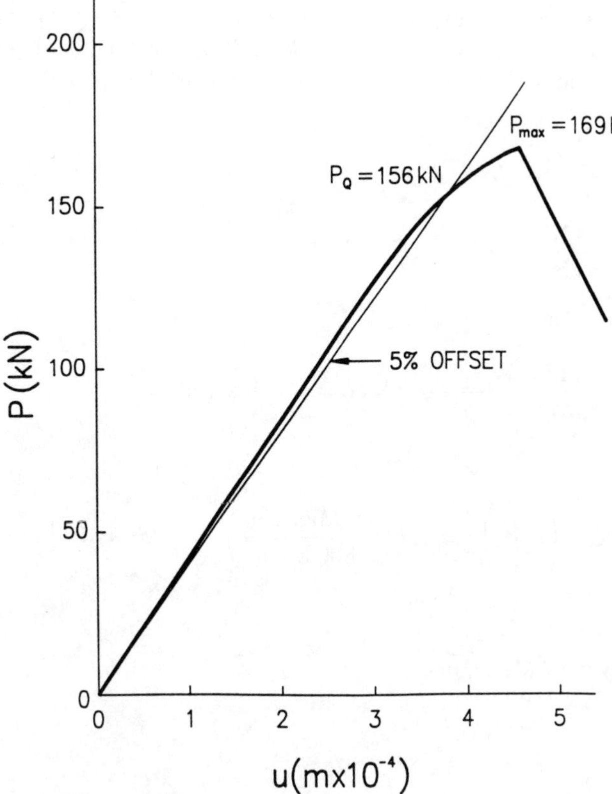

Fig. 5.15. Determination of P_Q on the load-displacement record of the compact tension specimen of Example 5.2.

a/W	$f(a/W)$	a/W	$f(a/W)$
0.450	8.34	0.500	9.66
0.455	8.46	0.505	9.81
0.460	8.58	0.510	9.96
0.465	8.70	0.515	10.12
0.470	8.83	0.520	10.29
0.475	8.96	0.525	10.45
0.480	9.09	0.530	10.63
0.485	9.23	0.535	10.80
0.490	9.37	0.540	10.98
0.495	9.51	0.545	11.17
–	–	0.550	11.36

We have $a/W = 0.5$. Thus, $f(a/W) = 9.66$. To determine P_Q in Equation (1) we draw a secant line through the origin of the load-displacement record of Figure 5.15 with slope equal to 0.95 of the initial slope of the record. We obtain

$$P_Q = 156 \text{ kN}$$

and

$$\frac{P_{\max}}{P_Q} = \frac{169 \text{ kN}}{156 \text{ kN}} = 1.08 < 1.10$$

K_Q is determined as

$$K_Q = \frac{(156 \text{ kN})}{(0.05 \text{ m})\,(0.12 \text{ m})^{1/2}} \times 9.66 = 87 \text{ MPa}\sqrt{\text{m}}\,.$$

We have

$$B, a, W > 2.5\left(\frac{K_Q}{\sigma_Y}\right)^2 = 2.5\left(\frac{87 \text{ MPa}\sqrt{\text{m}}}{800 \text{ MPa}}\right)^2 = 2.96 \times 10^{-2} \text{ m}$$

and therefore

$$K_{IC} = K_Q = 87 \text{ MPa}\sqrt{\text{m}}\,.$$

Example 5.3.

The stress intensity factor of a crack of length $2a$ in a plate subjected to a tension field is given by

$$K_{\text{I}} = \sigma(a)\, f(a/b)\,, \quad f(a/b) > 0\,. \tag{1}$$

The crack growth resistance curve of the plate is expressed by

$$K_R = m[a_0 + (a - a_0)^{1/2}] \tag{2}$$

where a_0 is the initial crack length and m is a constant.

Determine the critical stress σ_c and the critical length $2a_c$ at the point of instability.

Solution: At the point of unstable crack growth we have from Equation (5.11a) and the third Equation (4.40)

$$K_{\text{I}} = K_R\,, \tag{3}$$

$$\frac{\partial K_{\text{I}}}{\partial a} = \frac{\partial K_R}{\partial a}\,. \tag{4}$$

We have from Equations (1) and (2)

$$\frac{\partial K_{\mathrm{I}}}{\partial a} = \frac{\partial \sigma}{\partial a} f + \sigma \frac{\partial f}{\partial a}, \quad \sigma = \sigma(a), \quad f = f(a/b) \tag{5}$$

$$\frac{\partial K_R}{\partial a} = \frac{m}{2(a-a_0)^{1/2}}. \tag{6}$$

Equation (4) renders

$$a - a_0 = \left(\frac{m}{2}\right)^2 \left(\frac{\partial \sigma}{\partial a} f + \sigma \frac{\partial f}{\partial a}\right)^2. \tag{7}$$

Equation (3) becomes

$$\sigma f = m[a + (a-a_0)^{1/2}]. \tag{8}$$

Equations (7) and (8) give the critical stress σ_c and the critical crack length $2a_c$ at instability.

Example 5.4.

Determine the stability condition for a double cantilever beam (DCB) (Figure 4.14) subjected to an end load in a soft (load-controlled) or a hard (displacement-controlled) testing machine.

Solution: The compliance of the DCB is (Example 4.2)

$$C = \frac{u}{P} = \frac{2\eta a^3}{3EI}, \quad I = \frac{Bh^3}{12}. \tag{1}$$

From Equation (1) we obtain

$$\frac{\mathrm{d}C}{\mathrm{d}A} = \frac{3\eta a^2}{8EB^2h^3}, \quad \frac{\mathrm{d}^2 C}{\partial A^2} = \frac{3\eta a}{4EB^3h^3}. \tag{2}$$

For stability in a soft (load-controlled) testing machine Equation (5.18) becomes

$$\frac{1}{R}\frac{\mathrm{d}R}{\mathrm{d}A} \geq \frac{2}{A}. \tag{3}$$

From Equation (1) we obtain

$$\frac{\mathrm{d}C^{-1}}{\mathrm{d}A} = -\frac{24Eh^3}{\eta a^4}, \quad \frac{\mathrm{d}^2 C^{-1}}{\mathrm{d}A^2} = \frac{96Eh^3}{\eta a^5 B}. \tag{4}$$

For stability in a hard (displacement-controlled) testing machine Equation (5.19) becomes

$$\frac{1}{R}\frac{\mathrm{d}R}{\mathrm{d}A} \geq -\frac{4}{A}. \tag{5}$$

Equations (3) and (5) express the stability conditions in a soft or a hard testing machine. Note that Equation (5) is always satisfied for constant R. Equations (3) and (5) show that stability is achieved more easily with a hard than with a soft testing machine.

Example 5.5.

A cylindrical pressure vessel with closed ends has a radius $R = 1$ m and thickness $t = 40$ mm and is subjected to internal pressure p. The vessel must be designed safely against failure by yielding (according to the von Mises yield criterion) and fracture. Three steels with the following values of yield stress σ_Y and fracture toughness K_{Ic} are available for constructing the vessel.

Steel	σ_Y(MPa)	K_{Ic}(MPa $\sqrt{\text{m}}$)
A: 4340	860	100
B: 4335	1300	70
C: 350 Maraging	1550	55

Fracture of the vessel is caused by a long axial surface crack of depth a. The vessel should be designed with a factor of safety $S = 2$ against yielding and fracture. For each steel:

(a) Plot the maximum permissible pressure p_c versus crack depth a_c;

(b) Calculate the maximum permissible crack depth a_c for an operating pressure $p = 12$ MPa;

(c) Calculate the failure pressure p_c for a minimum detectable crack depth $a = 1$ mm.

Solution: Just as in Example 2.5, a material element of the vessel is subjected to a hoop σ_θ and a longitudinal σ_z stress given by

$$\sigma_\theta = \frac{pR}{t}\,, \quad \sigma_z = \frac{pR}{2t}\,. \tag{1}$$

The von Mises yield criterion expressed by Equation (1.3) takes the form

$$\sigma_\theta^2 - \sigma_\theta\sigma_z + \sigma_z^2 = (\sigma_Y/S)^2\,, \quad S = 2\,. \tag{2}$$

From Equations (1) and (2) we obtain

$$p_c = \frac{\sigma_Y t}{\sqrt{3}\,R} \tag{3}$$

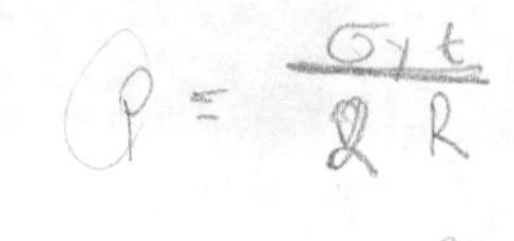

or

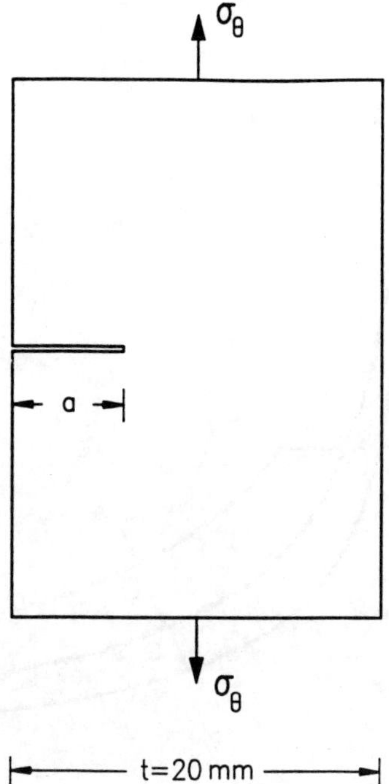

Fig. 5.16. A material element for the determination of stress intensity factor for a long axial surface crack of depth a.

$$p_c = 23.2 \times 10^{-3}\,\sigma_Y\,. \tag{4}$$

Equation (3) gives the maximum pressure the vessel can withstand without failure by yielding. For the three steels available we obtain

p_c(MPa)		
A	B	C
19.9	30.0	35.8

For a long axial surface crack of depth a the stress intensity factor is (Figure 5.16, Case 2 of Appendix 2.1).

$$K_{\mathrm{I}} = 1.12\sigma_\theta\,\sqrt{\pi a}\,. \tag{5}$$

The fracture condition is expressed by

$$K_{\mathrm{I}} = K_{\mathrm{Ic}}/S\,, \quad S = 2\,. \tag{6}$$

From Equations (1) to (3) we obtain

$$p = \frac{tK_{\mathrm{Ic}}}{2.24\sqrt{\pi}\,R\,\sqrt{a}} \tag{7}$$

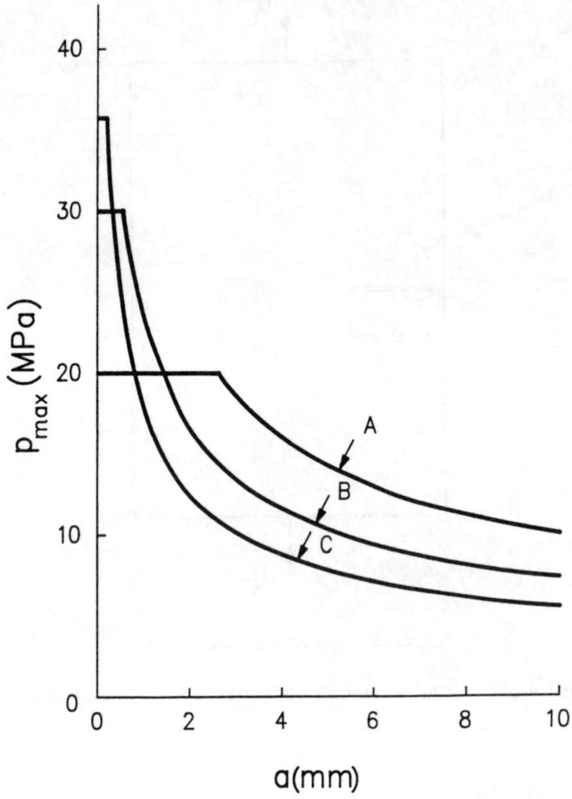

Fig. 5.17. Maximum pressure p_{max} versus crack depth a for the three steels of Example 5.5.

or

$$p = \frac{10.08 \times 10^{-3} K_{Ic}}{\sqrt{a}} \tag{8}$$

(a) Based on the results of the previous table and Equation (8) the maximum pressure versus crack depth curves for the three steels are shown in Figure 5.17.

(b) For $p = 12$ MPa the maximum permisible crack depths are calculated by Equation (8) for the three steels as

a_c(mm)		
A	B	C
7.04	3.64	2.12

Note that at $p = 12$ MPa material A failed by yielding.

(c) The failure pressure for a minimum detectable crack depth of $a = 1$ mm is calculated from Equation (8) for the three steels as

p_c(MPa)		
A	B	C
19.90(31.90)	22.96	17.60

Note that material A although withstands a pressure $p_{cr} = 31.90$ MPa for $a = 1$ mm, it fails at $p_c = 19.90$ MPa by yielding.

Example 5.6.

A cylindrical pipe with inner radius $b = 10$ cm and outer radius $c = 20$ cm is thermally stressed due to a temperature difference ΔT across the wall. Positive ΔT indicates that the outside wall temperature is higher than the inside. The pipe contains an initial crack of length $a = 1$ mm emanating from its inner radius. The material of the pipe has yield stress $\sigma_Y = 1000$ MPa, Poisson's ratio $\nu = 0.3$, modulus of elasticity $E = 210$ GPa, coefficient of thermal expansion $\alpha = 6.6 \times 10^{-6}$ °F and fracture toughness $K_{\mathrm{Ic}} = 100$ MPa $\sqrt{\mathrm{m}}$. Determine the maximum temperature difference $(\Delta T)_c$ the pipe can withstand without failure, with a factor of safety $S = 2$ against yielding and $S = 3$ against fracture.

Solution: The maximum stress at the rim of the inner radius of the pipe is given by (*Theory and Elasticity*, by S.P. Timoshenko and J.N. Goodier, Third Edition, p. 449)

$$\sigma_{\max} = \frac{\alpha E \Delta T}{2(1-\nu)} \left[\frac{2}{1-(b/c)^2} - \frac{1}{\log(c/b)} \right] . \tag{1}$$

The condition of failure by yielding is expressed by

$$\sigma_{\max} = \sigma_Y / S \,, \quad S = 2 \,. \tag{2}$$

From Equations (1) and (2) we obtain

$$\frac{1000}{2} = \frac{(6.6 \times 10^{-6}) \times 210 \times 10^3 (\Delta T)_c}{2(1-0.3)} \left[\frac{2}{1-0.5^2} - \frac{1}{\log 2} \right] . \tag{3}$$

Thus,

$$(\Delta T)_c = 413 \text{ °F} \,. \tag{4}$$

Equation (4) gives the maximum temperature the pipe can withstand without failure by yielding.

The stress intensity factor K_{I} at the crack tip is (Case 2 of Appendix 2.1)

$$K_{\mathrm{I}} = 1.12 \sigma_{\max} \sqrt{\pi a} \tag{5}$$

or

$$K_{\mathrm{I}} = 1.12 \times (1.21(\Delta T)_c)\sqrt{\pi \times 10^{-3}}$$

or

$$K_{\mathrm{I}} = 0.076(\Delta T)_c\,. \tag{6}$$

The condition of failure by fracture is expressed by

$$K_{\mathrm{I}} = K_{\mathrm{Ic}}/S\,,\ S = 3\,. \tag{7}$$

From Equations (6) and (7) we obtain

$$(\Delta T)_c = 438\ {}^{\circ}\mathrm{F}\,. \tag{8}$$

Equation (8) gives the maximum temperature the pipe can withstand without failure by fracture.

Comparing the values of $(\Delta T)_c$ from Equations (4) and (8) we see that there is little difference between the prediction based on the maximum stress criterion and that obtained by fracture mechanics.

Example 5.7.

A center cracked large plate of steel with thickness 10 mm is subjected to a uniform tension $\sigma = 300$ MPa perpendicular to the crack plane. Calculate the maximum crack length the plate can withstand without failure. $\sigma_Y = 860$ MPa, $K_{\mathrm{Ic}} = 100$ MPa $\sqrt{\mathrm{m}}$.

Solution: The applied stress $\sigma = 300$ MPa is smaller than the yield stress $\sigma_Y = 860$ MPa of the material. Thus, the plate does not fail by yielding.

The plate thickness $B = 10$ mm is smaller than the minimum thickness required for plane strain $B_c = 2.5\ (K_{\mathrm{Ic}}/\sigma_Y)^2 = 34$ mm. Thus, the critical stress intensity factor K_c of the plate is not the plane strain stress intensity factor K_{Ic}. It is calculated from Equation (5.7) as

$$K_c = 100\sqrt{1 + \frac{1.4}{(10 \times 10^{-3})^2}\left(\frac{100}{860}\right)^4} = 189\ \mathrm{MPa}\sqrt{\mathrm{m}}\,. \tag{1}$$

The stress intensity factor of a crack of length $2a$ in the plate is calculated as

$$K_{\mathrm{I}} = \sigma\sqrt{\pi a} = 532\sqrt{a}\,. \tag{2}$$

The fracture condition is expressed by

$$K_{\mathrm{I}} = K_c\,. \tag{3}$$

Equations (1) to (3) show that the maximum crack length the plate can withstand is

$$2a = 25.2 \text{ cm} . \tag{4}$$

Problems

5.1. A three-point bend specimen was tested according to the ASTM E399 procedure. The 0.2 percent offset yield stress of the material is $\sigma_Y = 1$ GPa and the modulus of elasticity is $E = 210$ GPa. The specimen was tested at a loading rate of 60 kN/min. A chevron starter notch was machined and the specimen was subjected to 20,000 cycles at $P_{max} = 20$ kN and $P_{min} = 0$. The final stage of fatigue crack growth was conducted for 40,000 cycles at $P_{max} = 15$ kN and $P_{min} = 0$. The specimen dimensions were measured as

$S = 40$ cm, $W = 10$ cm, $B = 5$ cm
$a_1 = 4.993$ cm
$a_2 = 5.008$ cm
$a_3 = 5.999$ cm
$a(\text{surface}) = 4.925$ cm
$a(\text{surface}) = 4.916$ cm.

The maximum load and the secant load of the test record were measured as $P_{max} = 108$ kN, $P_Q = 100$ kN. Calculate K_Q and comment on the validity of the test.

5.2. A compact tension specimen was tested according to the ASTM E399 procedure. The 0.2 per cent offset yield stress of the material is $\sigma_Y = 900$ MPa and the modulus of elasticity is $E = 210$ GPa. The specimen dimensions were measured as $W = 10$ cm, $B = 5$ cm, while the crack lengths at equal locations across the crack front were measured as 4.90, 4.93, 5.05, 4.95, 4.85 cm. The maximum load and the secant load of the test record were measured as $P_{max} = 70$ kN and $P_Q = 65$ kN. Calculate K_Q and comment on the validity of the test.

5.3. The load-displacement test record of the compact tension specimen of Problem 5.2 is shown in Figure 5.18. Calculate K_Q and comment on the validity of the test.

5.4. A three-point bend specimen with $S = 40$ cm, $W = 10$ cm, $a = 5$ cm and $B = 5$ cm was tested according to the ASTM E399 procedure. The 0.2 offset yield stress of the material is 400 MPa. The secant load is $P_Q = 60$ kN and the maximum load is $P_{max} = 65$ kN. Calculate K_{Ic} and comment on its validity.

5.5. A compact tension specimen of a steel with $W = 20$ cm, $a = 9$ cm and $B = 8$ cm was tested according to the ASTM E399 procedure. The 0.2 offset yield stress of the material is $\sigma_Y = 900$ MPa and the secant load P_Q was measured to be

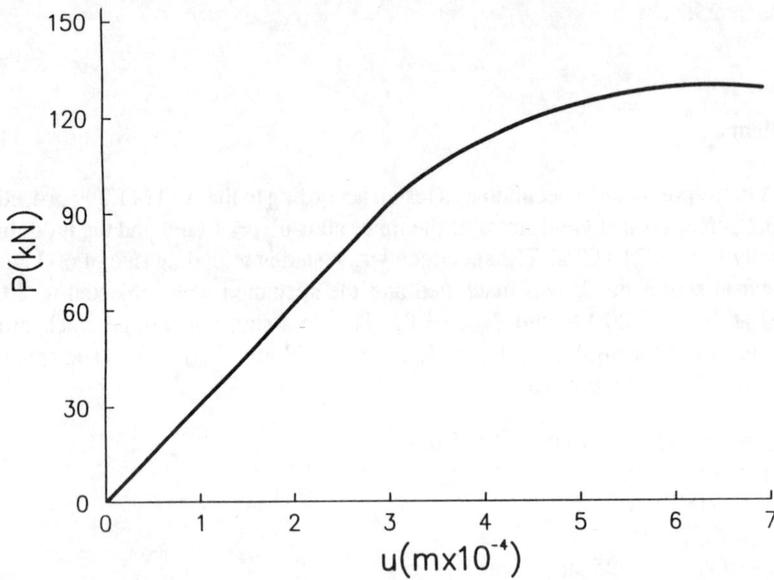

Fig. 5.18. Load-displacement $(P-u)$ record of the compact tension specimen of Problem 5.3.

300 kN. Calculate K_{Ic} and comment on its validity.

5.6. Determine the maximum K_{Ic} value that may be determined according to ASTM standards on a 30 mm thick plate with 0.2 offset yield stress (a) 400 MPa, (b) 2000 MPa.

5.7. Assume that the extension Δa of a crack of initial length a_0 is equal to the plane strain plastic zone radius calculated according to the Irwin model. For plane strain show that

$$\frac{\Delta a}{a_0} \leq 0.0212 \simeq 0.02\,.$$

5.8. Use the result of the previous problem to justify the determination of P_Q using a 5 percent secant offset line on the load-displacement record according to ASTM E399 procedure. For this reason use the result that when the load-displacement $(P-u)$ record takes the form

$$\frac{uEB}{P} = f(a/W)$$

where B and W are the specimen thickness and width, and E the modulus of elasticity, the quantity

$$H = \frac{a_0}{W}\,\frac{1}{f}\,\frac{\mathrm{d}f}{\mathrm{d}(a_0/W)}\,, \quad f = f(a/W)$$

takes an average value of 2.5 for the recommended range of values a_0/W lying between 0.45 and 0.55.

5.9. The crack growth resistance curve of a certain material at a thickness 2 mm is expressed by

$$R = \frac{K_{\mathrm{Ic}}^2}{E} + \frac{1}{2}(\Delta a)^{0.5}\ \mathrm{MJ/m^2}\,, \quad K_{\mathrm{Ic}} = 95\ \mathrm{MPa}\sqrt{\mathrm{m}}\,, \quad E = 210 \times 10^3\ \mathrm{MPa}\,.$$

Consider a center cracked plate of width 10 cm and thickness 2 mm with a crack of length 1 cm. Calculate the length of stable crack growth, the critical crack length, and the critical stress at instability.

5.10. As in Problem 5.9 for a center cracked plate of width 5 cm.

5.11. Consider a double cantilever beam (DCB, Figure 4.14) with dimensions: $B = 2$ cm, $a = 50$ mm, $h = 10$ mm. For the material of Problem 5.9 calculate the length of stable crack growth, the critical length, and the critical load at instability.

5.12. Determine the stability condition for the double cantilever beam (DCB) of Problem 4.7 tested in a soft (load-controlled) or a hard (displacement-controlled) testing machine.

5.13. As in Problem 5.12 for the strip of Problem 4.10.

5.14. As in Problem 5.12 for the DCB of Problem 4.24.

5.15. As in Problem 5.12 for a center cracked plate.

5.16. Calculate the load required for crack growth of a double cantilever beam (Figure 4.14) with $a = 20$ cm, $h = 4$ cm, $B = 1$ cm, $E = 210$ GPa and $G_c = 200$ kJ/m^2.

5.17. Calculate the critical stress for growth of a crack of length 2 mm in a large thin plate of steel, loaded by a uniform stress perpendicular to the crack plane. Take $E = 210$ GPa, $\gamma = 2$ Jm^{-2}, $\gamma_p = 2 \times 10^4$ Jm^{-2}.

5.18. A sheet of glass with width 50 cm and thickness 1 mm is subjected to a tensile stress of 10 MPa. Determine the minimum crack length that would lead to fracture. Take the following properties of glass: $E = 60$ GPa, $\nu = 0.25$, $\gamma = 0.5$ J/m^2, $\sigma_f = 150$ MPa.

5.19. A center cracked plate of steel with width 10 cm is subjected to a uniform tension 200 MPa perpendicular to the crack plane. Calculate the maximum crack length the plate can withstand without failure. $K_{\mathrm{Ic}} = 55$ MPa $\sqrt{\mathrm{m}}$.

5.20. Consider an edge cracked plate of steel with width 5 cm subjected to a uniform tension 200 MPa perpendicular to the crack plane. Calculate the maximum crack length the plate can withstand without failure. $K_{Ic} = 55$ MPa $\sqrt{\text{m}}$.

5.21. A crack of length 2 mm emanates from a circular hole of radius 50 mm in a large plate of steel subjected to a uniform stress perpendicular to the crack plane. Calculate the maximum stress the plate can withstand without failure. $\sigma_Y = 860$ MPa, $K_{Ic} = 100$ MPa $\sqrt{\text{m}}$.

5.22. A spherical vessel of steel with radius 1 m and thickness 40 mm contains a through crack of length 1 mm. Calculate the maximum internal pressure the vessel can withstand without failure. $\sigma_Y = 860$ MPa, $K_{Ic} = 100$ MPa $\sqrt{\text{m}}$.

5.23. A crack of length 2 mm emanates from an elliptical hole with axes 100 mm and 50 mm in a large plate of steel subjected to a uniform stress perpendicular to the crack plane. Calculate the maximum stress the plate can withstand without failure. $\sigma_Y = 860$ MPa, $K_{Ic} = 100$ MPa $\sqrt{\text{m}}$.

5.24. A center cracked plate of width 10 cm contains a crack of length 4 cm. The plate is subjected to a uniform stress perpendicular to the crack plane. Calculate the maximum stress the plate can withstand without failure. $\sigma_Y = 450$ MPa and $K_{Ic} = 25$ MPa $\sqrt{\text{m}}$.

5.25. Calculate the maximum load the three-point bend specimen of Problem 2.32 (Figure 2.30) can withstand. $\sigma_Y = 1500$ MPa, $K_{Ic} = 60$ MPa $\sqrt{\text{m}}$. The factor of safety for failure by yielding is 3 and by fracture is 2.

5.26. Calculate the maximum radial interference the shaft of Problem 2.36 (Figure 2.33) can withstand. $c = 5$ cm, $b = 2$ cm, $a = 1$ mm; $E = 210$ GPa, $\nu = 0.3$, $\sigma_Y = 900$ MPa, $K_{Ic} = 100$ MPa $\sqrt{\text{m}}$. The factor of safety for failure by yielding and fracture is 2.

5.27. Calculate the maximum crack length the cylindrical pipe of Problem 2.37 can withstand when it is subjected to a temperature difference $\Delta T = 200$ °F, across the wall. $c = 15$ cm, $b = 10$ cm; $E = 210$ GPa, $\nu = 0.3$, $\alpha = 6.6 \times 10^{-6}$ °F, $\sigma_Y = 900$ MPa, $K_{Ic} = 100$ MPa $\sqrt{\text{m}}$. The factor of safety for failure by yielding and fracture is 2.

5.28. Plot the allowable stress versus crack length curves for a large plate with a through crack subjected to a uniform stress perpendicular to the crack plane for the three steels of Example 5.5.

5.29. Consider a circular disk of radius $c = 20$ cm with a hole of radius $b = 10$ cm rotating at an angular velocity $\omega = 2\pi N/60$, where N is the number of revolutions per minute. The disk contains an initial crack of length $a = 1$ cm emanating from its hole. The material of the disk has mass density $\rho = 7800$ kg/m^3, yield stress $\sigma_Y = 1000$ MPa, Poisson's ratio $\nu = 0.3$, and fracture toughness $K_{Ic} = 100$ MPa $\sqrt{\text{m}}$. Determine the maximum number of revolutions per minute N_{max} that the disk can rotate without failure, when the factor of safety is $S = 2$ and $S = 3$ against failure by yielding and fracture, respectively.

5.30. The circular disk of Problem 5.29 may be constructed from the three steels of Example 5.5. For each steel:

(a) Plot the maximum permissible number of revolutions per minute N_{max} versus crack length;

(b) Calculate the maximum permissible crack length for $N = 5,000$ rpm;

(c) Calculate N_{max} for a minimum detectable crack $a = 0.05$ mm.

5.31. The spherical vessel of Problem 5.22 may be constructed from the three steels of Example 5.5. The factor of safety against failure by yielding is 3 and by fracture is 2. For each steel:

(a) Plot the maximum permissible pressure p_{max} versus crack length a;

(b) Calculate the maximum permissible crack length a_{cr} for an operating pressure $p_c = 10$ MPa;

(c) Calculate p_{max} for a minimum detectable crack $a = 0.05$ mm.

5.32. The cylindrical pipe of Example 5.6 may be constructed from the three steels of Example 5.5. For each steel:

(a) Plot the maximum permissible temperature T_{cr} versus crack length;

(b) Calculate the maximum permissible crack length a_{cr} for an operating temperature difference $\Delta T = 200$ °F;

(c) Calculate the critical temperature difference $(\Delta T)_{cr}$ for a minimum detectable crack length $a = 0.05$ mm.

5.33. A rectangular panel of width 1 m is subjected to a tensile load of 100 MN perpendicular to the width. The smallest crack size that can be detected is 1 mm. Two steels with the following values of yield stress and fracture toughness are available for constructing the panel: A. $\sigma_Y = 860$ MPa, $K_{Ic} = 100$ MPa $\sqrt{\text{m}}$ and

B. $\sigma_Y = 1550$ MPa, $K_{Ic} = 55$ MPa $\sqrt{m}$. Find which material gives the smallest thickness of the panel.

5.34. Calculate the maximum stress the plate of Problem 5.19 can withstand without failure when it has a thickness of 5 mm.

5.35. Calculate the maximum stress the plate of Problem 5.24 can withstand without failure when it has a thickness of 20 mm.

5.36. A long steel plate of width 20 cm and thickness 2 cm is subjected to a load 1.2 MN. Calculate the maximum crack length of a central or edge crack the plate can withstand without failure. $\sigma_Y = 860$ MPa, $K_{Ic} = 100$ MPa $\sqrt{m}$.

Appendix 5.1

Values of critical stress intensity factor K_{Ic} and 0.2 per cent offset yield stress σ_Y at room temperature for various alloys.

Material	σ_Y		K_{Ic}	
	MPa	ksi	MPa $\sqrt{m}$	ksi $\sqrt{in}$
300 Maraging Steel	1669	242	93.4	85
350 Maraging Steel	2241	325	38.5	35
D6AC Steel	1496	217	66.0	60
AISI 4340 Steel	1827	265	47.3	43
A533B Reactor Steel	345	50	197.8	180
Carbon Steel	241	35	219.8	200
Al 2014–T4	448	65	28.6	26
Al 2024–T3	393	57	34.1	31
Al 7075–T651	545	79	29.7	27
Al 7079–T651	469	68	33.0	30
Ti 6Al–4V	1103	160	38.5	35
Ti 6Al–6V–2Sn	1083	157	37.4	34
Ti 4Al–4Mo–2Sn–0.5Si	945	137	70.3	64

References

5.1. Krafft, J.M., Sullivan, A.M. and Boyle, R.W. (1961) 'Effect of dimensions on fast fracture instability of notched sheets', *Proceedings of Crack Propagation Symposium*, College of Aeronautics, Cranfield (England), Vol. 1, pp. 8–28.

5.2. *Standard Practice for R-Curve-Determination. ASTM Annual Book of Standards*, Part 10, American Society for Testing and Materials, E561–81, pp. 680–699 (1981).

5.3. Irwin, G.R. (1948) 'Fracture dynamics', in *Fracture of Metals*, American Society for Metals, Cleveland, U.S.A., pp. 147–166.

5.4. Orowan, E. (1948) 'Fracture and strength of solids', in *Reports on Progress in Physics* XII, pp. 185–232.

Chapter 6

J-Integral and Crack Opening Displacement Fracture Criteria

6.1. Introduction

A number of investigators have proposed the mathematical formulation of elastostatic conservation laws as path independent integrals of some functionals of the elastic field over the bounding surface of a closed region. For notch problems, Rice [6.1] introduced the two-dimensional version of the conservation law, a path independent line integral, known as the J-integral. The present chapter is devoted to the theoretical foundation of the path independent J-integral and its use as a fracture criterion. The critical value of the opening of the crack faces near the crack tip is also introduced as a fracture criterion.

First we present the definition of the J-integral in two-dimensional crack problems, and its physical interpretation in terms of the rate of change of potential energy with respect to an incremental extension of the crack. The path independent nature of the integral allows the integration path to be taken close or far away from the crack tip. Although the J-integral is based on purely elastic (linear or nonlinear) analysis, its use for plasticity-type materials has been supported by experimentation or numerical analysis. We present experimental methods for the evaluation of the integral. We introduce a failure criterion based on the J-integral and describe the standard ASTM method for the experimental determination of the critical value J_{Ic} of the J-integral. The chapter concludes with a brief presentation of the crack opening displacement fracture criterion.

6.2. Path-independent integrals

First we introduce some path-independent surface integrals in three-dimensional space. We assume that the solid body is linearly or nonlinearly elastic, homogeneous, anisotropic and in a state of static equilibrium under the action of a system of traction T_k. Denote by Σ the bounding surface of the region R occupied by the body and refer all quantities to a fixed Cartesian coordinate system $Ox_1x_2x_3$. To simplify the analysis we assume that the deformation is small. The stress tensor σ_{ij} is obtained from the elastic strain energy ω by

$$\sigma_{ij} = \sigma_{ji} = \frac{\partial \omega}{\partial \varepsilon_{ij}}\,, \quad \omega(0) = 0 \tag{6.1}$$

where ε_{ij} denotes the strain tensor. The strain energy density ω is considered to be continuously differentiable with respect to strain. For elastic behavior Equation (6.1) becomes

$$\omega = \int_0^{\varepsilon_{kl}} \sigma_{ij}\,\mathrm{d}\varepsilon_{ij} \tag{6.2}$$

where the integral is path independent in the strain space.

In the absence of body forces the equations of equilibrium take the form

$$\sigma_{ij,i} = \frac{\partial \sigma_{ij}}{\partial x_i} = 0 \tag{6.3}$$

and the traction vector T_j on surface S is given by

$$T_j = \sigma_{ij} n_i \tag{6.4}$$

where n_i denotes the normal vector.

For small deformation the strain tensor is derived from the displacement field by

$$\varepsilon_{ij} = \varepsilon_{ji} = \frac{1}{2}\left(u_{i,j} + u_{j,i}\right). \tag{6.5}$$

Consider the integrals

$$Q_j = \int_\Sigma \left(\omega n_j - T_k u_{k,j}\right)\mathrm{d}\Sigma\,, \quad j,k = 1,2,3 \tag{6.6}$$

where Σ is a closed surface bounding a region R which is assumed to be free of singularities.

Equation (6.6) can be put in the form

$$Q_j = \int_\Sigma \left(\omega n_j - \sigma_{lk} n_l u_{k,j}\right)\mathrm{d}\Sigma = \int_\Sigma \left(\omega \delta_{jl} - \sigma_{lk} u_{k,j}\right) n_l\,\mathrm{d}\Sigma\,. \tag{6.7}$$

and Gauss's divergence theorem gives

$$Q_j = \int_R \left(\omega \delta_{jl} - \sigma_{lk} u_{k,j}\right)_{,l}\,\mathrm{d}V\,. \tag{6.8}$$

The integrand takes the form

$$
\begin{aligned}
(\omega\delta_{jl} - \sigma_{lk}u_{k,j})_{,l} &= \omega_{,j} - \sigma_{lk,l}u_{k,j} - \sigma_{lk}u_{k,jl} \\
&= \frac{\partial\omega}{\partial\varepsilon_{lk}}\,\varepsilon_{lk,j} - \sigma_{lk}u_{k,lj} = \sigma_{lk}(\varepsilon_{lk,j} - u_{k,lj}) \\
&= \sigma_{lk}(\varepsilon_{lk} - u_{k,l})_{,j} = \sigma_{kl}(\varepsilon_{kl} - u_{k,l})_{,j} \\
&= -\sigma_{kl}r_{kl,j} = 0\,.
\end{aligned}
$$

In these derivations we used Equations (6.1) to (6.2). r_{kl} denotes the nonsymmetrical rotation tensor for small deformation given by

$$r_{ij} = -r_{ji} = \frac{1}{2}(u_{i,j} - u_{j,i})\,. \tag{6.9}$$

Thus, we have

$$Q_j = 0\,. \tag{6.10}$$

We can prove in a similar manner that Equation (6.10) holds when the elastic strain energy density ω in Equation (6.6) is replaced by the complementary elastic strain energy density Ω given by

$$\Omega = \int_0^{\sigma_{kl}} \varepsilon_{ij}\mathrm{d}\sigma_{ij} \tag{6.11}$$

or

$$\Omega(\sigma_{ij}) = \sigma_{ij}\varepsilon_{ij} - \omega(\varepsilon_{ij})\,, \quad \Omega(0) = 0\,. \tag{6.12}$$

The integral in Equation (6.11) is path independent in the stress space. The strain is obtained from Ω by

$$\varepsilon_{ij} = \frac{\partial\Omega}{\partial\sigma_{ij}} \tag{6.13}$$

in the same way that the stress is obtained from Ω by Equation (6.1).

6.3. *J*-integral

(a) Definition

For the particular case of the two-dimensional plane elastic problem, consider the integral

$$J = Q_1 = \int_\Gamma (\omega n_1 - T_k u_{k},1)\,\mathrm{d}s \tag{6.14}$$

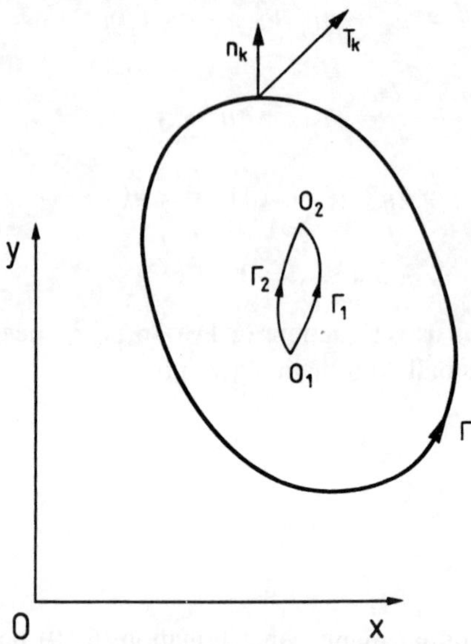

Fig. 6.1. A closed contour Γ and paths Γ_1 and Γ_2 between two points O_1 and O_2 in a continuum.

where Γ is a closed contour bounding a region R (Figure 6.1). With

$$n_1 = \frac{\mathrm{d}y}{\mathrm{d}s} \,. \tag{6.15}$$

Equation (6.14) becomes

$$J = \int_{\Gamma} \omega \,\mathrm{d}y - T_k \frac{\partial u_k}{\partial x} \,\mathrm{d}s \quad (k = 1, 2) \tag{6.16}$$

and defines the J-integral along a closed contour in the two-dimensional space. From Equation (6.10) it follows that $J = 0$. Since J is zero for any closed paths, the J-integrals along any two paths Γ_1, Γ_2 connecting any two points O_1 and O_2, are equal (Figure 6.1). We have

$$J_1 = J_{\Gamma_1}[\ldots] = J_2 = \int_{\Gamma_2} [\ldots] \,. \tag{6.17}$$

(b) Application to notches and cracks

Consider a notch or a crack, with flat surfaces parallel to the x-axis, which may have an arbitrary root radius (Figure 6.2). The J-integral defined from Equation (6.16) is calculated along a path Γ in a counterclockwise sense starting from an arbitrary point on the flat part of the lower notch surface and ending at an arbitrary point on the flat part of the upper surface of the notch. The region R bounded by the closed

? How to introduce 'c' into J-integral formular

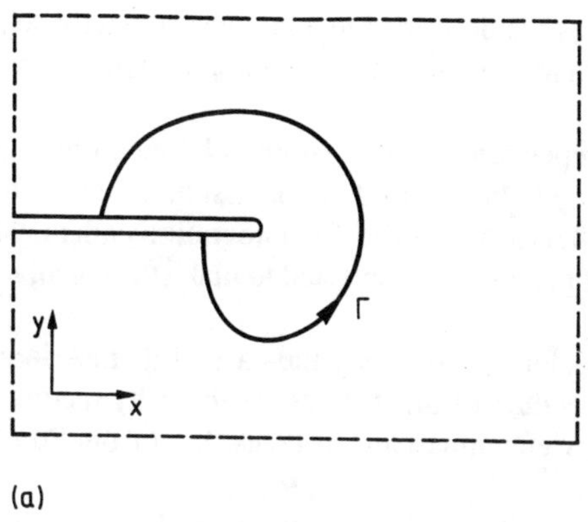

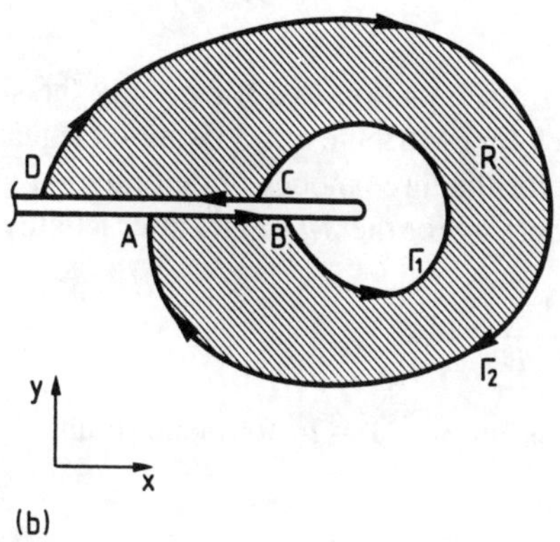

Fig. 6.2. (a) Path Γ starting from the lower and ending up to the upper face of a notch in a two-dimensional body. The flat notch surfaces are parallel to the x-axis. (b) Paths Γ_1 and Γ_2 around a notch tip.

contour $AB\Gamma_1 CD\Gamma_2 A$ is free of singularities and the J-integral calculated along $AB\Gamma_1 CD\Gamma_2 A$ is zero. We have

$$J_{AB\Gamma_1 CD\Gamma_2 A} = J_{B\Gamma_1 C} + J_{CD} + J_{D\Gamma_2 A} + J_{AB} = 0\,. \tag{6.18}$$

When parts AB and CD of the notch surfaces that are parallel to the x-axis are traction free, $\mathrm{d}y = 0$ and $T_k = 0$, and Equation (6.16) implies that

$$J_{CD} = J_{AB} = 0\,.$$

Equation (6.18) takes the form

$$J_{B\Gamma_1 C} + J_{D\Gamma_2 A} = 0$$

or

$$J_{B\Gamma_1 C} = J_{A\Gamma_2 D} \tag{6.19}$$

when the contour $A\Gamma_2 D$ is described in a counterclockwise sense. Equation (6.19) establishes the path independence of the J-integral defined by Equation (6.16) for notch problems.

Note that path independence of the J-integral defined by Equation (6.16) for an arbitrary path Γ (Figure 6.2a) is based on the assumption that the notch surfaces are traction free and parallel to the x-axis. The integration path may be taken close or far away from the crack tip, and can be selected to make the calculation of the J-integral easy.

As an example problem, let us study the case of a crack in a mixed-mode stress field governed by the values of the three stress intensity factors K_{I}, K_{II}, K_{III}. For a circular path of radius r encompassing the crack tip, J from Equation (6.16) becomes

$$J = r \int_{-\pi}^{\pi} \left[\omega(r,\theta)\, \cos\,\theta - T_k(r,\theta)\, \frac{\partial u_k(r,\theta)}{\partial x} \right] \mathrm{d}\theta\,. \tag{6.20}$$

Letting $r \to 0$ we see from Equation (6.20) that only the singular terms of the stress field around the crack tip contribute to J. Using the equations of the singular stresses and displacement for the three modes of deformation given in Section 2.4, we obtain the following equation for the J-integral after lengthy algebra.

$$J = \frac{\eta K_{\mathrm{I}}^2}{E} + \frac{\eta K_{\mathrm{II}}^2}{E} + \frac{1+\nu}{E}\, K_{\mathrm{III}}^2 \tag{6.21}$$

where $\eta = 1$ for plane stress and $\eta = 1 - \nu^2$ for plane strain.

6.4. Relationship between the J-integral and potential energy

We look for a physical interpretation of the J-integral in terms of the rate of change of potential energy with respect to incremental change of crack size. The derivation concerns a linear or nonlinear elastic plane body, with a crack of length a subjected to prescribed tractions and displacements along parts of its boundary. Tractions and displacements are assumed to be independent of crack length. The body is referred to a fixed system of Cartesian coordinates $x_1 x_2$ with the x_1-axis parallel to the crack faces (Figure 6.3). It is also assumed that the crack extends in a self-similar manner.

The potential energy $\Pi(a)$ of the body is given by

$$\Pi(a) = \int_A \omega\, \mathrm{d}A - \int_\Gamma T_k u_k\, \mathrm{d}s \tag{6.22}$$

where A is the area of the body and Γ its boundary.

Under the previous assumptions differentiation of Equation (6.22) with respect to crack length a yields

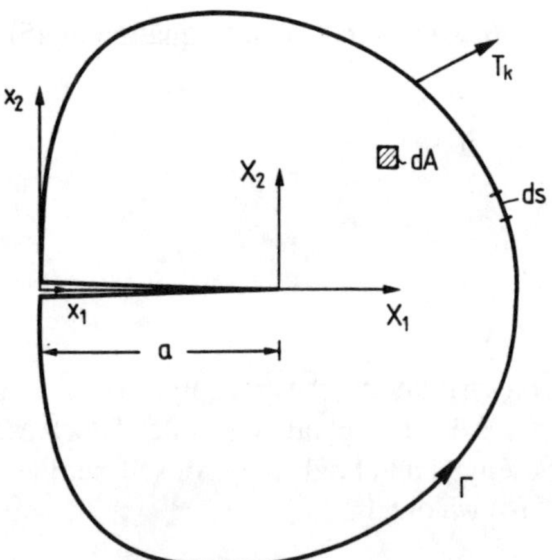

Fig. 6.3. A two-dimensional cracked body.

$$\frac{\mathrm{d}\Pi}{\mathrm{d}a} = \int\limits_A \frac{\mathrm{d}\omega}{\mathrm{d}a}\,\mathrm{d}A - \int\limits_\Gamma T_k\,\frac{\mathrm{d}u_k}{\mathrm{d}a}\,\mathrm{d}s\,. \tag{6.23}$$

A new coordinate system X_1X_2 attached to the crack tip is introduced:

$$X_1 = x_1 - \alpha\,;\quad X_2 = x_2 \tag{6.24a}$$

and

$$\frac{\mathrm{d}}{\mathrm{d}a} = \frac{\partial}{\partial a} + \frac{\partial}{\partial X_1}\,\frac{\partial X_1}{\partial a} = \frac{\partial}{\partial a} - \frac{\partial}{\partial x_1}\,. \tag{6.24b}$$

Thus, Equation (6.23) takes the form

$$\frac{\mathrm{d}\Pi}{\mathrm{d}a} = \int\limits_A \left(\frac{\partial\omega}{\partial a} - \frac{\partial\omega}{\partial x_1}\right)\mathrm{d}A - \int\limits_\Gamma T_k\left(\frac{\partial u_k}{\partial a} - \frac{\partial u_k}{\partial x_1}\right)\mathrm{d}s\,. \tag{6.25}$$

We have

$$\frac{\partial\omega}{\partial a} = \frac{\partial\omega}{\partial\varepsilon_{ij}}\,\frac{\partial\varepsilon_{ij}}{\partial a} = \sigma_{ij}\,\frac{\partial\varepsilon_{ij}}{\partial a}$$

and by applying the principle of virtual work we obtain

$$\int\limits_A \frac{\partial\omega}{\partial a}\,\mathrm{d}A = \int\limits_A \sigma_{ij}\,\frac{\partial\varepsilon_{ij}}{\partial a}\,\mathrm{d}A = \int\limits_\Gamma T_k\,\frac{\partial u_k}{\partial a}\,\mathrm{d}s\,. \tag{6.26}$$

Furthermore, the divergence theorem yields

$$\int\limits_A \frac{\partial\omega}{\partial x_1}\,\mathrm{d}A = \int\limits_\Gamma \omega\,\mathrm{d}x_2\,. \tag{6.27}$$

Introducing Equations (6.26) and (6.27) into Equation (6.25) we get

$$-\frac{\mathrm{d}\Pi}{\mathrm{d}a} = \int_{\Gamma} \omega \,\mathrm{d}x_2 - T_k \frac{\partial u_k}{\partial x_1} \,\mathrm{d}s \tag{6.28}$$

or

$$J = -\frac{\mathrm{d}\Pi}{\mathrm{d}a} \tag{6.29}$$

for any path of integration surrounding the crack tip.

Equation (6.29) expresses the J-integral as the rate of decrease of potential energy with respect to the crack length, and holds only for self-similar crack growth. From Equations (6.29) and (4.10) we obtain

$$J = G \tag{6.30}$$

for "fixed-grips" or "dead-load" loading during crack growth.

For a crack in a mixed-mode stress field, the value of the J-integral given by Equation (6.30) represents the elastic strain energy release rate only for self-similar crack growth. However, a crack under mixed-mode loading does not extend along its own plane and therefore the value of energy release rate G given from Equation (6.30) is physically unrealistic.

6.5. J-integral fracture criterion

The J-integral can be viewed as a parameter which characterizes the state of affairs in the region around the crack tip. This argument is supported by the following fundamental properties of J:

(i) J is path independent for linear or nonlinear elastic material response;

(ii) J is equal to $-\mathrm{d}\Pi/\mathrm{d}a$ for linear or nonlinear elastic material response;

(iii) J is equal to G;

(iv) J can easily be determined experimentally;

(v) J can be related to the crack-tip opening displacement δ by a simple relation of the form $J = M\sigma_Y\delta$ (for the Dugdale model $M = 1$ (Example 6.2)).

Because of these characteristic properties J has been proposed [6.2] as an attractive candidate for a fracture criterion. Under opening-mode loading, the criterion for crack initiation takes the form

$$J = J_c \tag{6.31}$$

where J_c is a material property for a given thickness under specified environmental conditions. Under plane strain conditions, the critical value of J, J_{Ic}, is related to the plane strain fracture toughness K_{Ic} by (Equation (6.21))

$$J_{\mathrm{Ic}} = \frac{1-\nu^2}{E} K_{\mathrm{Ic}}^2 \,. \tag{6.32}$$

The above properties of the J-integral, which support its use as a fracture criterion, were derived under elastic material response. Attempts have been made to extend the realm of applicability of the J-integral fracture criterion to ductile fracture where extensive plastic deformation and possibly stable crack growth precede fracture instability.

Strictly speaking, the presence of plastic enclaves nullifies the path independence property of the J-integral. For any closed path surrounding the crack tip and taken entirely within the plastic zone or within the elastic zone, the necessary requirements for path independence (Equations (6.1) and (6.2)) are not satisfied. The stress is not uniquely determined by the strain, and the stress-strain constitutive equations relate strain increments to stresses and stress increments. In an effort to establish path independence for the J-integral the deformation theory of plasticity has been invoked. This theory is a nonlinear elasticity theory, and no unloading is permitted. Any solution based on the deformation theory of plasticity coincides exactly with a solution based on the flow (incremental) theory of plasticity under proportional loading (the stress components change in fixed proportion to one another). No unloading is permitted at any point of the plastic zone. Although, strictly speaking, the condition of proportional loading is not satisfied in practice it is argued that in a number of stationary problems, under a single monotonically applied load, the loading condition is close to proportionality. A number of finite element solutions have supported this proposition [6.3, 6.4].

J is used today as a fracture criterion in situations of appreciable plastic deformation. J-dominance conditions were formulated for such circumstances. For the special cases of the bend and center cracked specimen they take the form [6.5–6.7]

$$\frac{b\sigma_Y}{J} > 25 \tag{6.33}$$

for the bend specimen, and

$$\frac{b\sigma_Y}{J} > 175 \tag{6.34}$$

for the center cracked specimen, where b is the uncracked ligament.

6.6. Experimental determination of the J-integral

This section deals with the experimental determination of the J-integral and its critical value J_{Ic}. We present the multiple-specimen method, the one-specimen method and

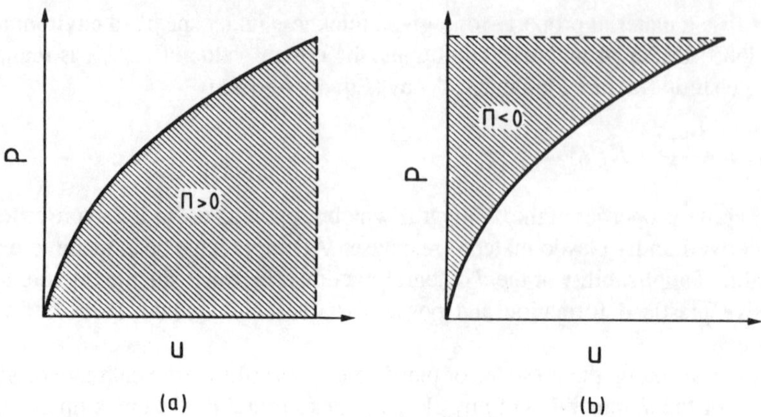

Fig. 6.4. Potential energy shown as shaded area for (a) "fixed-grips" and (b) "dead-load" conditions.

the standard test method according to the ASTM specifications. Before proceeding to the details of these methods we derive some general equations inherent in the test methods.

(a) General equations

The experimental determination of J follows from Equation (6.29). This states that J is equal to the rate of decrease of potential energy (defined from Equation (6.22)) with respect to the crack length. Experiments are usually performed under either "fixed-grips" (prescribed displacement) or "dead-load" (prescribed load) conditions. In the load-displacement diagram the potential energy is equal to the area included between the load-displacement curve and the displacement axis or the load axis, for fixed-grips or dead-load conditions, respectively (shaded areas of Figure 6.4(a) and 6.4(b)). Observe that the potential energy is positive for fixed-grips, and negative for dead-load conditions.

Consider the load-displacement curves corresponding to the crack lengths a and $a + \Delta a$ for "fixed-grips" or "dead-load" conditions in Figures 6.5(a) and 6.5(b). The area included between the two curves represents the value of $J\,\Delta a$. Equation (6.29) and Figures 6.5(a) and 6.5(b) show that for crack growth under fixed grips

$$J = -\left(\frac{\partial \Pi}{\partial a}\right)_u = -\int_0^{u_0} \left(\frac{\partial P}{\partial a}\right)_u \mathrm{d}u \tag{6.35}$$

and under dead-load

$$J = -\left(\frac{\partial \Pi}{\partial a}\right)_P = -\int_0^{P_0} \left(\frac{\partial u}{\partial a}\right)_P \mathrm{d}P\,. \tag{6.36}$$

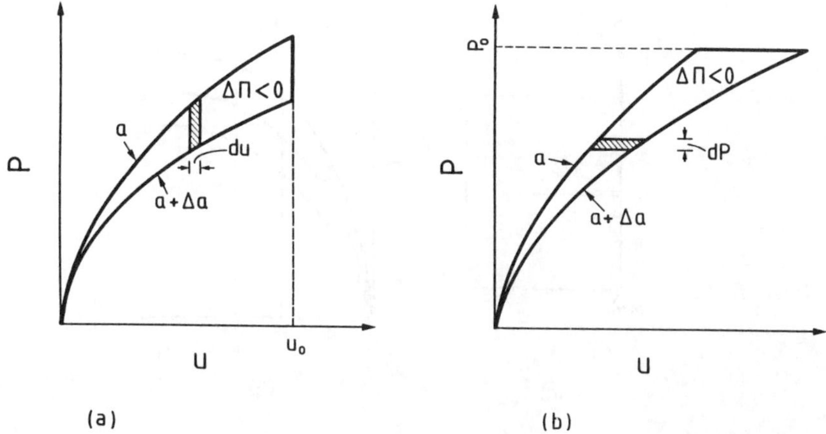

Fig. 6.5. Load-displacement curves for crack lengths a and $a + \Delta a$ under (a) "fixed-grips" and (b) "dead-load" conditions.

Equations (6.29), (6.35) and (6.36) form the basis for the experimental determination of J.

(b) Multiple-specimen method

This method is based on Equation (6.29) and was first introduced by Begley and Landes [6.2]. A number of identically loaded specimens with neighboring crack lengths is used (Figure 6.6(a)). The procedure is as follows:

(i) Load-displacement $(P - u)$ records, under fixed-grips, are obtained for several precracked specimens, with different crack lengths (Figure 6.6(b)). For given values of displacement u, we calculate the area underneath the load-displacement record, which is equal to the potential energy Π of the body at that displacement.

(ii) Π is plotted versus crack length for the previously selected displacements (Figure 6.6(c)).

(iii) The negative slopes of the $\Pi - a$ curves are determined and plotted versus displacement for different crack lengths (Figure 6.6(d)). Thus the $J - u$ curves are obtained for different crack lengths.

The critical value J_{Ic} of J is determined from the displacement at the onset of crack extension. Since J_{Ic} is a material constant, the values of J_{Ic} obtained from different crack lengths should be the same.

The multiple-specimen method presents the disadvantage that several specimens are required to obtain the J versus displacement relation. Furthermore, accuracy

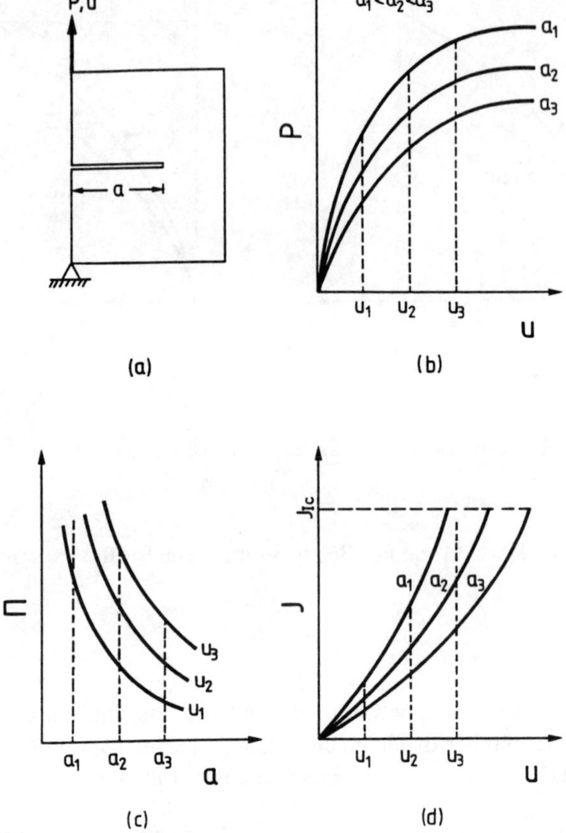

Fig. 6.6. Multiple-specimen method for calculating the J-integral.

problems enter in the numerical differentiation of the $\Pi - a$ curves. A technique for determining J from a single test becomes very attractive and is described next.

(c) Single-specimen method

This method was first proposed by Rice *et al.* [6.8] and is based on Equation (6.35) or (6.36). We consider a deeply cracked bend specimen, a compact specimen and a three-point bend specimen.

For the cracked bend specimen shown in Figure 6.7(a), Equation (6.36) becomes

$$J = \int_0^M \left(\frac{\partial \theta}{\partial a}\right)_M \mathrm{d}M . \tag{6.37}$$

Here M is the applied moment per unit thickness and θ is the angle of relative rotation

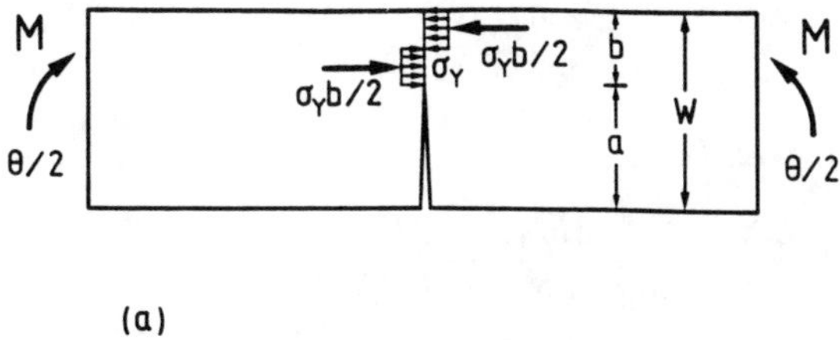

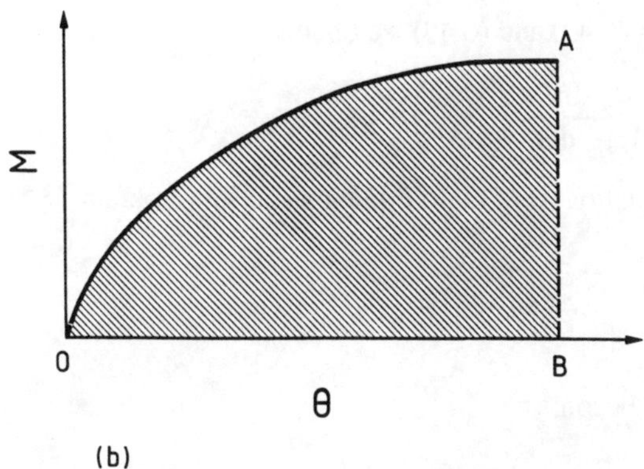

Fig. 6.7. (a) A deeply cracked bend specimen and (b) bending moment M versus angle θ of relative rotation of the specimen end sections.

of the end sections of the specimen. The angle θ can be put in the form

$$\theta = \theta_{nc} + \theta_c \tag{6.38}$$

where θ_{nc} represents the relative rotation of the uncracked specimen and θ_c is the additional rotation caused by the presence of the crack. It is now assumed that the ligament b is small compared to W, so that the rotation θ_c is mainly due to deformation of the ligament. Equation (6.37) takes the form

$$J = \int_0^M \left(\frac{\partial \theta_c}{\partial a}\right)_M \mathrm{d}M \tag{6.39}$$

since θ_{nc} is independent of a. When L is large compared to W we can assume that θ_c depends only on M/M_0, where M_0 is the plastic limit moment. We have

$$\theta_c = f\left(\frac{M}{M_0}\right) \tag{6.40}$$

where

$$M_0 = \frac{\sigma_Y b^2}{4} \,. \tag{6.41}$$

Equation (6.40) gives

$$\left(\frac{\partial \theta_c}{\partial a}\right)_M = -\frac{\partial \theta_c}{2b} = \frac{M}{M_0^2}\frac{\mathrm{d}M_0}{\mathrm{d}b}\frac{\partial f}{\partial (M/M_0)} \tag{6.42}$$

and

$$\left(\frac{\partial \theta_c}{\partial M}\right)_a = \frac{1}{M_0}\frac{\partial f}{\partial (M/M_0)} \,. \tag{6.43}$$

From Equations (6.42) and (6.43) we obtain

$$\left(\frac{\partial \theta_c}{\partial a}\right)_M = \frac{M}{M_0}\frac{\mathrm{d}M_0}{\mathrm{d}b}\left(\frac{\partial \theta_c}{\partial M}\right)_a \,. \tag{6.44}$$

Substituting Equation (6.44) into Equation (6.39) we obtain

$$J = \frac{1}{M_0}\frac{\mathrm{d}M_0}{\mathrm{d}b}\int_0^{\theta_c} M \,\mathrm{d}\theta_c \,. \tag{6.45}$$

Equation (6.41) becomes

$$\frac{1}{M_0}\frac{\mathrm{d}M_0}{\mathrm{d}b} = \frac{2}{b} \,. \tag{6.46}$$

From Equations (6.45) and (6.46) we obtain

$$J = \frac{2}{b}\int_0^{\theta_c} M \,\mathrm{d}\theta_c \tag{6.47}$$

where the integral in Equation (6.47) is the area underneath the $M = M(\theta_c)$ curve and the θ_c-axis (Figure 6.7(b)).

It is not possible to apply Equation (6.47) directly to determine the J-integral because that is the total angle θ which is measured in an experiment. The critical value of J, J_{Ic}, can be obtained by determining the area under the M versus angle θ curve up to the point of crack extension, and subtracting the area of a similar, but uncracked, specimen. The second area is usually very small compared to the first and Equation (6.47) can be approximated as

$$J = \frac{2}{b}\int_0^{\theta} M \,\mathrm{d}\theta \,. \tag{6.48}$$

Equation (6.48) allows J to be determined from a single experiment.

For a compact specimen (Figure 6.8) we find [6.9]

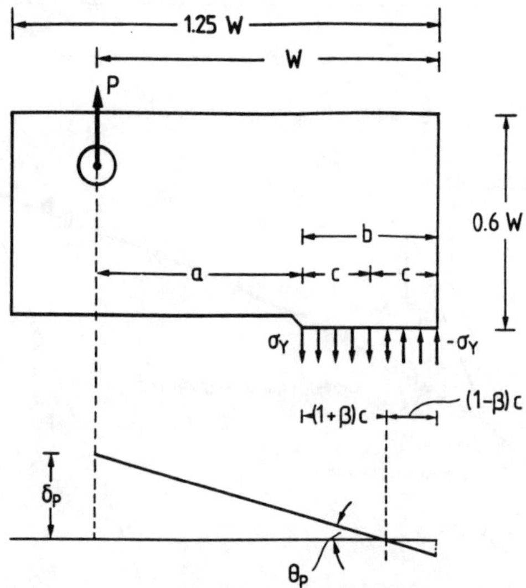

Fig. 6.8. A deeply cracked compact specimen at plastic collapse.

$$J = \frac{2}{b}\frac{1+\beta}{1+\beta^2}\int_0^{\delta_p} P\,\mathrm{d}\delta_p + \frac{2}{b}\frac{\beta(1-2\beta-\beta^2)}{(1+\beta^2)^2}\int_0^{P} \delta_p\,\mathrm{d}P \tag{6.49}$$

where δ_p is the plastic contribution to the load-point displacement and β is given by

$$\beta = 2\left[\left(\frac{a}{b}\right)^2 + \frac{a}{b} + \frac{1}{2}\right]^{1/2} - 2\left(\frac{a}{b} + \frac{1}{2}\right). \tag{6.50}$$

For $a/W > 0.5$ it has been found that one can use the total displacement δ instead of δ_p in Equation (6.49). Furthermore, $\beta = 0$ for deeply cracked specimens and Equation (6.49) becomes

$$J = \frac{2}{b}\int_0^{\delta} P\,\mathrm{d}\delta \tag{6.51}$$

which is similar to Equation (6.48).

For the deeply cracked three-point bend specimen J is again given by Equation (6.51).

(d) Standard test method

ASTM [6.10] issued a standard test method for determining J_{Ic}, the plane strain value of J at initiation of crack growth for metallic materials. The recommended

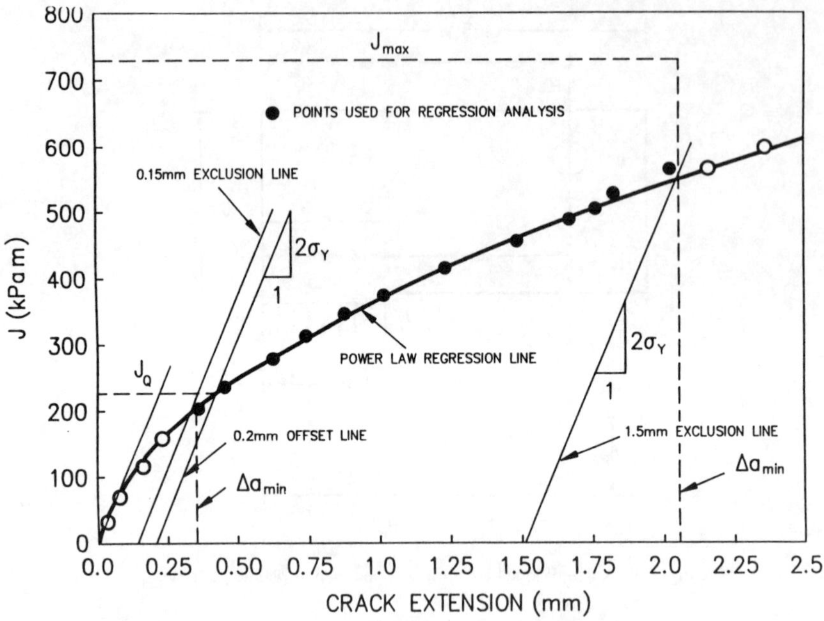

Fig. 6.9. Determination of J_{Ic} according to ASTM standards [6.10].

specimens are the three-point bend specimen and the compact specimen that contain deep initial cracks. The specimens are loaded to special fixtures, and applied loads and load-point displacements are simultaneously recorded during the test. For a valid J_{Ic} value, the crack ligament b and the specimen thickness B must be greater than $25J_{\mathrm{Ic}}/\sigma_Y$. The initial crack length for the three-point bend specimen must be at least $0.5W$, but not greater than $0.75W$, where W is the specimen width. The overall specimen length is $4.5W$, and the specimen thickness is $0.5W$. The geometry of the compact specimen is shown in Figure 6.8, where the initial crack length a is taken to satisfy $0.5W < a < 0.75W$, and the specimen thickness is $0.5W$.

To determine the value of J_{Ic} that corresponds to the onset of slow stable crack propagation we use the following procedure. We determine the J-integral for the bend specimen from Equation (6.48) and for the compact specimen from Equation (6.49). The latter can be approximated by

$$J = \frac{2}{b}\frac{1+\beta}{1+\beta^2}\int_0^\delta P\,\mathrm{d}\delta \tag{6.52}$$

where β is given from Equation (6.50). J is plotted against physical crack growth length, using at least four data points within specified limits of crack growth (Figure 6.9). We fit a power law expression of the form

$$J = C_1(\Delta a)^{C_2} \tag{6.53}$$

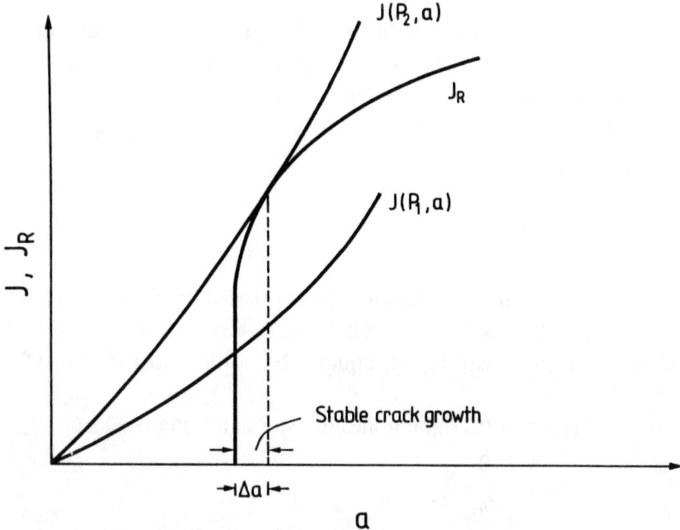

Fig. 6.10. Stable crack growth by J-resistance curve analysis.

to the experimental points and find the point at which it intersects the line originating at $\Delta a = 0.2$ mm with slope $2\sigma_Y$ (parallel to the blunting line $J = 2\sigma_Y \Delta a$). The value of J which corresponds to the point of intersection is J_{Ic}.

The blunting line approximates the apparent crack advance due to crack tip blunting when there is no slow stable crack tearing. We choose this line because we assume that, before tearing, the crack advance is equal to one half of the crack-tip opening displacement ($\Delta a = 0.5\delta$).

We draw two additional offset lines parallel to the blunting line and starting from the points $\Delta a = 0.15$ and 1.5 mm. For a valid test all data should be placed inside the area enclosed by the two parallel offset lines and the line $J = J_{\max} = b_0 \sigma_Y / 15$. Data outside these limits are not valid. The valid data points are used to determine the final regression curve of Equation (6.53).

The value of J_{Ic} can also be used to obtain an estimate of K_{Ic} from Equation (6.21) which, for opening-mode, takes the form

$$K_{\mathrm{Ic}}^2 = J_{\mathrm{Ic}} E / (1 - \nu^2) \,. \tag{6.54}$$

Equation (6.54) is used in situations where large specimen dimensions are required for a valid K_{Ic} test according to the ASTM specifications (Section 5.4).

6.7. Stable crack growth studied by the J-integral

The crack growth resistance curve method for the study of crack growth under small-scale yielding developed in Section 5.5 has been extended to large-scale yielding, using J instead of G or K (Figure 6.10). The J_R-resistance curve is assumed to be a geometry-independent material property for given thickness and environmental conditions. During crack growth the J-integral, which is interpreted as the crack driving force, must be equal to the material resistance to crack growth. Stability of crack growth requires

$$J(P,a) = J_R(\Delta a) \tag{6.55a}$$

$$\frac{\mathrm{d}J(P,a)}{\mathrm{d}a} < \frac{\mathrm{d}J_R(\Delta a)}{\mathrm{d}a} \tag{6.55b}$$

where P represents the applied loading. Crack growth becomes unstable when the inequality (6.55b) is reversed. The methodology developed in Section 5.5 for the study of slow crack propagation can equally be applied to the J-resistance curve method.

Paris *et al.* [6.11] introduced the nondimensional tearing moduli

$$T = \frac{E}{\sigma_0^2}\frac{\mathrm{d}J}{\mathrm{d}a}\,, \quad T_R = \frac{E}{\sigma_0^2}\frac{\mathrm{d}J_R}{\mathrm{d}a} \tag{6.56}$$

where σ_0 is an appropriate yield stress in tension when the material has strain hardening. σ_0 is usually taken equal to $(\sigma_Y + \sigma_u)/2$, with σ_u being the ultimate stress of the material in tension. The tearing modulus T_R is a material parameter that we can reasonably assume to be temperature independent. Using the tearing moduli we may write the stability condition as

$$T < T_R\,. \tag{6.57}$$

We can question the validity of this analysis which uses the J-integral for the study of slow stable crack growth. Crack growth involves some elastic unloading and, therefore, nonproportional plastic deformation near the crack tip. However, the J-integral is based on deformation plasticity theory which is incapable of adequately modeling both of these aspects of plastic crack propagation. The conditions for J-controlled crack growth were first studied by Hutchinson and Paris [6.12].

The argument for J-controlled crack growth requires that the region of elastic unloading and nonproportional plastic loading be well contained within the J-dominance zone (Figure 6.11). The condition for J-controlled crack growth may be expressed as

$$\omega = \frac{b}{J}\frac{\mathrm{d}J}{\mathrm{d}a} \gg 1 \tag{6.58}$$

where according to [6.12], ω should be of the order of 40.

6.8. Crack opening displacement (COD) fracture criterion

(a) Outline of the method

Wells [6.13] and Cottrell [6.14] independently introduced the concept of a critical crack opening displacement as a fracture criterion for the study of crack initiation

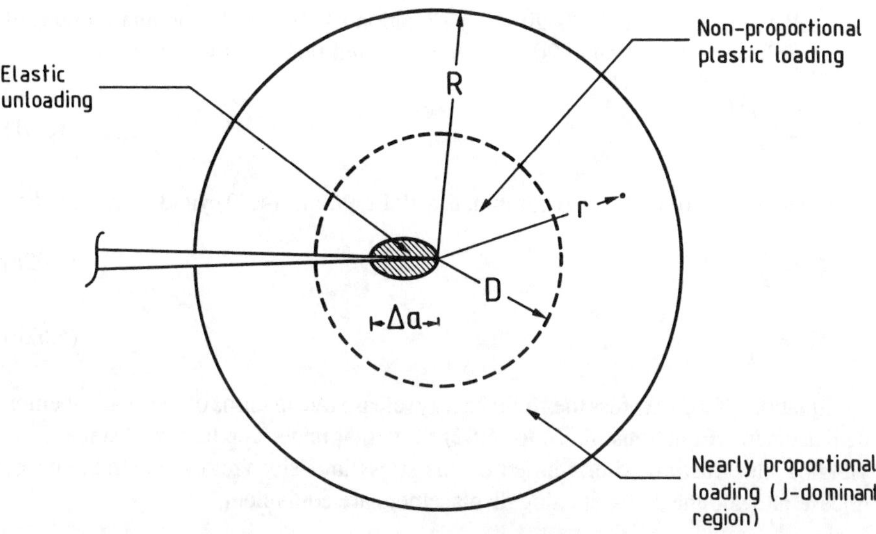

Fig. 6.11. Elastic unloading, nonproportional plastic loading and J-dominant region around the tip of a growing crack.

in situations where significant plastic deformation precedes fracture. Under such conditions they argued that the stresses around the crack tip reach the critical value and therefore fracture is controlled by the amount of plastic strain. Crack extension takes place by void growth and coalescence with the original crack tip, a mechanism for which the crack-tip strain is responsible. A measure of the amount of crack-tip plastic strain is the separation of the crack faces or crack opening displacement (COD), especially very close to the crack tip. It is thus expected that crack extension will begin when the crack opening displacement reaches some critical value which is characteristic of the material at a given temperature, plate thickness, strain rate and environmental conditions. The criterion takes the form

$$\delta = \delta_c \tag{6.59}$$

where δ is the crack opening displacement and δ_c is its critical value. It is assumed that δ_c is a material constant independent of specimen configuration and crack length. This assumption has been confirmed by experiments which indicate that δ has the same value at fracture.

We use the Irwin or Dugdale models to obtain an analytical expression for Equation (6.59) in terms of applied load, crack length, specimen geometry and other fracture parameters. In both models δ is taken as the separation of the faces of the effective crack at the tip of the physical crack. According to the Irwin model, δ is given by Equation (3.10) for plane stress. δ_c is expressed by

$$\delta_c = \frac{4}{\pi}\frac{K_c^2}{E\sigma_Y} . \tag{6.60}$$

For the Dugdale model δ is given by Equation (3.19) which, for small values of σ/σ_Y, reduces to Equation (3.20). Under such conditions δ_c is given by

$$\delta_c = \frac{K_c^2}{E\sigma_Y} \,. \tag{6.61}$$

Equations (3.10) and (3.19) combined with Equations (4.22) yield, respectively,

$$G = \frac{\pi}{4}\,\sigma_Y \delta \tag{6.62a}$$

$$G = \sigma_Y \delta \,. \tag{6.62b}$$

Equations (6.62) express the strain energy release rate in terms of the crack opening displacement. Equations (6.60) to (6.62) show that under conditions of small-scale yielding, the fracture criteria based on the stress intensity factor, the strain energy release rate and the crack opening displacement are equivalent.

(b) COD design curve

The objective of the COD design curve is to establish a relationship between the crack opening displacement and the applied load and crack length. Knowing the critical crack opening displacement we can determine the maximum permissible stress, or the maximum allowable crack length, in a structure. For determining the COD design curve we should select an analytical model. Burdekin and Stone [6.15] used the Dugdale model (Section 3.4) to obtain the following equation which expresses the overall strain $\varepsilon(\varepsilon = u/2y)$ of two equidistant points P from the crack (Figure 6.12)

$$\frac{\varepsilon}{\varepsilon_Y} = \frac{2}{\pi}\left[2n \coth^{-1}\left[\frac{1}{n}\sqrt{\frac{k^2+n^2}{1-k^2}}\right] + (1-\nu)\cot^{-1}\sqrt{\frac{k^2+n^2}{1-k^2}} + \nu \cos^{-1} k\right] \tag{6.63a}$$

with

$$n = \frac{a}{y}\,, \quad k = \cos\left(\frac{\pi\sigma}{2\sigma_Y}\right)\,, \quad \varepsilon_Y = \frac{\sigma_Y}{E}\,. \tag{6.63b}$$

In this equation a is half the crack length, y is the distance of point P from the crack, σ_Y is the yield stress of the material in tension and σ is the applied stress.

Define the dimensionless crack opening displacement ϕ by

$$\phi = \frac{\delta}{2\pi\,\varepsilon_Y a} \tag{6.64}$$

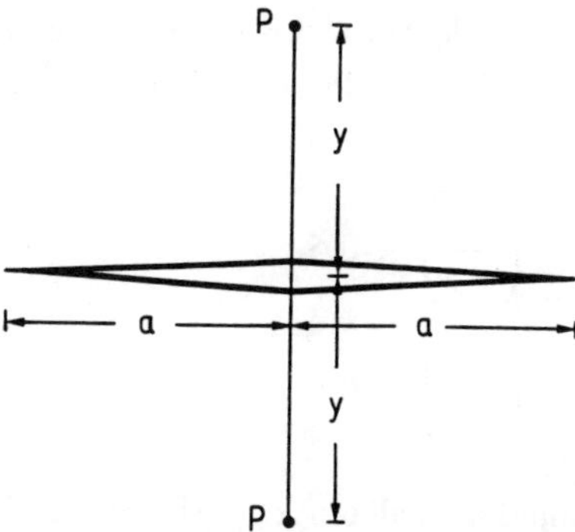

Fig. 6.12. Points P at equal distances y from a crack of length $2a$.

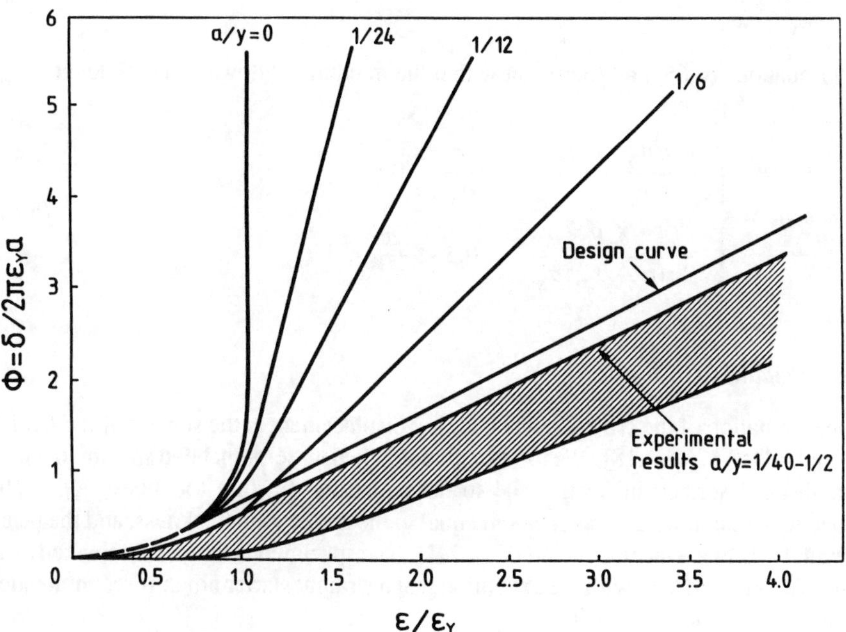

Fig. 6.13. Design curves according to crack opening displacement criterion [6.17].

where δ is given by Equation (3.19) for the Dugdale model. Eliminating stress σ from Equations (6.63) and (6.64), we obtain the design curves shown in Figure 6.13. This figure enables us to determine the maximum allowable overall strain in a cracked structure at position a/y, when we know the critical crack opening displacement δ_c and the crack length a.

Experimental data relating δ_c and the maximum strain at fracture, fall into a single scatter band of Figure 6.13 for a wide range of a/y values. It is thus shown that the design curve based on the Dugdale model is far from reality. An empirical equation was obtained [6.16] to describe the experimental data of Figure 6.13. This equation has the form

$$\phi = \begin{cases} \left(\dfrac{\varepsilon}{\varepsilon_Y}\right)^2, & \left(\dfrac{\varepsilon}{\varepsilon_Y}\right) < 0.5 \\ \dfrac{\varepsilon}{\varepsilon_Y} - 0.25\,, & \left(\dfrac{\varepsilon}{\varepsilon_Y}\right) > 0.5\,. \end{cases} \tag{6.65}$$

Dawes [6.17] argued that for small cracks ($a/W < 0.1$, W being the plate width) and applied stresses below the yield value,

$$\frac{\varepsilon}{\varepsilon_Y} = \frac{\sigma}{\sigma_Y}\,. \tag{6.66}$$

Equations (6.65) and (6.66) show that the maximum allowable crack length $a_{\max}$ is

$$a_{\max} = \begin{cases} \dfrac{\delta_c E \sigma_Y}{2\pi\sigma^2}\,, & \dfrac{\sigma}{\sigma_Y} < 0.5 \\ \dfrac{\delta_c E}{2\pi(\sigma - 0.25\sigma_Y)}\,, & 0.5 < \dfrac{\sigma}{\sigma_Y} < 1\,. \end{cases} \tag{6.67}$$

(d) Standard COD test

Determination of the critical crack opening displacement is the subject of the British Standard BS 5762 [6.18]. We use the edge-notched three-point bend specimen which has been described in Section 5.4 to determine the fracture toughness K_{Ic}. The specimen thickness B is taken about equal to the application thickness, and the beam width W is twice the thickness ($W = 2B$). The specimen is fatigue precracked as in the K_{Ic} standard test, with the exception that a straight starter notch is recommended rather than a chevron notch.

The load versus crack mouth displacement is recorded from the experiment. Clip gages are usually installed at a distance z from the specimen surface. The load-displacement records fall into the five cases shown in Figure 6.14. Four categories of crack-tip opening displacement are defined in relation to Figure 6.14: δ_c at the onset of unstable crack growth (case I) or pop-in (case II), when no stable crack growth is observed; δ_u at the onset of unstable crack growth (case III) or pop-in (case IV) when stable crack growth takes place before instability; δ_i at the commencement of stable crack growth (cases III, IV and V); δ_m at the maximum load P_m (case V) when it is preceded by stable crack growth.

The critical crack-tip opening displacement $\delta(\delta_i, \delta_c, \delta_u$ or $\delta_m)$ is determined from the test record by

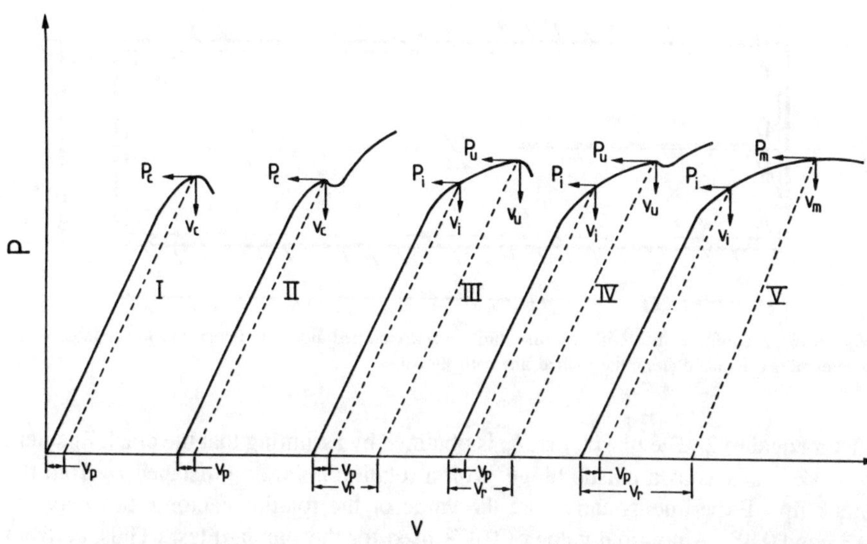

Fig. 6.14. Different types of load-clip gauge displacement records according to British standards [6.18].

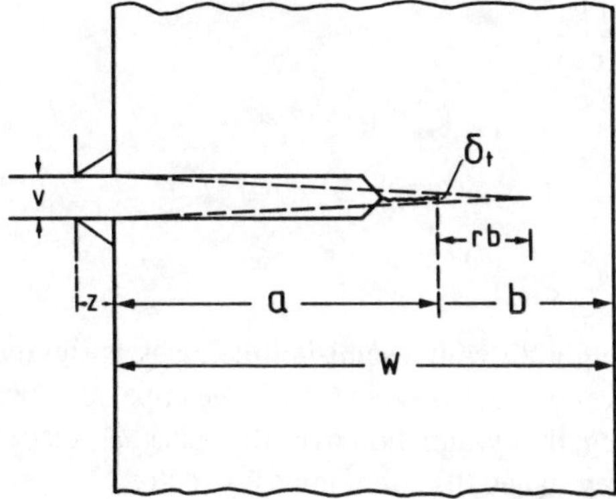

Fig. 6.15. Definition of δ_t and its relation to V.

$$\delta = \delta_e + \delta_p \tag{6.68a}$$

where

$$\delta_e = \frac{K_I^2(1-\nu^2)}{2\sigma_Y E}\,, \quad \delta_p = \frac{V_p rb}{rb + a + z}\,. \tag{6.68b}$$

Here V_p is the plastic component of the measured displacement V (Figure 6.14) and the quantities r, b, a and z are shown in Figure 6.15.

In Equation (6.68a) the crack opening displacement δ is equal to elastic δ_e plus the plastic δ_p contribution. The elastic part δ_e is calculated from Equation (3.20) of the Dugdale model which is modified for plane strain and by a plastic constraint

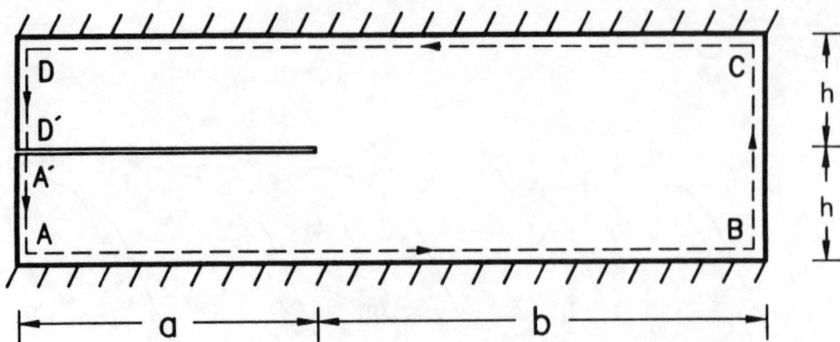

Fig. 6.16. An infinite strip with a semi-infinite crack clamped along its upper and lower faces. The J-integral is calculated along the dashed-line path shown.

factor equal to 2. The plastic part δ_p is obtained by assuming that the crack ligament $b = W - a$ acts as a plastic hinge, with a rotation point at a distance rb from the crack tip. Experiments show that the value of the rotation factor r lies between 0.33 and 0.48. A nominal value of 0.4 is used for the standard test. Thus, δ_p from Equation (6.78b) with $z = 0$ becomes

$$\delta_p = \frac{0.4(W - a)\,V_p}{0.4W + 0.6a}\,. \tag{6.69}$$

Examples

Example 6.1.

An infinite strip of height $2h$ with a semi-infinite crack is rigidly clamped along its upper and lower faces at $y = \pm h$ (Figure 6.16). The upper and lower faces are moved in the positive and negative y-direction over distances u_0, respectively. Determine the value of the J-integral and the stress intensity factor.

Solution: To determine the value of the J-integral we consider the path $A'ABCDD'$ extended along the upper and lower surfaces of the strip up to infinity and traversing the strip perpendicularly to the crack. J is calculated from

$$J = J_{A'ABCDD'} = J_{AB} + J_{BC} + J_{CD} + J_{DD'} + J_{A'A}\,. \tag{1}$$

We have: for path $AB, CD:\ \mathrm{d}y = 0, \partial u_{1,2}/\partial x = 0$, implying that

$$J_{AB} = J_{CD} = 0 \tag{2}$$

for path $DD', A'A$: the stresses vanish, $\partial u_{1,2}/\partial x = 0$, implying that

$$J_{DD'} = J_{A'A} = 0 \tag{3}$$

and for path $BC:\ \partial u_{1,2}/\partial x = 0$, implying that

$$J_{BC} = \int_{-h}^{h} \omega|_{x\to\infty}\, \mathrm{d}y\, . \tag{4}$$

For linear elastic material

$$\omega|_{x\to\infty} = \frac{1}{2}\,\sigma_y \varepsilon_y\, . \tag{5}$$

Putting

$$\varepsilon_y = \frac{u_0}{h}\, , \quad \sigma_y = \beta E \varepsilon_y = \frac{\beta E u_0}{h} \tag{6}$$

where

$$\beta = \frac{1}{1-\nu^2} \tag{7}$$

for plane stress, and

$$\beta = \frac{1-\nu}{(1+\nu)\,(1-2\nu)} \tag{8}$$

for plane strain, we obtain

$$J = \frac{\beta E u_0^2}{h} \tag{9}$$

K_{I} is computed from Equation (6.21) as

$$K_{\mathrm{I}} = \left(\frac{JE}{\eta}\right)^{1/2} = \left(\frac{\beta}{\eta h}\right)^{1/2} E u_0 \tag{10}$$

where $\eta = 1$ for plane stress and $\eta = 1 - \nu^2$ for plane strain.

Example 6.2.

Determine the value of J-integral for the Dugdale model.

Solution: Referring to Figure 3.10 we take the integration path along the line ABC, around the yield strip boundary from the lower side at $x = a$, through the tip of the effective crack. Along ABC, $\mathrm{d}y = 0$, and J is computed as

$$J = -\int_{a}^{a+c} \sigma_Y \,\frac{\partial}{\partial x_1}\,(u_2^+ - u_2^-)\, \mathrm{d}x_1 = \int_{0}^{\delta} \sigma_Y\, \mathrm{d}\delta \tag{1}$$

or

$$J = \sigma_Y \delta \tag{2}$$

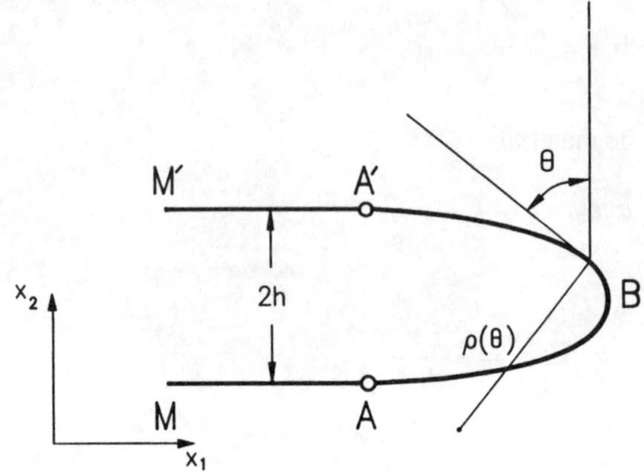

Fig. 6.17. A narrow notch and coordinates used to describe its surface.

where δ is the opening of the effective crack at the tip of the physical crack.

Substituting the value of δ from Equation (3.19) into Equation (2) we obtain

$$J = \frac{8\sigma_Y^2 a}{\pi E} \ln \left[\sec\left(\frac{\pi a}{2\sigma_Y}\right)\right]. \tag{3}$$

By expanding Equation (3) into a series and retaining the first term for small values of σ/σ_Y (small scale yielding) we obtain

$$J_{ssy} = \frac{K_I^2}{E} \tag{4}$$

which is the value of J for plane stress (Equation (6.21)).

From Equations (3) and (4) with $K_I = \sigma\sqrt{\pi a}$ we obtain

$$\frac{J}{J_{ssy}} = \frac{8}{\pi^2}\left(\frac{\sigma_Y}{\sigma}\right)^2 \ln\left[\sec\left(\frac{\pi\sigma}{2\sigma_Y}\right)\right]. \tag{5}$$

For $\sigma/\sigma_Y \ll 1$, J/J_{ssy} approaches unity, while for $\sigma \to \sigma_Y$ it becomes unbounded.

Example 6.3.

Consider a narrow notch whose lower and upper surfaces are flat and parallel up to the points A and A', so that the arc $A'A$ forms the curved tip of the notch (Figure 6.17). The opening $2h$ of the notch is small compared to its length. Assume that the surface stress may be approximated as

$$\sigma = \sigma(\theta) = \sigma_{\max} \cos^2 \theta \tag{1}$$

where θ is the tangent angle and σ_{max} is the maximum stress. For a semicircular tip, show that

$$\sigma_{max} = \sqrt{\frac{15}{8}} \frac{K_I}{\sqrt{\rho}} \tag{2}$$

where ρ is the radius of curvature at the semicircular notch tip and K_I is the stress intensity factor for a similarly loaded body with a crack of the same length as the notch.

Solution: Consider the J-integral and compute it for the path $MABA'M'$ that coincides with the notch surface. We have

$$J = J_{MA} + J_{ABA'} + J_{A'M'} \,. \tag{3}$$

The notch is traction-free. Since $\mathrm{d}y = 0$ along the flat surfaces MA and $A'M'$ of the notch, we have

$$J_{MA} = J_{A'M'} = 0 \tag{4}$$

so that

$$J = J_{ABA'} = \int_{ABA'} \omega \, \mathrm{d}y \,. \tag{5}$$

For conditions of plane strain we have

$$\omega = \frac{1-\nu^2}{2E} \sigma^2 \tag{6}$$

and Equation (5) becomes

$$J = \frac{1-\nu^2}{2E} \int_{ABA'} \sigma^2 \, \mathrm{d}y \tag{7}$$

or

$$J = \frac{1-\nu^2}{2E} \int_{-\pi/2}^{\pi/2} \sigma^2(\theta)\, \rho(\theta)\, \cos\theta \, \mathrm{d}\theta \,. \tag{8}$$

For a semicircular notch tip $\rho(\theta) = \rho$, and Equation (1) gives

$$J = \frac{(1-\nu^2)\,\sigma_{max}^2 \rho}{E} \int_{-\pi/2}^{\pi/2} \cos^5\theta \, \mathrm{d}\theta = \frac{8(1-\nu^2)\,\sigma_{max}^2 \rho}{15E} \,. \tag{9}$$

On the other hand, if the path of integration of J is selected far from the notch we have (Equation (6.21))

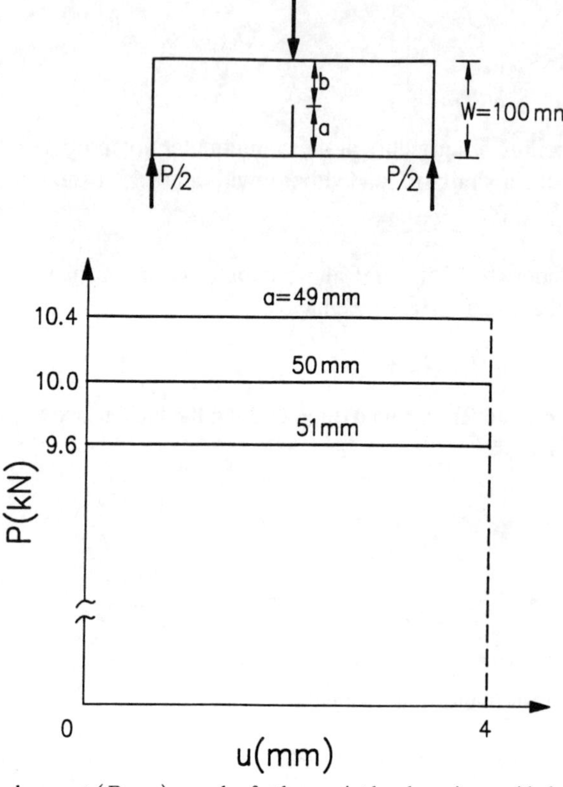

Fig. 6.18. Load-displacement $(P - u)$ records of a three-point bend specimen with three different crack lengths.

$$J = \frac{1 - \nu^2}{E} K_\mathrm{I}^2 \tag{10}$$

where K_I is the stress intensity factor for a similarly loaded body with a crack of the same length as the notch.

From Equations (9) and (10) we obtain Equation (2).

Example 6.4.

The load-displacement records of a three-point bend specimen of width $W = 100$ mm and thickness $B = 50$ mm containing three cracks of length $a = 49, 50$ and 51 mm are shown in Figure 6.18. The specimen with $a = 50$ mm fails at a displacement $u = 4$ mm. Determine the value of J using Equation (6.51) and Equation (6.36).

Solution: Equation (6.51) becomes

$$J = \frac{2U}{Bb} \tag{1}$$

where U is the area of the load-displacement record of the specimen with $a = 50$ mm up to displacement $u = 4$ mm. We have

$$U = (10\ \text{kN}) \times (4 \times 10^{-3}\ \text{m}) = 40 \times 10^{-3}\ \text{kN m} \tag{2}$$

and Equation (1) becomes

$$J = \frac{2 \times (40 \times 10^{-3})\ \text{kN m}}{(50 \times 10^{-3}\ \text{m}) \times (50 \times 10^{-3}\ \text{m})} = 32\ \text{kN/m}\,. \tag{3}$$

Equation (6.36) becomes

$$J = -\frac{1}{B}\frac{\Delta U}{\Delta a}\,. \tag{4}$$

We have:

$$\Delta U = -(0.4\ \text{kN}) \times (4 \times 10^{-3}\ \text{m}) = -1.6 \times 10^{-3}\ \text{kN m} \tag{5}$$

and Equation (4) gives

$$J = \frac{1.6 \times 10^{-3}\ \text{kN m}}{(50 \times 10^{-3}\ \text{m}) \times (1 \times 10^{-3}\ \text{m})} = 32\ \text{kN/m}\,. \tag{6}$$

Example 6.5.

Determine the minimum thickness $B_{\min}$ and/or crack ligament $b_{\min}$ of a three-point bend specimen required for a valid J_{Ic} test according to the ASTM standards for a material with $K_{\text{Ic}} = 100$ MPa $\sqrt{\text{m}}$, $\sigma_Y = 400$ MPa, $E = 210$ GPa and $\nu = 0.3$. Compare the results with those for a valid K_{Ic} test.

Solution: J_{Ic} is calculated according to Equation (6.32) as

$$J_{\text{Ic}} = \frac{(1-\nu^2)\,K_{\text{Ic}}^2}{E} = \frac{(1-0.3^2) \times (100\ \text{MPa}\sqrt{\text{m}})^2}{120 \times 10^3\ \text{MPa}} = 0.0433\ \text{MPa m}\,.$$

The minimum thickness $B_{\min}$ and/or crack ligament $b_{\min}$ required for a valid J_{Ic} test according to ASTM standards for the three-point bend specimen is calculated according to Equation (6.33) as

$$B_{\min}, b_{\min} = \frac{25 J_{\text{Ic}}}{\sigma_Y} = \frac{25 \times (0.0433\ \text{MPa m})}{400\ \text{MPa}} = 2.7\ \text{mm}\,.$$

The minimum thickness $B_{\min}$ required for a valid K_{Ic} test is calculated according to Equation (5.8a) as

$$B_{\min} = 2.5\left(\frac{K_{\text{Ic}}}{\sigma_Y}\right)^2 = 2.5\left(\frac{100\ \text{MPa}\sqrt{\text{m}}}{400\ \text{MPa}}\right)^2 = 156.2\ \text{mm}\,.$$

Observe that the minimum thickness required for a valid J_{Ic} test is nearly two orders of magnitude smaller than that required for a valid K_{Ic} test. This indicates that J_{Ic} testing may be used to estimate K_{Ic} in situations where large specimen dimensions are required for a valid K_{Ic} test.

Example 6.6.

A steel structural member with a stress-concentration factor of 3 is subjected to a nominal design stress $\sigma_Y/2$, where σ_Y is the yield stress of the material in tension. Using the crack opening displacement design method determine the maximum crack length the member can withstand without failure. The modulus of elasticity of steel is $E = 210$ GPa, the yield stress is $\sigma_Y = 1$ GPa and the critical crack opening displacement is $\delta_c = 0.5$ mm.

Solution: For a stress concentration factor of 3 and an applied stress of $\sigma_Y/2$ we have

$$\frac{\varepsilon}{\varepsilon_Y} = 3\,\frac{\varepsilon_Y}{2}\,\frac{1}{\varepsilon_Y} = 1.5\,.$$

For this case we obtain from the second Equation (6.65)

$$\phi = 1.5 - 0.25 = 1.25\,.$$

The yield strain ε_Y is

$$\varepsilon_Y = \frac{\sigma_Y}{E} = \frac{1\text{ GPa}}{210\text{ GPa}} = 0.0048\,.$$

The critical half crack length a_c is calculatd from Equation (6.64) as

$$a_c = \frac{(0.5 \times 10^{-3}\text{ m})}{2\pi \times (0.0048) \times 1.25} = 13.26\text{ mm}\,.$$

The critical crack length is

$$2a_c = 26.5\text{ mm}\,.$$

Example 6.7.

A three-point bend specimen with $S = 25$ cm, $W = 6$ cm, $a = 3$ cm, and $B = 3$ cm is used to determine the critical crack opening displacement δ_c of a steel plate according to the British Standard BS 5762. The load versus crack mouth displacement $(P - V)$ record of the test is shown in Figure 6.19. Determine δ_c when $E = 210$ GPa, $\nu = 0.3$ and $\sigma_Y = 800$ MPa for steel.

Solution: The critical crack opening displacement δ_c is calculated from Equations (6.68) and (6.69) (for $z = 0$ and $r = 0.4$). We have

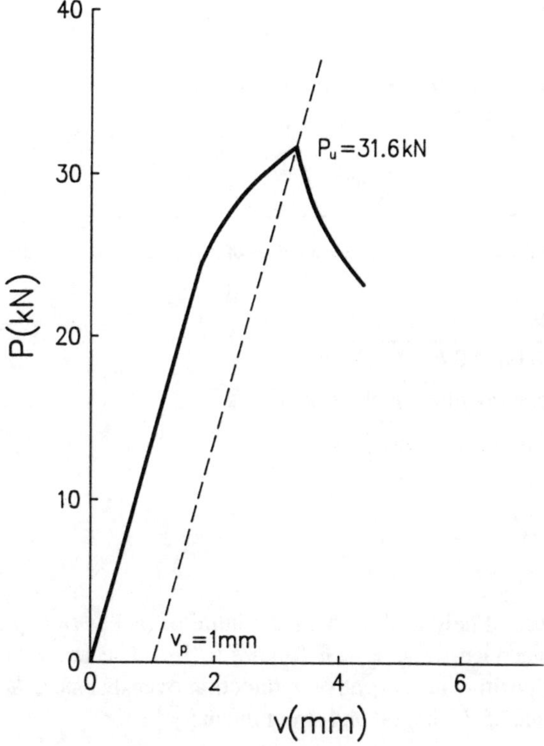

Fig. 6.19. Load-crack mouth displacement $(P - V)$ record of a three-point bend specimen.

$$\delta_c = \delta_e + \delta_p \,, \tag{1}$$

where

$$\delta_e = \frac{K_\text{I}^2(1-\nu^2)}{2\sigma_Y E} \,, \quad \delta_p = \frac{0.4(W-a)\,V_p}{0.4W + 0.6a} \,. \tag{2}$$

For the three-point bend specimen, K_I is calculated from Equations (1) and (2) of Example 5.1. Table of Example 5.1 gives for $a/W = 0.5$, $f(a/W) = 2.66$. K_I is computed as

$$K_\text{I} = \frac{(31.6\text{ kN}) \times (0.25\text{ m})}{(0.03\text{ m}) \times (0.06\text{ m})^{3/2}} \times 2.66 = 47.7\text{ MPa}\sqrt{\text{m}} \,. \tag{3}$$

We have

$$\delta_e = \frac{(47.7\text{ MPa}\sqrt{\text{m}})^2 \times (1 - 0.3^2)}{2 \times (800\text{ MPa}) \times (210 \times 10^3\text{ MPa})} = 0.006\text{ mm} \,. \tag{4}$$

The plastic component of the crack mouth displacement V_p is determined from the test record $P - V$ (Figure 6.19) by drawing a line from the maximum load parallel to the linear portion of the curve. We have $V_p = 1$ mm. δ_p is determined as

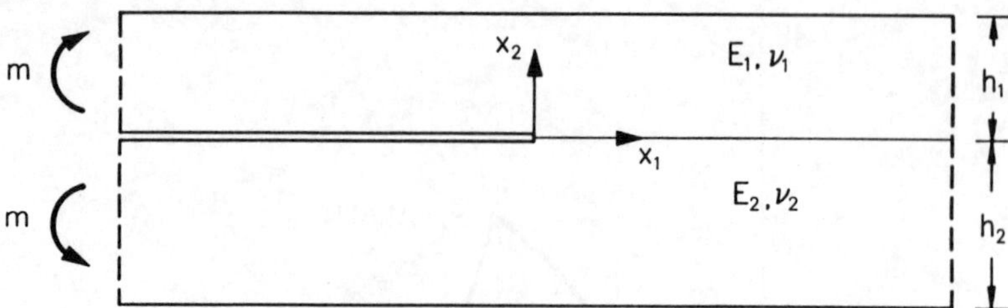

Fig. 6.20. A semi-infinite crack along the interface of two infinite layers of different heights.

$$\delta_p = \frac{0.4(0.06 - 0.03)\text{ (m)} \times 1\text{ mm}}{(0.4 \times 0.06 + 0.6 \times 0.03)\text{ mm}} = 0.286\text{ mm}\,. \tag{5}$$

The critical crack opening displacement δ_c is

$$\delta_c = 0.006 + 0.286 = 0.292\text{ mm}\,. \tag{6}$$

Problems

6.1. An infinite strip of height $2h$ with a semi-infinite crack is rigidly clamped along its upper and lower faces at $y = \pm h$ (Figure 6.16). The upper and lower faces are moved in the positive and negative x-direction over distances u_0, respectively. Determine the value of J-integral and stress intensity factor.

6.2. An infinite strip of height $2h$ with a semi-infinite crack is rigidly clamped along its upper and lower faces at $y = \pm h$ (Figure 6.16). The upper and lower faces are moved in the positive and negative z-direction over distances w_0, respectively. Determine the value of J-integral and stress intensity factor.

6.3. Two infinite layers of heights h_1 and h_2 and large thicknesses are made of different materials with moduli of elasticity and Poisson's ratios E_1, ν_1 and E_2, ν_2 respectively (Figure 6.20). The layers are joined across their interface forming a semi-infinite crack, and are subjected to bending moments m at $x \to -\infty$. Determine the value of J-integral.

6.4. Two infinite layers of heights h_1 and h_2 and large thicknesses are made of different materials with moduli of elasticity and Poisson's ratios E_1, ν_1, and E_2, ν_2 respectively (Figure 6.21). The layers are joined across their interface forming a semi-infinite crack, and are rigidly clamped along their bases at $y = h_1$ and $y = -h_2$. The upper and lower bases are moved in the positive and negative x-direction over distances u_0 respectively. Determine the value of J-integral.

6.5. Show that for a crack in a mixed-mode stress field governed by the values of stress intensity factors K_{I}, K_{II} and K_{III}, the value of J-integral is given by Equation (6.21).

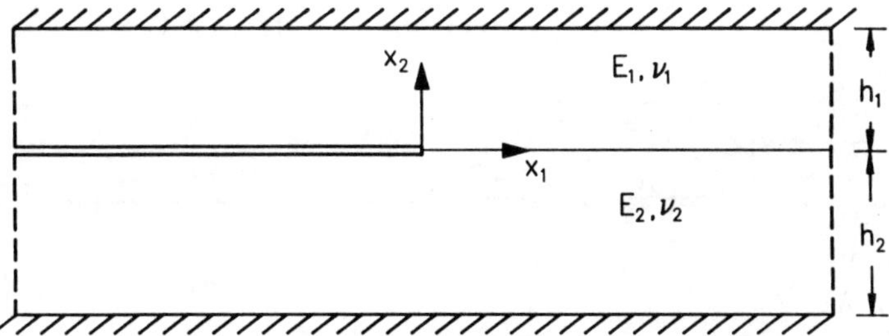

Fig. 6.21. A semi-infinite crack along the interface of two infinite layers of different heights with rigidly clamped bases.

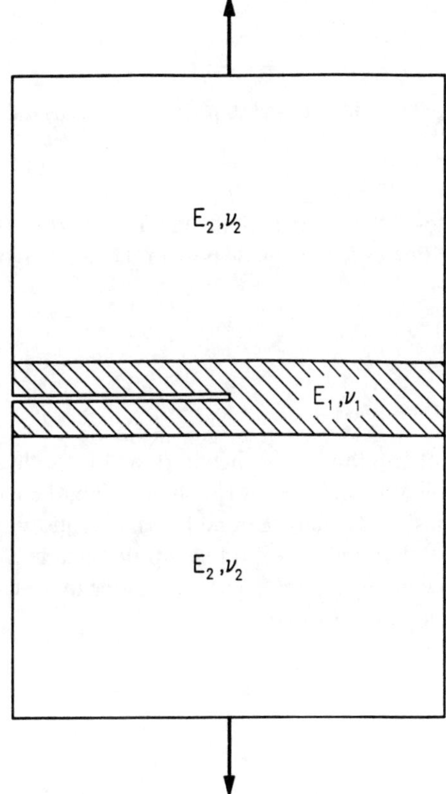

Fig. 6.22. A thin cracked layer perfectly bonded to two large layers of a different material.

6.6. A thin layer containing a crack, is made of a material with modulus of elasticity E_1 and Poisson's ratio ν_1. It is perfectly bonded to two large layers of the same material with modulus of elasticity E_2 and Poisson's ratio ν_2 (Figure 6.22). If k_{I} is

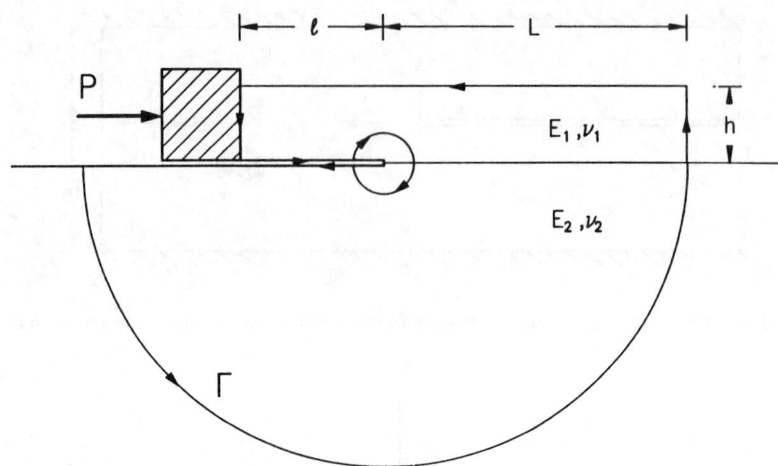

Fig. 6.23. A beam with an end force bonded to a half-plane and integration path for the determination of J-integral.

the stress intensity factor at the crack tip of the thin layer and K_{I} is the stress intensity factor for the same loading of a plate made only of the large layers show that

$$k_{\mathrm{I}} = \sqrt{\frac{(1-\nu_2^2)\,E_1}{(1-\nu_1^2)\,E_2}}\;K_{\mathrm{I}}\,.$$

6.7. An elastic beam of length $(l + L)$, height h, and large thickness has modulus of elasticity E_1 and Poisson's ratio ν_1. It is bonded along the length L to an elastic half-plane with modulus of elasticity E_2 and Poisson's ratio ν_2 (Figure 6.23). The beam is subjected at its left end to a rigid stamp that exerts a force P. Take the integration path Γ shown in Figure 6.23, where the circle in the half-plane has a large radius. Show that J-integral is given by

$$J = \frac{(1-\nu_1^2)\,P^2}{2hE_1}\,.$$

6.8. According to the atomic or molecular cohesive force theory introduced by Barenblatt (*Advances in Applied Mechanics*, Academic Press, Vol. 7, pp. 55–129, 1962) large forces of atomic or molecular attraction act in small zones at the ends of the crack (Figure 6.24a). These forces pull the crack faces together and induce stress singularities at the crack ends which cancel out the stress singularities introduced by the applied loads. Consequently, the stresses are bounded, and the faces of the crack join smoothly in cusp form at the ends. If δ is the separation distance between the

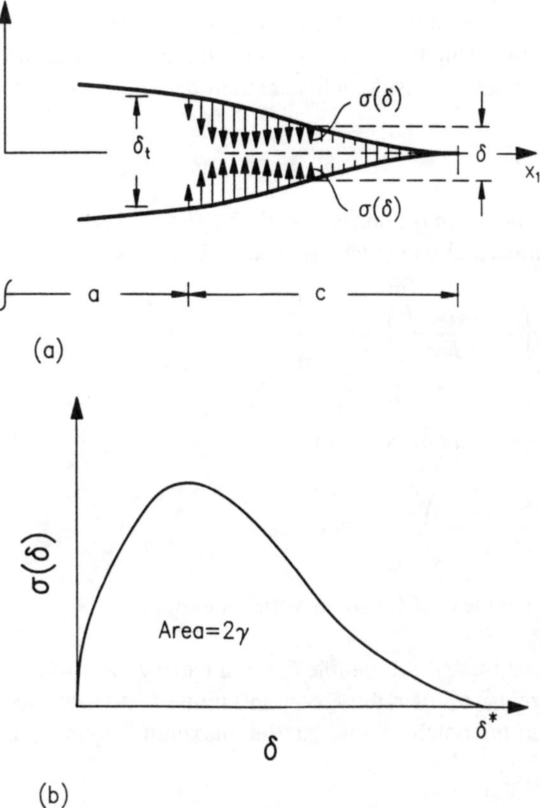

Fig. 6.24. (a) Cohesive forces at the end of a brittle crack and (b) cohesive stress versus separation relationship.

upper and lower crack faces and $\sigma(\delta)$ is the cohesive stress, show that

$$J = \int_0^{\delta_t} \sigma(\delta)\,\mathrm{d}\delta$$

where δ_t is the separation distance at the crack tip.

Suppose that δ_* is the separation distance when the atoms are pulled so that there is no attraction among them (Figure 6.24b). The critical value of J at crack growth is obtained from the above equation when $\delta_t = \delta_*$. Note that the area under the $\sigma(\delta) - \delta$ curve (Figure 6.24b) is twice the surface energy γ. Use this observation to establish that the Barenblatt theory and the Griffith theory are identical for crack growth with small cohesive zones.

6.9. The Barenblatt cohesive force theory is, in many ways, similar to the Dugdale model (Section 3.4) for the study of the plastic zones at the crack tip, although it has a completely different physical meaning. The equation in the previous problem applies to a modified Dugdale model in which the stress distribution $\sigma(\delta)$ inside the plastic zone is variable. Consider a linear strain-hardening behavior

$$\sigma(\delta) = \sigma_Y + E_w \frac{\delta}{h}$$

in which the plastic strain is approximated by δ/h, where σ_Y is the yield stress, E_w the tangent modulus and h the plate thickness. Show that

$$\delta_t = \frac{\sigma_Y h}{E_w}\left[\left(1 + \frac{2E_w J}{h\sigma_Y^2}\right)^{1/2} - 1\right].$$

6.10. For the Dugdale model show that

$$\frac{J}{J_{ssy}} = 1 + \frac{\pi^2}{24}\left(\frac{\sigma}{\sigma_Y}\right)^2$$

where J_{ssy} is the value of J for small-scale yielding.

6.11. Apply Equation (2) of Example 6.3 to a narrow flat-surfaced notch of length $2a$, with semicircular tip of radius ρ, in an infinite plate subjected to a stress σ at infinity normal to the notch. Show that the maximum stress $\sigma_{\max}$ along the notch surface is given by

$$\sigma_{\max} = 2.43\sigma\left(\frac{a}{\rho}\right)^{1/2}.$$

6.12. Use Example 6.3 to show that, for a narrow flat-surfaced notch of thickness $2h$, the maximum stress $\sigma_{\max}$ along the notch surface satisfies the inequality

$$\sigma_{\max} \geq \frac{K_{\mathrm{I}}}{h^{1/2}}.$$

Here K_{I} is the stress intensity factor for a similarly loaded body with a crack of the same length as the notch.

6.13. Figure 6.25 shows the load-displacement records of a three-point bend specimen of width $W = 50$ mm and thickness $B = 20$ mm containing three cracks of length $a = 24$, 25 and 26 mm. The specimen with $a = 25$ mm fails at a displacement $u = 2$ mm. Determine the value of J using Equation (6.51) and Equation (6.36).

6.14. A number of compact tension specimens of thickness $B = 2$ mm with different

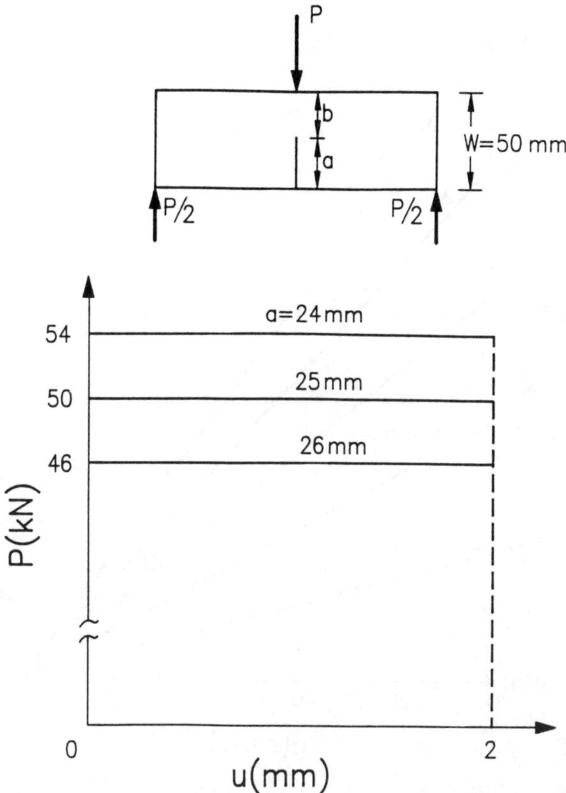

Fig. 6.25. Load-displacement $(P - u)$ records of a three-point bend specimen with three different crack lengths.

crack lengths were tested, and the load-displacement records were obtained. From these records the absorbed potential energy Π at different selected displacements was calculated and is plotted in Figure 6.26 versus crack length. Plot the variation of J-integral versus displacement.

6.15. The load-displacement $(P - \delta)$ diagram of a compact tension specimen of dimensions $W = 40$ mm, $a = 25$ mm, $B = 20$ mm can be approximated as

$$\delta = 10^{-22} P^5$$

where δ is measured in meters and P in Newtons. Initiation of crack growth occurred at $\delta = 10$ mm. Estimate J.

6.16. What is the maximum K_{Ic} value that may be determined on a 20 mm thick plate with: $\sigma_Y = 500$ MPa, $E = 210$ GPa and $\nu = 0.3$ (a) according to ASTM standard E399 for estimating K_{Ic} directly and (b) according to ASTM standard E813–87 for estimating J_{Ic} and then calculating K_{Ic}.

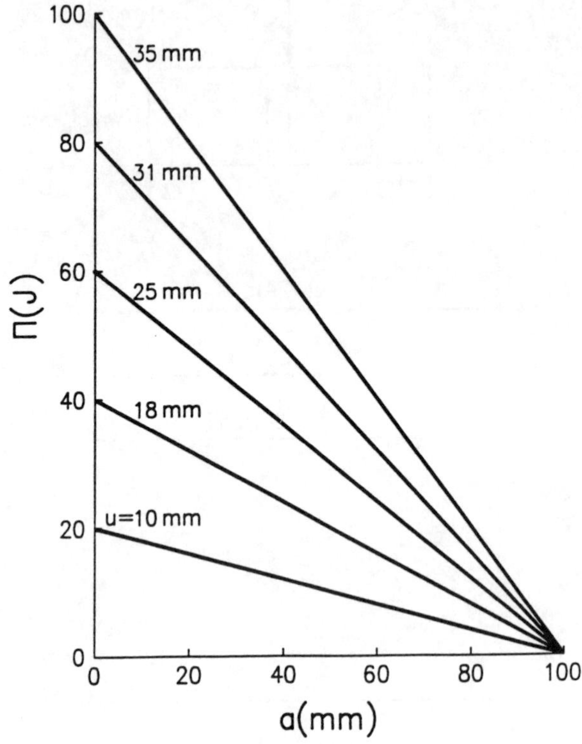

Fig. 6.26. Potential energy versus crack length at different displacements of a number of compact tension specimens.

6.17. The following data were obtained from a series of tests on three-point bend specimens with thickness $B = 30$ mm and crack ligament $b = 30$ mm made of a steel with 0.2 offset yield stress $\sigma_Y = 450$ MPa and ultimate stress $\sigma_u = 550$ MPa.

$J(\text{kJ/m}^2)$	$\Delta a(\text{mm})$
120	0.2
150	0.3
180	0.5
220	0.7
260	1.0
280	1.2
310	1.5
340	1.8
360	2.0

Estimate the provisional value J_Q according to ASTM standard E813–87 and check whether $J_{Ic} = J_Q$.

6.18. Consider a crack in a mixed-mode stress field governed by the values of stress intensity factors K_{I} and K_{II}. Show that the path-independent line integral

$$J_2 = Q_2 = \int_\Gamma (\omega n_2 - T_k u_{k,2})\,\mathrm{d}s$$

defined from Equation (6.6) with $j = 2$ is given by

$$J_2 = -\frac{\kappa + 1}{4\mu} K_{\mathrm{I}} K_{\mathrm{II}}\,.$$

6.19. Consider a crack in a mixed-mode stress field governed by the values of stress intensity factors K_{I} and K_{II}. Consider the vector

$$\mathbf{Q} = Q_1\mathbf{i} + Q_2\mathbf{j} = \mathbf{J} = J_1\mathbf{i} + J_2\mathbf{j}$$

where $J_1 = J$ and J_2 as was defined in Problem 6.18. Take the projection of this vector along the crack growth direction given by

$$J(\theta) = J_1 \cos\theta + J_2 \sin\theta\,.$$

Crack growth is governed by the following hypotheses

(i) The crack extends along the radial direction $\theta = \theta_c$ on which $J(\theta)$ becomes maximum.

(ii) Fracture starts when that maximum of $J(\theta)$ reaches the value 2γ, where γ is the energy required to form a unit of new surface in the Griffith theory.

Under these assumptions determine the angle θ_c and the critical load for initiation of crack growth.

6.20. Use Problem 6.19 to determine the crack extension angle θ_c and the critical stress σ_c for initiation of growth of a crack of length $2a$ in an infinite plate. Suppose that the crack subtends an angle β with the direction of applied uniform stress σ. Plot the curves $\theta_c = \theta_c(\beta)$ and $\sigma_c = \sigma_c(\beta)$.

6.21. A welded tension member of width 20 mm contains an edge crack at the weld toe and is subjected to a tensile stress $\sigma = 200$ MPa perpendicular to the crack axis. There is a residual stress 400 MPa at the weld. Using the crack opening displacement design method, determine the maximum length of crack the member can withstand without failure. The yield stress of the weld is 500 MPa, the modulus of elasticity is $E = 210$ GPa and the critical crack opening displacement is $\delta_c = 0.2$ mm.

6.22. A cylindrical pressure vessel of diameter 2 m and wall-thickness 20 mm is subjected to a pressure of 1 MPa. The material of the vessel is a low carbon steel with modulus of elasticity 210 GPa and yield stress 400 MPa. The critical crack opening displacement of a specimen taken from the vessel is $\delta_c = 0.1$ mm. Determine the maximum crack length the vessel can withstand without failure using the crack opening displacement design method. Assume that there are residual stresses of magnitude equal to the yield stress of the material of the vessel.

6.23. A compact tension specimen (Figure 5.7) with $W = 14$ cm, $B = 7$ cm and $a = 7$ cm is used to determine the critical crack opening displacement δ_c of a steel plate according to the British Standard BS 5762. From the load versus crack mouth displacement $(P - V)$ record of the test it is obtained that the maximum load is $P_u = 50$ kN and the plastic component of the crack mouth displacement is $V_p = 1.5$ mm. Determine δ_c when $E = 210$ GPa, $\nu = 0.3$ and $\sigma_Y = 500$ MPa.

References

6.1. Rice, J.R. (1968) 'A path independent integral and the approximate analysis of strain concentration by notches and cracks', *Journal of Applied Mechanics, Trans. ASME* **35**, 379–386.

6.2. Begley, J.A. and Landes, J.D. (1972) 'The J-integral as a fracture criterion', *Fracture Toughness, ASTM STP 514*, American Society for Testing and Materials, Philadelphia, pp. 1–23 and 24–39.

6.3. Bucci, R.J., Paris, P.C., Landes, J.D., and Rice, J.R. (1972) 'J-integral estimation procedures', *Fracture Toughness, ASTM STP 514*, American Society for Testing and Materials, Philadelphia, pp. 40–69.

6.4. McMeeking, R.M. (1977) 'Finite deformation analysis of crack opening in elastic-plastic materials and implications for fracture', *Journal of the Mechanics and Physics of Solids* **25**, 357–381.

6.5. Shih, C.F. and German, M.D. (1981) 'Requirements for a one parameter characterization of crack tip fields by the HRR singularity', *International Journal of Fracture* **17**, 27–43.

6.6. McMeeking, R.M. and Parks, D.M. (1979) 'On criteria for J-dominance of crack tip fields in large scale yielding', in *Elastic-Plastic Fracture, ASTM STP 668*, American Society for Testing and Materials, Philadelphia, pp. 175–194.

6.7. Hutchinson, J.W. (1983) 'Fundamentals of the phenomenological theory of nonlinear fracture mechanics', *Journal of Applied Mechanics, Trans. ASME* **50**, 1042–1051.

6.8. Rice, J.R., Paris, P.C., and Merkle, J.G. (1973) 'Some further results of J-integral analysis and estimates', in *Progress in Flaw Growth and Fracture Toughness Testing, ASTM STP 536*, American Society for Testing and Materials, Philadelphia, pp. 213–245.

6.9. Merkle, J.G. and Corten, H.T. (1974) 'A J-integral analysis for the compact specimen, considering axial force as well as bending effects', *Journal of Pressure Vessel Technology* **96**, 286–292.

6.10. Standard test method for J_{Ic}, a measure of fracture toughness, *ASTM Annual Book of Standards*, Part 10, E813–87, American Society for Testing and Materials, Philadelphia, pp. 968–990 (1987).

6.11. Paris, P.C., Tada, H., Zahoor, A., and Ernst, H. (1979) 'The theory of instability of the tearing mode of elastic-plastic crack growth', in *Elastic-Plastic Fracture, ASTM STP 668*, American Society for Testing and Materials, Philadelphia, pp. 5–36 and 251–265.

6.12. Hutchinson, J.W. and Paris, P.C. (1979) 'Stability analysis of J-controlled crack growth', in *Elastic-Plastic Fracture, ASTM STP 668*, American Society for Testing and Materials, Philadelphia, pp. 37–64.

6.13. Wells, A.A. (1961) 'Unstable crack propagation in metals: cleavage and fracture', *Proceedings of the Crack Propagation Symposium*, College of Aeronautics, Cranfield, Vol. 1, pp. 210–230.

6.14. Cottrell, A.H. (1961) 'Theoretical aspects of radiation damage and brittle fracture in steel pressure vessels', *Iron Steel Institute Special Report No. 69*, pp. 281–296.

6.15. Burdekin, F.M. and Stone, D.E.W. (1966) 'The crack opening displacement approach to fracture mechanics in yielding materials', *Journal of Strain Analysis* **1**, 145–153.

6.16. Dawes, M.G. (1974) 'Fracture control in high yield strength weldments', *Welding Journal Research Supplement* **53**, 369S-379S.

6.17. Dawes, M.G. (1980) 'The COD design curve', in *Advances in Elastic-Plastic Fracture Mechanics* (ed. L.H. Larsson), Applied Science Publishers, pp. 279–300.

6.18. BS 5762, Methods for crack opening displacement (COD) testing, British Standards Institution, London (1979).

Chapter 7

Strain Energy Density Failure Criterion: Mixed-Mode Crack Growth

7.1. Introduction

So far, we have studied growth of a crack only for the case when the load is applied normal to the crack and such that the crack propagates in a self-similar manner. However, often the loads are not aligned to the orientation of the crack. In such cases, the crack-tip stress field is no longer governed by a single opening-mode stress intensity factor K_{I} but by a combination of the three stress intensity factors K_{I}, K_{II}, and K_{III}. Moreover, the direction of crack initiation is not known a priori but depends on a failure criterion involving some combination of K_{I}, K_{II}, and K_{III}. As a rule the crack follows a curved path.

The need of a criterion that can predict crack growth under mixed-mode loading in a simple and unified manner, led to the development by Sih [7.1–7.3] of the strain energy density (SED) criterion. The fundamental quantity in the SED criterion is the strain energy $\mathrm{d}W/\mathrm{d}V$ contained in a unit volume of material at a given instant of time. The quantity $\mathrm{d}W/\mathrm{d}V$ serves as a useful failure criterion and has successfully been applied in the solution of a host of engineering problems of major interest. These include: two- and three-dimensional crack problems; cracked nonhomogeneous and composite materials; plates and shells with cracks; dynamic crack problems; failure initiating from notches; ductile fracture involving the prediction of crack initiation; slow stable crack growth and final separation; fatigue crack growth, etc. Many of the results have been published in the introductory chapters of the seven volumes of the series "Mechanics of Fracture" edited by Sih [7.4–7.10]. The author has demonstrated the usefulness and versatility of the SED criterion for determining the allowable load and crack growth direction corresponding to a variety of engineering problems of practical interest in the book *Problems of Mixed Mode Crack Propagation* [7.11].

In this chapter we review the SED theory and apply it to problems of fundamental importance. Following the definition of volume strain energy density and the basic hypotheses of the theory, we use the SED criterion to solve the general two-dimensional linear elastic crack problem, placing emphasis on the uniaxial extension of an inclined crack. Furthermore, we use the SED criterion to study ductile fracture.

The chapter concludes with a brief presentation of the maximum stress criterion for non-self-similar crack growth.

7.2. Volume strain energy density

SED theory is based on the idea that a continuum may be viewed as an assembly of small building blocks, each of which contains a unit volume of material and can store a finite amount of energy at a given instant of time. The energy per unit volume will be referred to as the (volume) strain energy density function, $\mathrm{d}W/\mathrm{d}V$, and is expected to vary from one location to another.

The strain energy density function can be computed from

$$\frac{\mathrm{d}W}{\mathrm{d}V} = \int_0^{\varepsilon_{ij}} \sigma_{ij}\,\mathrm{d}\varepsilon_{ij} + f(\Delta T, \Delta C) \tag{7.1}$$

where σ_{ij} and ε_{ij} are the stress and strain components and ΔT and ΔC are the changes in temperature and moisture concentration, respectively. Equation (7.1) shows that a material element can contain energy even when the stresses are zero.

Historically, the energy quantity has been used for the description of failure of a material element by yielding. There are two separate theories: the total energy or Beltrami-Haigh theory; the distortional energy or Hubert-von Mises-Hencky theory. According to these theories, failure in a material by yielding occurs when the total, or the distortional, strain energy per unit volume absorbed by the material equals the energy per unit volume stored in the material loaded in uniaxial tension at yield. This quantity corresponds to the limiting strain energy and is regarded as a material constant. Extensive experimental evidence is available on the use of the strain energy quantity to describe failure by yielding.

For linear elastic material behavior, the strain energy density function $\mathrm{d}W/\mathrm{d}V$ can be written as

$$\begin{aligned}\frac{\mathrm{d}W}{\mathrm{d}V} = {} & \frac{1}{2E}\left(\sigma_x^2 + \sigma_y^2 + \sigma_z^2\right) - \frac{\nu}{E}\left(\sigma_x\sigma_y + \sigma_y\sigma_z + \sigma_z\sigma_x\right) \\ & + \frac{1}{2\mu}\left(\tau_{xy}^2 + \tau_{yz}^2 + \tau_{zx}^2\right)\end{aligned} \tag{7.2}$$

where σ_x, σ_y, σ_z, τ_{xy}, τ_{yz} and τ_{zx} are the stress components, E the Young modulus, ν the Poisson ratio and μ the shear modulus of elasticity such that $E = 2\mu(1+\nu)$.

For the plane elasticity problems the quantity $\mathrm{d}W/\mathrm{d}V$ takes the form

$$\frac{\mathrm{d}W}{\mathrm{d}V} = \frac{1}{4\mu}\left[\frac{\kappa+1}{4}\left(\sigma_x + \sigma_y\right)^2 - 2\left(\sigma_x\sigma_y - \tau_{xy}^2\right)\right] \tag{7.3}$$

where $\kappa = 3 - 4\nu$ for plane strain and $\kappa = (3-\nu)\,/\,(1+\nu)$ for generalized plane stress.

The strain energy per unit volume, $\mathrm{d}W/\mathrm{d}V$, can be further decomposed into two parts:

$$\frac{\mathrm{d}W}{\mathrm{d}V} = \left(\frac{\mathrm{d}W}{\mathrm{d}V}\right)_d + \left(\frac{\mathrm{d}W}{\mathrm{d}V}\right)_\nu \tag{7.4}$$

in which

$$\left(\frac{\mathrm{d}W}{\mathrm{d}V}\right)_d = \frac{1+\nu}{6E}\Big[(\sigma_x-\sigma_y)^2+(\sigma_y-\sigma_z)^2+(\sigma_z-\sigma_x)^2+ +6(\tau_{xy}^2+\tau_{yz}^2+\tau_{zx}^2)\Big] \tag{7.5}$$

represents the distortional strain energy per unit volume, corresponding to the deviatoric stress tensor that is associated with distortion of an element undergoing no volume change.

The quantity

$$\left(\frac{\mathrm{d}W}{\mathrm{d}V}\right)_\nu = \frac{1-2\nu}{6E}\,(\sigma_x+\sigma_y+\sigma_z)^2 \tag{7.6}$$

represents the part of the strain energy per unit volume associated with volume change and no shape change, i.e. with dilatation.

By means of the plane strain condition

$$\sigma_3 = \nu(\sigma_1+\sigma_2) \tag{7.7}$$

where $\sigma_1, \sigma_2, \sigma_3$ are the principal stresses, we find that the ratio $(\mathrm{d}W/\mathrm{d}V)_\nu \,/\, (\mathrm{d}W/\mathrm{d}V)_d$ obtained from Equations (7.5) and (7.6) takes the form

$$\frac{(\mathrm{d}W/\mathrm{d}V)_\nu}{(\mathrm{d}W/\mathrm{d}V)_d} = \frac{(1-2\nu)\,(1+\nu)\,[(\sigma_1/\sigma_2)+1]^2}{[(\sigma_1/\sigma_2)-1]^2+[(1-\nu)-\nu(\sigma_1/\sigma_2)]^2+[(1-\nu)\,(\sigma_1/\sigma_2)-\nu]^2} \tag{7.8}$$

in terms of the principal stresses.

Figure 7.1 gives the variations of the ratio $(\mathrm{d}W/\mathrm{d}V)_\nu \,/\, (\mathrm{d}W/\mathrm{d}V)_d$ with the ratio of the principal stresses σ_1/σ_2 for $\nu = 0, 0.1, 0.2, 0.3$, and 0.4. This plot shows that the greatest volume change takes place for $\sigma_1 = \sigma_2$, corresponding to a two-dimensional hydrostatic stress state. For most metals, with ν ranging from 0.2 to 0.3, $(\mathrm{d}W/\mathrm{d}V)_\nu \,/\, (\mathrm{d}W/\mathrm{d}V)_d$ varies from 4 to 6.5. From the relative magnitudes of $(\mathrm{d}W/\mathrm{d}V)_\nu$ and $(\mathrm{d}W/\mathrm{d}V)_d$, we see that both quantities should be taken into account when considering the failure of material elements either by yielding and/or fracture.

Failure of material elements in a solid is caused by permanent deformation or fracture which can be related to shape change (distortion) and volume change (dilatation). In general, a material element is subjected to both distortion and dilatation and the corresponding energies for linear elastic material response can be computed from

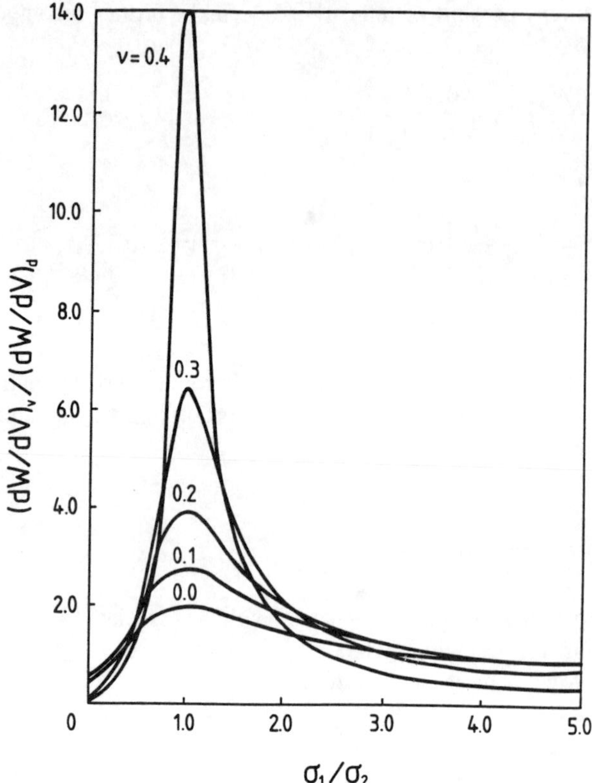

Fig. 7.1. Variation of the ratio of the dilatational and distortional strain energy density $(\mathrm{d}W/\mathrm{d}V)_\nu\,/\,(\mathrm{d}W/\mathrm{d}V)_d$ versus the ratio σ_1/σ_2 of the principal stresses for a state of plane strain. The Poisson ratio ν takes the values 0, 0.1, 0.2, 0.3 and 0.4.

Equations (7.5) and (7.6). Realistic modeling of material failure requires knowledge of damage at both the microscopic and macroscopic levels. X-ray examination of the problem of brittle fracture of a tensile crack in a low carbon steel reveals a thin layer of highly distorted material along the fracture surface. Consider a macrocrack in a tensile stress field (Figure 7.2). The elastic zone ahead of the crack contains shear planes, and the plastic zone out of the plane of the macrocrack contains cleavage planes. The continuum mechanics solution of the stress problem shows that the principal stresses σ_1 and σ_2 are equal in the macroelement ahead of the crack. Equation (7.8) shows that for a Poisson's ratio $\nu = 0.3$, $(\mathrm{d}W/\mathrm{d}V)_\nu$, is 6.5 times larger than $(\mathrm{d}W/\mathrm{d}V)_d$. Even though the macroelement ahead of the crack fractures due to the dilatational component of the strain energy density, the distortional component is not negligible and is responsible for the creation of slip planes or microcracks. Similarly, maximum distortion of the macroelement off the axis of the macrocrack takes place while micro cleavage planes appear perpendicular to the direction of tension.

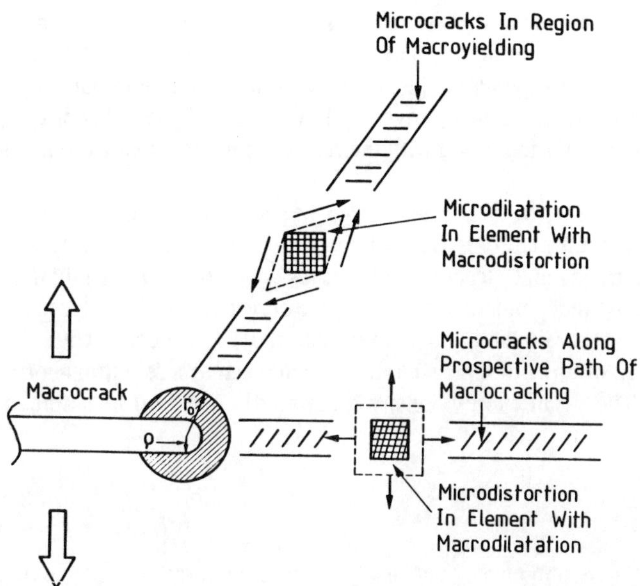

Fig. 7.2. Schematic of macro- and micro-damage in the region ahead of the crack tip.

Stress analysis shows that the distortional component of the strain energy density becomes maximum in this element. The dilatation component is not negligible and is responsible for the creation of the cleavage planes.

These arguments suggest that we must consider both components, the dilatational and the distortional, of strain energy density for a complete description of material damage. They both play a role in the material damage process. Microyielding may lead to macrofracture and microfracture to macroyielding. Thus, the two processes of yielding and fracture are inseparable; they are unique features of material damage and should be treated simultaneously by a single failure criterion.

For the macrocrack under tension, the model of Figure 7.2 suggests that macrofracture coincides with the direction in which $(\mathrm{d}W/\mathrm{d}V)_v > (\mathrm{d}W/\mathrm{d}V)_d$ and macroyielding with the direction in which $(\mathrm{d}W/\mathrm{d}V)_d > (\mathrm{d}W/\mathrm{d}V)_v$. These directions are determined by appealing to physical hypotheses.

7.3. Basic hypotheses

The strain energy density (SED) criterion provides a complete description of material damage by including both the distortional and dilatational effects. The previous arguments show that both components of strain energy density should be included. Distortion and dilatation vary in proportion, depending on the load history, location, and nonuniformity in stress or energy fields. Their contributions to distortion or dilatation of a macroelement are weighted automatically by taking the stationary values of the total strain energy density with respect to appropriate space variables referred to the site of failure initiation. The relative local minimum of $\mathrm{d}W/\mathrm{d}V$ corresponds to large volume change and is identified with the region dominated by macrodilatation leading to fracture; the relative local maximum of $\mathrm{d}W/\mathrm{d}V$ corresponds to large shape change and is identified with the region dominated by macrodistortion leading to yielding.

The strain energy density criterion may be stated in terms of three basic hypotheses, and applied to all materials (reversible or irreversible), loading types (monotonic, cyclic or fatigue) and structure geometries, with or without initial defects. The hypotheses are independent of material type or restrictions introduced by constitutive equations. The strain energy density function $\mathrm{d}W/\mathrm{d}V$ decays with distance r from the crack tip, or any other possible failure site such as a re-entrant corner, inclusion, void, etc. The strain energy density function $\mathrm{d}W/\mathrm{d}V$ will be assumed to have the form

$$\frac{\mathrm{d}W}{\mathrm{d}V} = \frac{S}{r} \tag{7.9}$$

where S is the strain energy density factor and r the radial distance measured from the site of possible failure initiation. The singular dependency $1/r$ is a fundamental character of the Newtonian potential and is independent of the constitutive relation. The three hypotheses of the strain energy density criterion are:

Hypothesis (1)

The location of fracture coincides with the location of relative minimum strain energy density, $(\mathrm{d}W/\mathrm{d}V)_{\mathrm{min}}$, and yielding with relative maximum strain energy density, $(\mathrm{d}W/\mathrm{d}V)_{\mathrm{max}}$.

Hypothesis (2)

Failure by fracture or yielding occurs when $(\mathrm{d}W/\mathrm{d}V)_{\mathrm{min}}$ or $(\mathrm{d}W/\mathrm{d}V)_{\mathrm{max}}$ reach their respective critical values.

Hypothesis (3)

The crack growth increments $r_1, r_2, \ldots, r_j, \ldots, r_c$ during stable crack growth satisfy the equation

$$\left(\frac{\mathrm{d}W}{\mathrm{d}V}\right)_c = \frac{S_1}{r_1} = \frac{S_2}{r_2} = \ldots = \frac{S_j}{r_j} = \ldots = \frac{S_c}{r_c}\,. \tag{7.10}$$

There is unstable fracture or yielding when the critical ligament size r_c is reached.

These hypotheses of the SED criterion will be used later for the solution of many problems of fundamental importance.

7.4. Two-dimensional linear elastic crack problems

This section deals with the general problem of crack extension in a mixed-mode stress field governed by the values of the stress intensity factors K_{I} and K_{II}.

The fundamental quantity is the strain energy density factor S, the amplitude of the energy field that possesses an r^{-1}-type of singularity. We consider the situation of brittle crack growth where crack initiation coincides with final instability. The strain energy density factor S is direction sensitive, unlike the stress intensity factor K, which is a measure of the local stress amplitude. K is scalar, while S is somewhat like a vector. The factor S, defined as $r(\mathrm{d}W/\mathrm{d}V)$ in Equation (7.9), represents the local energy release for a segment of crack growth r. There is unstable crack growth when the critical ligament size r_c is reached.

For crack growth in a two-dimensional stress field the first hypothesis of the SED criterion can be expressed mathematically by the relations

$$\frac{\partial S}{\partial \theta} = 0\,, \quad \frac{\theta^2 S}{\partial \theta^2} > 0 \tag{7.11}$$

where θ is the polar angle. Crack initiation occurs when

$$S(\theta_c) = S_c\,. \tag{7.12}$$

Here θ is defined by (7.11) and S_c is the critical value of the strain energy density factor which is a material constant. S_c represents the fracture toughness of the material.

Consider a crack in a mixed-mode stress field governed by the values of the opening-mode K_{I} and sliding-mode K_{II} stress intensity factors. The singular stress field near the crack tip is expressed by (see Equations (2.28) and (2.43))

$$\begin{aligned}
\sigma_x &= \frac{K_{\mathrm{I}}}{\sqrt{2\pi r}}\cos\frac{\theta}{2}\left(1-\sin\frac{\theta}{2}\sin\frac{3\theta}{2}\right)-\frac{K_{\mathrm{II}}}{\sqrt{2\pi r}}\sin\frac{\theta}{2}\left(2+\cos\frac{\theta}{2}\cos\frac{3\theta}{2}\right)\\
\sigma_y &= \frac{K_{\mathrm{I}}}{\sqrt{2\pi r}}\cos\frac{\theta}{2}\left(1+\sin\frac{\theta}{2}\sin\frac{3\theta}{2}\right)+\frac{K_{\mathrm{II}}}{\sqrt{2\pi r}}\sin\frac{\theta}{2}\cos\frac{\theta}{2}\cos\frac{3\theta}{2}\\
\tau_{xy} &= \frac{K_{\mathrm{I}}}{\sqrt{2\pi r}}\cos\frac{\theta}{2}\sin\frac{\theta}{2}\cos\frac{3\theta}{2}+\frac{K_{\mathrm{II}}}{\sqrt{2\pi r}}\cos\frac{\theta}{2}\left(1-\sin\frac{\theta}{2}\sin\frac{3\theta}{2}\right).
\end{aligned} \tag{7.13}$$

Introducing these equations into Equation (7.3), we obtain the following quadratic form for the strain energy density factor S:

$$S = a_{11}k_{\mathrm{I}}^2 + 2a_{12}k_{\mathrm{I}}k_{\mathrm{II}} + a_{22}k_{\mathrm{II}}^2 \tag{7.14}$$

where the coefficients a_{ij} $(i,j=1,2)$ are given by

$$\begin{aligned}
16\mu a_{11} &= (1+\cos\theta)\,(\kappa-\cos\theta)\\
16\mu a_{12} &= \sin\theta\,[2\cos\theta-(\kappa-1)]\\
16\mu a_{22} &= (\kappa+1)\,(1-\cos\theta)+(1+\cos\theta)\,(3\cos\theta-1)\,.
\end{aligned} \tag{7.15}$$

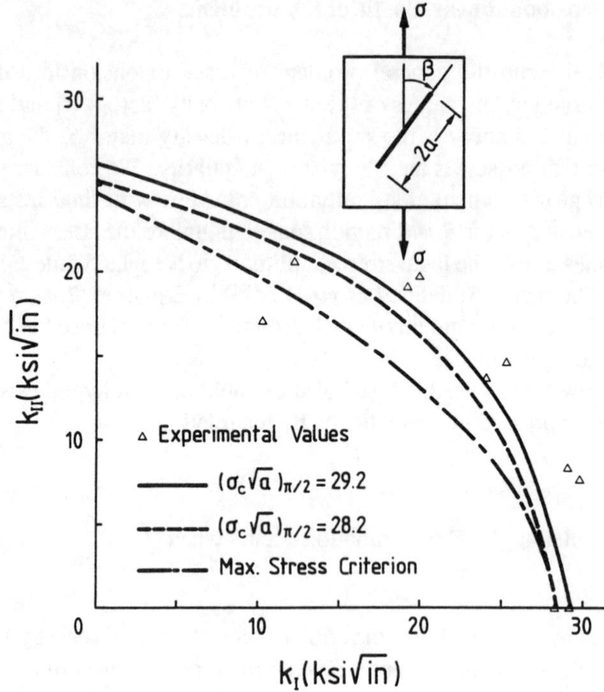

Fig. 7.3. Mixed-mode fracture criterion for cracks under tension.

In these equations $k_j = K_j/\sqrt{\pi}$ $(j = \mathrm{I}, \mathrm{II})$.
Substituting S from Equation (7.14) into relations (7.11) we obtain

$$[2\cos\theta - (\kappa - 1)] \sin\theta\, k_{\mathrm{I}}^2 + 2[2\cos 2\theta - (\kappa - 1)\cos\theta]\, k_{\mathrm{I}} k_{\mathrm{II}} +$$

$$+[(\kappa - 1 - 6\cos\theta)\sin\theta]\, k_{\mathrm{II}}^2 = 0 \tag{7.16a}$$

$$[2\cos 2\theta - (\kappa - 1)\cos\theta]\, k_{\mathrm{I}}^2 + 2[(\kappa - 1)\sin\theta - 4\sin 2\theta]\, k_{\mathrm{I}} k_{\mathrm{II}} +$$

$$+[(\kappa - 1)\cos\theta - 6\cos 2\theta]\, k_{\mathrm{II}}^2 > 0\,. \tag{7.16b}$$

Relations (7.16a) and (7.16b) represent the general formulas of the strain energy density criterion for a crack in a two-dimensional stress field under mixed-mode loading conditions. Suppose k_{I}, k_{II} stress intensity factors are known for a particular problem. Introducing these values into equation (7.16a) we obtain the values of the crack extension angle θ_c as the roots of the equation which satisfy the inequality (7.16b). Substituting these roots, θ_c, into Equation (7.14) we obtain the minimum values $S_{\min}$ of the strain energy density factor. Then, we obtain the critical values of the applied loads corresponding to the onset of rapid crack propagation by equating $S_{\min}$ to the material constant critical strain energy density factor S_c.

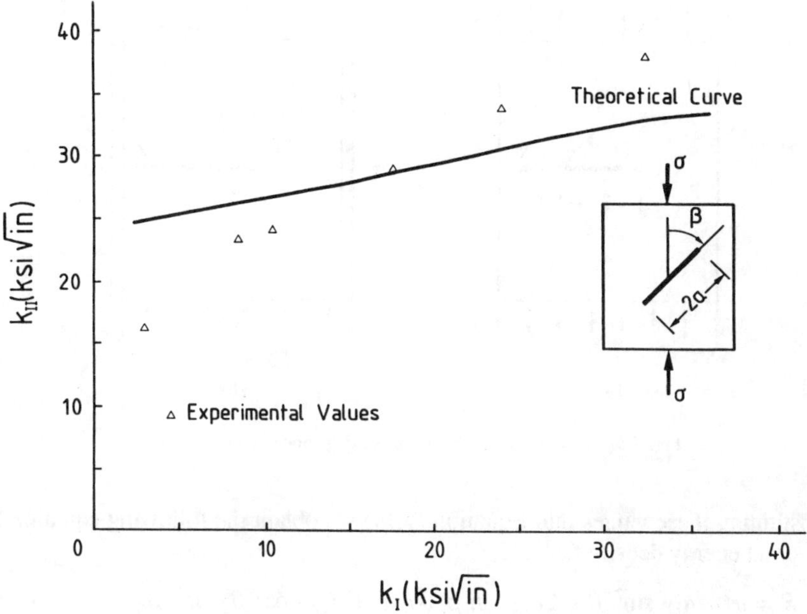

Fig. 7.4. Mixed-mode fracture criterion for cracks under compression.

The crack growth condition expressed by Equations (7.12), (7.14), (7.15) and (7.16a) defines the fracture locus in the $k_I - k_{II}$ plane. It is shown in Figure 7.3 for two aluminum alloys with $4.8(\mu S_c)^{1/2} = 28.2$ kip/in$^{5/2}$ and 29.2 kip/in$^{5/2}$, and for tensile applied loads. The third curve represents the prediction based on the maximum stress criterion (Section 7.7). The experimental results relate to the uniaxial tension plate with an inclined crack. Observe that prediction based on the SED criterion is closer to the experimental results. Figure 7.4 shows the $k_I - k_{II}$ locus for compressive applied stresses for glass, together with experimental results. The $k_I - k_{II}$ curve for compression is basically different from that in tension. The curve does not intersect the k_I-axis because a crack under mode-I does not extend in compression.

7.5. Uniaxial extension of an inclined crack

Consider a central crack of length $2a$ in an infinite plate subjected to a uniform uniaxial stress σ at infinity; suppose the axis of the crack makes an angle β with the direction of stress σ. Mixed-mode conditions predominate near the crack tip, and the values of the k_I, k_{II} stress intensity factors are given by (Problem 2.22):

$$k_I = \sigma a^{1/2} \sin^2 \beta \,, \quad K_{II} = \sigma a^{1/2} \sin \beta \cos \beta \,. \tag{7.17}$$

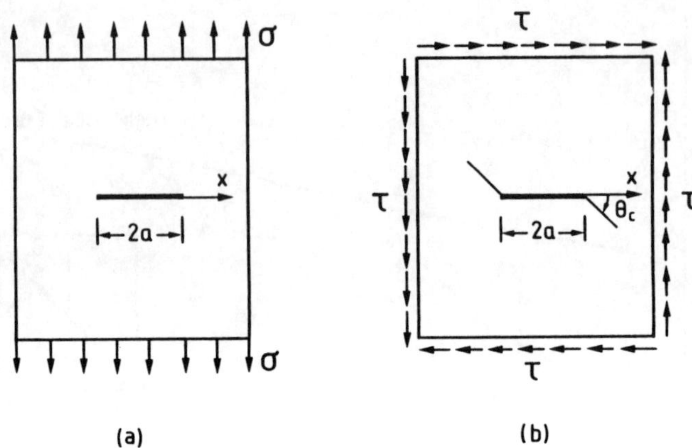

Fig. 7.5. (a) Opening-mode and (b) sliding-mode crack extension.

Substituting these values into Equation (7.14) we obtain the following equation for the strain energy density factor S:

$$S = \sigma^2 a (a_{11} \sin^2 \beta + 2a_{12} \sin \beta \cos \beta + a_{22} \cos^2 \beta) \sin^2 \beta \tag{7.18}$$

where the coefficients a_{ij} are given by Equations (7.15).

Equation (7.16a) for the calculation of the angle θ_c of initial crack extension takes the form:

$$(\kappa - 1) \sin (\theta_c - 2\beta) - 2 \sin [2(\theta_c - \beta)] - \sin 2\theta_c = 0\,, \quad \beta \neq 0\,. \tag{7.19}$$

Before proceeding to the general case of a crack of any inclination with respect to the loading direction, we will consider separately the two common cases of opening-mode and sliding-mode crack extension.

(a) Opening-mode crack extension

This case corresponds to the trivial Griffith crack configuration consisting of an infinite body with a central crack of length $2a$ subjected to a uniform uniaxial stress σ at infinity (Figure 7.5(a)). Because of load symmetry the crack propagates in its own plane. Let us suppose that this is an unsolved problem and analyze it using the strain energy density theory. Inserting the values of the stress intensity factors k_{I}, k_{II}:

$$k_{\text{I}} = \sigma a^{1/2}\,, \quad k_{\text{II}} = 0 \tag{7.20}$$

into Equation (7.14) we obtain the equation

$$S = \frac{\sigma^2 a}{16\mu} (1 + \cos \theta)\,(\kappa - \cos \theta)\,. \tag{7.21}$$

Furthermore, Equation (7.16a), which gives the stationary values of S, takes the form

$$[2\cos\theta - (\kappa - 1)]\sin\theta = 0 \tag{7.22}$$

while inequality (7.16b) becomes

$$2\cos 2\theta - (\kappa - 1)\cos\theta > 0\,. \tag{7.23}$$

Equation (7.22) is satisfied when $\theta_c = 0$ or $\theta_c = \arccos[(\kappa - 1)/2]$. The second root θ_c does not satisfy inequality (7.22) because for the elastic constant κ is $1 \leq \kappa \leq 3$, and thus is ignored. Hence, the minimum value of S corresponds to an initial crack extension angle $\theta_c = 0$, which means that the crack extends in its own plane. The plane ($\theta_c = 0$) corresponds to the direction of maximum potential energy, a position of unstable equilibrium. For $\theta_c = 0$, Equation (7.21) gives the minimum value S_{min} of S

$$S_{min} = \frac{(\kappa - 1)\,\sigma^2 a}{8\mu}\,. \tag{7.24}$$

Equating S_{min} with the material constant S_c, we obtain the following expression for the critical stress σ_c corresponding to the onset of crack extension:

$$\sigma_c a^{1/2} = \left(\frac{8\mu S_c}{\kappa - 1}\right)^{1/2}\,. \tag{7.25}$$

The value of the stress intensity factor given by Equation (7.20) corresponds to an infinite plate. For the general case of a mode-I crack with stress intensity factor K_I, Equation (7.24) becomes

$$S_c = \frac{(1+\nu)\,(1-2\nu)\,K_{Ic}^2}{2\pi E} \tag{7.26}$$

when it refers to the critical state of unstable crack extension under plane strain conditions.

Equation (7.26) relates the critical strain energy density factor, S_c, to the critical stress intensity factor, K_{Ic}, which can be determined by the methods described in Section 5.4. S_c is a material constant and characterizes the fracture toughness of the material.

(b) Sliding-mode crack extension

This case corresponds to an infinite body containing a central crack of length $2a$ and subjected to a uniform shear stress τ at infinity (Figure 7.5b). The k_I, k_{II} stress intensity factors are given by:

$$k_I = 0\,, \quad k_{II} = \tau a^{1/2} \tag{7.27}$$

and Equation (7.14) gives

$$S = \frac{\tau^2 a}{16\mu}\,[(\kappa + 1)\,(1 - \cos\theta) + (1 + \cos\theta)\,(3\cos\theta - 1)]\,. \tag{7.28}$$

TABLE 7.1. Fracture angle $-\theta_0$ under pure shear and plane strain conditions

ν	0	0.1	0.2	0.3	0.4	0.5
$-\theta_0$	70.5°	74.5°	78.5°	82.3°	86.2°	90.0°

Working as in the previous case, we find that the angle of crack extension θ_c is given by

$$\theta_c = \arccos\left(\frac{\kappa - 1}{6}\right) . \tag{7.29}$$

Note that θ_c is a function of Poisson's ratio ν. Table 7.1 shows the values of the predicted fracture angle θ_c for ν ranging from 0 to 0.5 in plane strain conditions $(\kappa = 3 - 4\nu)$.

Introducing the value of the angle of initial crack extension θ_c into Equation (7.28) and equating the resulting value of $S_{\min}$ to S_c we derive the following expression for the critical shear stress τ_c:

$$\tau_c a^{1/2} = \left(\frac{192\mu S_c}{-\kappa^2 + 14k - 1}\right)^{1/2} . \tag{7.30}$$

(c) Inclined crack; tensile loads

From the resulting values of the angle θ, those which satisfy inequality (7.16b) provide the angles of initial crack extension θ_c. Figure 7.6 displays the variation of $-\theta_c$ versus the crack angle β for $\nu = 0, 0.1, 0.2, 0.3, 0.4$ and 0.5 under plane strain conditions. Results for plane stress can be obtained by replacing ν with $\nu/(1+\nu)$. In the same figure the dashed curve represents the results obtained by the maximum stress criterion (Section 7.7). We see that these results agree with those based on Equation (7.19) for large values of β, and represent a lower bound for small values of β. In general it can be taken as an average curve. It is worth noting that the crack extension angle θ_c is always negative for uniaxial tensile loads. The results of Figure 7.6 are in good agreement with experimental results obtained from plexiglas plates with a central crack.

Figure 7.6 gives the minimum values of the strain energy density factor S. By equating these values of S with the critical strain energy density factor S_c, we find the critical values of the tensile stress σ_c for crack propagation. Figure 7.7 shows the variation of the quantity $16\mu S_c/\sigma_c^2 a$ versus the crack inclination angle β for $\nu = 0$, 0.1, 0.2, 0.3, 0.333 and 0.4, and plane strain conditions. We see that the quantity $16\mu S_c/\sigma_c^2 a$ increases with the crack angle β, reaching a maximum for opening-mode crack extension. Furthermore, $16\mu S_c/\sigma_c^2 a$ increases as the Poisson's ratio ν of the plate decreases. Since S_c is a material constant, this means that the quantity $\sigma_c^2 a$ decreases as the crack angle β increases, and increases as Poisson's ratio increases.

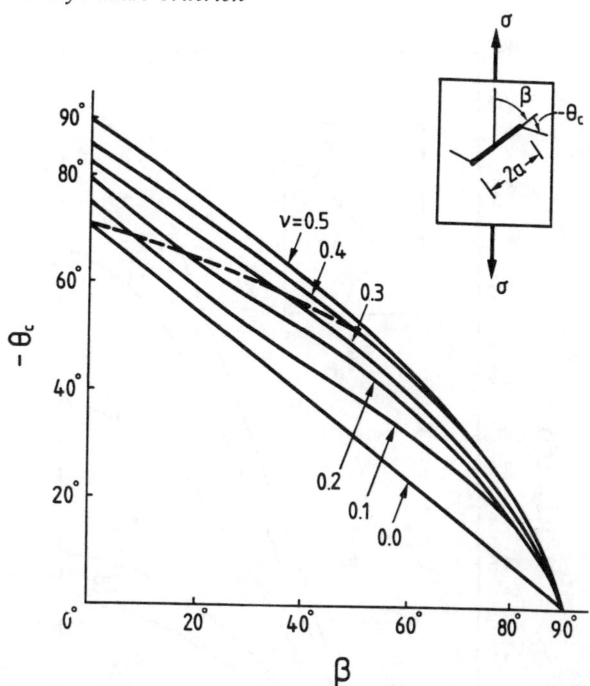

Fig. 7.6. Variation of the crack extension angle $-\theta_c$ versus the crack inclination angle β under plane strain conditions for tensile applied loads.

Thus, the lowest value of the applied stress σ_c that will initiate crack propagation occurs at $\beta = \pi/2$.

(d) Inclined crack; compressive loads

Equation (7.19) has negative roots and positive roots. The negative ones correspond to uniaxial tensile loads; the positive ones to uniaxial compression. Since S depends on σ^2, Equations (7.18) and (7.19) contain the solutions for uniaxial tension $+\sigma$ and compression $-\sigma$. We assume that there is no overlapping of crack surfaces if there is compression.

Figure 7.8 presents the variation of the positive crack extension angle θ_c versus the crack inclination angle β for compressive applied loads and various values of Poisson's ratio ν. We see that under uniaxial compression the crack path extends towards the direction of loading. This phenomenon was observed by Hoek and Bieniawski [7.12] who made tests on a number of glass plates with inclined cracks under uniaxial compression. Unfortunately, they did not report the angle of initial crack extension and therefore it is not possible to compare their experimental results with the theoretical results of Figure 7.8.

The stationary values of the strain energy density factor are obtained following the procedure used for tensile loading. Figure 7.9 presents the variation of the

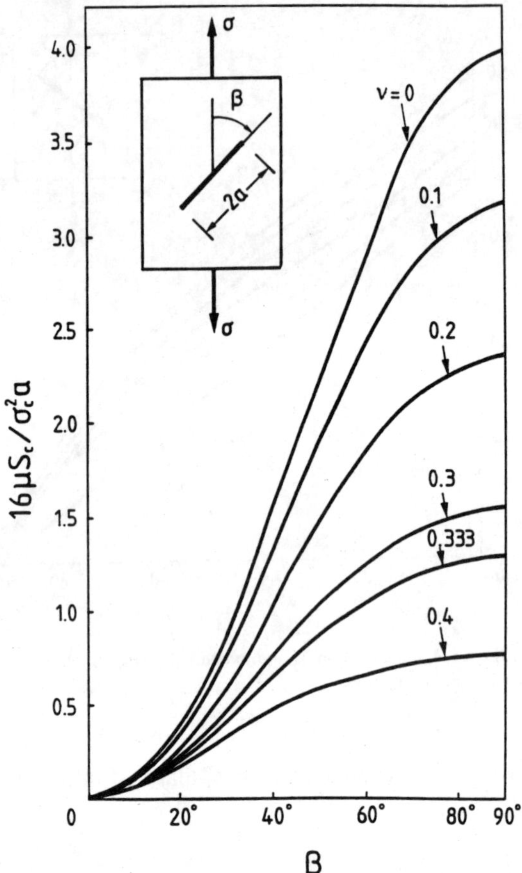

Fig. 7.7. Variation of the quantity $16\mu S_c/\sigma_c^2 a$ versus the crack angle β under plane strain conditions for tensile applied loads.

quantity $16\mu S_c/\sigma_c^2 a$ versus the crack inclination angle β for plane strain conditions. Figure 7.9 shows that the quantity $16\mu S_c/\sigma_c^2 a$ reaches a maximum in the interval $0 < \beta < 90°$ depending on the value of Poisson's ratio ν. We see also that the critical stress σ_c increases and tends to infinity as the crack becomes parallel ($\beta = 0$) or perpendicular ($\beta = 90°$) to the direction of loading. This result corresponds to the physical observation that a crack parallel or perpendicular to the direction of a compressive applied stress has no influence on the fracture behavior of the plate.

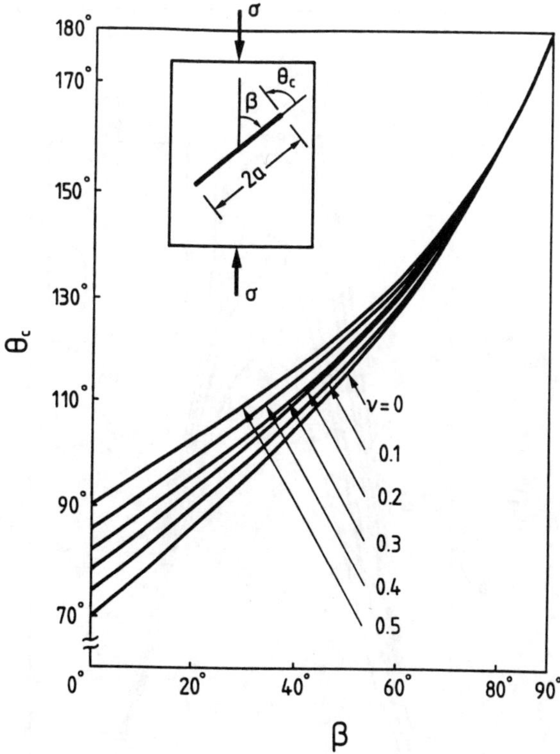

Fig. 7.8. Variation of the crack extension angle θ_c versus the crack inclination angle β under plane strain conditions for compressive applied loads.

7.6. Ductile fracture

(a) Introductory remarks

The term ductile fracture is generally used to indicate failure where unstable crack propagation is preceded by plastic deformation. (Ductile indicates the presence of stable deformation, while fracture designates load instability associated with the sudden creation of a macrocrack surface.) A characteristic feature of ductile fracture is that the crack grows slowly at first before the onset of unstable crack propagation. The crack growth process can be separated into the phases of crack initiation, and stable and unstable crack growth. Generally speaking, all fracture processes may be regarded as transitions from stable to unstable crack propagation. When the amount of stable crack growth is small, it is usually assumed that onset of crack initiation coincides with crack instability. The phenomenon of ductile fracture is associated with a nonlinear load versus deformation relation which is attributed to plastic deformation and slow stable crack growth. These two effects take place simultaneously

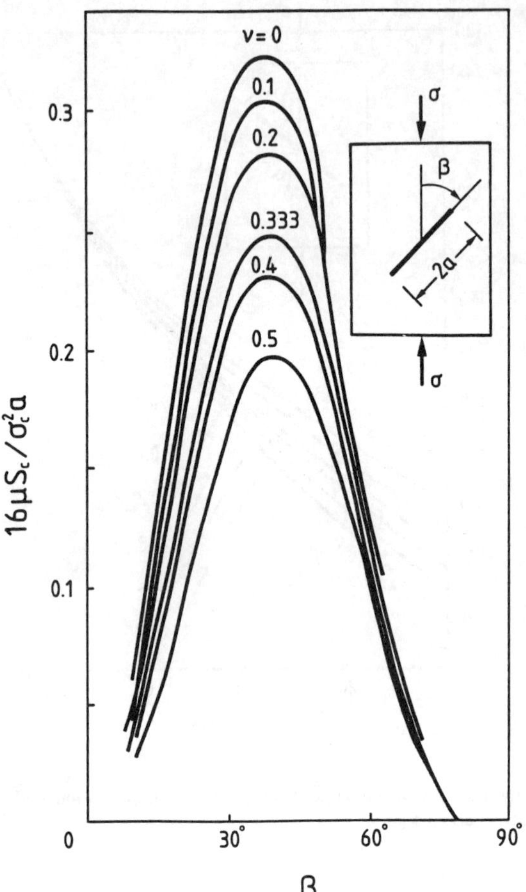

Fig. 7.9. Variation of the quantity $16\mu S_c/\sigma_c^2 a$ versus the crack angle β under plane strain conditions for compressive applied loads.

by known experimental methods.

Ductile fracture is controlled by the rate and history of loading, specimen size and geometry, material properties and environmental conditions. Thus, small metal specimens may exhibit high ductility, while large structures of the same material can behave in a brittle fashion. Furthermore, a substantial amount of subcritical crack growth may occur in a specimen made of a brittle material when it is slowly loaded. On the contrary, brittle fractures may occur in structures made of a ductile material when the load is applied suddenly or when they are subjected to low temperatures. The development of plastic zones during crack growth in ductile fracture, which corresponds to material damage at a microscopic scale, reduces the amount of energy available for macrocrack instability.

The strain energy density theory will be used in the following to address the entire history of crack growth, including the phases of initiation, stable and unstable crack growth. The theory has been applied successfully to the solution of a number of problems related to ductile fracture.

(b) Energy dissipation

During the process of crack growth not all the energy is consumed by the creation of new macroscopic surfaces. Dissipation of energy also takes place at the microscopic level; this is called yielding in continuum mechanics. Even though yielding occurs in a direction at some angle to the crack propagation (Figure 7.2) the amount of energy consumed at the microscopic level along the path of crack growth is not negligible and should be taken into account.

Consider the true stress-true strain diagram of the material in tension (Figure 7.10), and suppose that the stress exceeds the yield stress σ_Y; the unloading path will follow the line PM which is almost parallel to the direction OA of the elastic portion of the diagram. If the specimen is reloaded the new stress-strain curve will be the line MPF. During the unloading and reloading procedure the amount of energy dissipated is represented by the area $OAPM = (\mathrm{d}W/\mathrm{d}V)_p$. Thus, for a stress level σ the available strain energy $(\mathrm{d}W/\mathrm{d}V)_c^*$ for crack growth is represented by the area $MPFF'$, or

$$\left(\frac{\mathrm{d}W}{\mathrm{d}V}\right)_c^* = \left(\frac{\mathrm{d}W}{\mathrm{d}V}\right)_c - \left(\frac{\mathrm{d}W}{\mathrm{d}V}\right)_p . \tag{7.31}$$

This equation gives the total energy per unit volume required for failure of a material element.

Thus, Equation (7.10), which expresses the third hypothesis of the strain energy density theory, should be modified to read

$$\left(\frac{\mathrm{d}W}{\mathrm{d}V}\right)_c^* = \frac{S_1}{r_1} = \frac{S_2}{r_2} = \ldots \frac{S_j}{r_j} = \frac{S_c^*}{r_c^*} \text{ or } \frac{S_0^*}{r_0^*} . \tag{7.32}$$

The material damage process increases monotonically up to global instability when S or r increase during stable crack growth, that is,

$$S_1 < S_2 < \ldots < S_j < \ldots < S_c$$

$$r_1 < r_2 < \ldots < r_j < \ldots < r_c \tag{7.33}$$

and comes to rest when S or r decrease during stable crack growth, that is,

$$S_1 > S_2 > \ldots > S_j > \ldots > S_0$$

$$r_1 > r_2 > \ldots > r_j > \ldots > r_0 \tag{7.34}$$

where r_0 is the radius of the fracture core region.

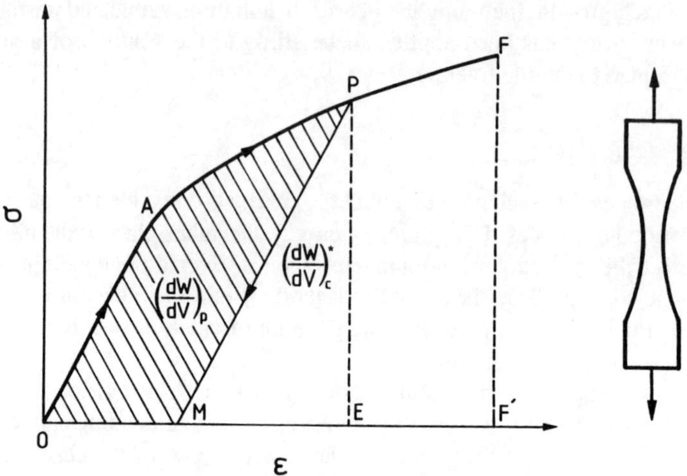

Fig. 7.10. Dissipated, $(dW/dV)_p$, and available, $(dW/dV)_c^*$, energy of a tension specimen.

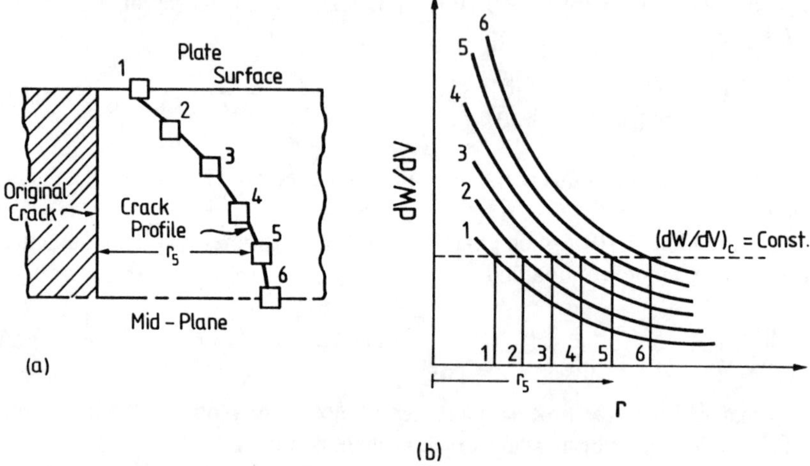

Fig. 7.11. Development of crack growth profile.

(d) Development of crack profile

Equation (7.10), expressing the third hypothesis of the strain energy density theory, can be used to determine crack profiles in three dimensions during slow stable crack growth. The procedure is illustrated in Figure 7.11, which refers to a crack in a tensile plate specimen. Since the configuration is symmetric we take six points along one-half of the plate thickness. Figure 7.11b shows the variation of the strain energy density function dW/dV versus distance r ahead of the crack front. The intersection of the line $(\mathrm{d}W/\mathrm{d}V)_c = \text{const}$ with curves $\mathrm{d}W/\mathrm{d}V - r$ for the six points determines the values of r_j $(j = 1, 2, 3, 4, 5, 6)$ which define the crack profile. The same procedure can be repeated to describe the crack growth profile during slow stable crack growth.

This procedure for constructing crack profiles has been applied to a straight crack in a tensile plate specimen [7.13, 7.14]. Larger intervals of crack growth have been obtained for material elements near the plate midsection than for those near the plate boundaries. The diagram illustrates the influence of plate thickness, material type, and loading step, on the shape and size of crack growth front during stable crack growth. These results explain the well-known crack tunneling effect and verify experimental observation.

7.7. The stress criterion

This chapter deals with mixed-mode crack growth; it is therefore appropriate to present the maximum circumferential stress criterion proposed by Erdogan and Sih [7.15]. Consider a crack in a mixed-mode stress field governed by the values of the opening-mode K_{I} and sliding-mode K_{II} stress intensity factors. The singular polar stress components near the crack tip are expressed by (see Equations (2.35) and Problem 2.15)

$$\sigma_r = \frac{K_{\mathrm{I}}}{\sqrt{2\pi r}}\left(\frac{5}{4}\cos\frac{\theta}{2} - \frac{1}{4}\cos\frac{3\theta}{2}\right) + \frac{K_{\mathrm{II}}}{\sqrt{2\pi r}}\left(-\frac{5}{4}\sin\frac{\theta}{2} + \frac{3}{4}\sin\frac{3\theta}{2}\right) \tag{7.35a}$$

$$\sigma_\theta = \frac{K_{\mathrm{I}}}{\sqrt{2\pi r}}\left(\frac{3}{4}\cos\frac{\theta}{2} + \frac{1}{4}\cos\frac{3\theta}{2}\right) + \frac{K_{\mathrm{II}}}{\sqrt{2\pi r}}\left(-\frac{3}{4}\sin\frac{\theta}{2} - \frac{3}{4}\sin\frac{3\theta}{2}\right) \tag{7.35b}$$

$$\tau_{r\theta} = \frac{K_{\mathrm{I}}}{\sqrt{2\pi r}}\left(\frac{1}{4}\sin\frac{\theta}{2} + \frac{1}{4}\sin\frac{3\theta}{2}\right) + \frac{K_{\mathrm{II}}}{\sqrt{2\pi r}}\left(\frac{1}{4}\cos\frac{\theta}{2} + \frac{3}{4}\cos\frac{3\theta}{2}\right). \tag{7.35c}$$

The assumptions made in the criterion for crack extension in brittle materials may be stated as

(i) The crack extension starts from its tip along the radial direction $\theta = \theta_c$ on which σ_θ becomes maximum.

(ii) Fracture starts when that maximum of σ_θ reaches a critical stress σ_c equal to the fracture stress in uniaxial tension.

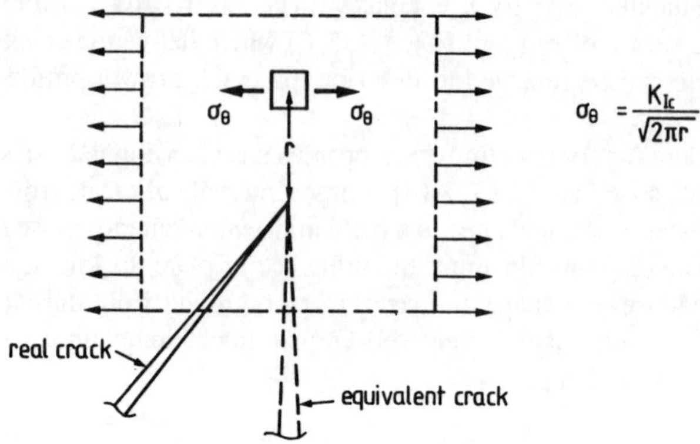

Fig. 7.12. Equivalent opening-mode crack model according to the maximum circumferential stress criterion.

The hypotheses can be expressed mathematically by the relations

$$\frac{\partial \sigma_\theta}{\partial \theta} = 0\,; \quad \frac{\partial^2 \sigma_\theta}{\partial \theta^2} < 0 \tag{7.36a}$$

$$\sigma_\theta(\theta_c) = \sigma_c\,. \tag{7.36b}$$

Observe that the circumferential stress σ_θ in the direction of crack extension is a principal stress, and the shear stress $\tau_{r\theta}$ for that direction vanishes. The crack extension angle θ_c is calculated (see Equation (7.35c)) by

$$K_{\mathrm{I}} \left(\sin \frac{\theta}{2} + \sin \frac{3\theta}{2} \right) + K_{\mathrm{II}} \left(\cos \frac{\theta}{2} + 3 \cos \frac{3\theta}{2} \right) = 0 \tag{7.37}$$

or

$$K_{\mathrm{I}} \sin \theta + K_{\mathrm{II}} \left(3 \cos \theta - 1 \right) = 0\,. \tag{7.38}$$

For the calculation of the stress σ_θ from Equation (7.35b) we must introduce a critical distance r_0 measured from the crack tip. To circumvent the determination of the core region radius r_0, the second hypothesis of the stress criterion is often referred to as follows:

Fracture starts when σ_θ has the same value as in an equivalent opening-mode (Figure 7.12), that is

$$\sigma_\theta = \frac{K_{\mathrm{Ic}}}{\sqrt{2\pi r}}\,. \tag{7.39}$$

The fracture condition following from Equations (7.39) and (7.35b) takes the form

$$K_{\mathrm{I}} \left(3 \cos \frac{\theta_c}{2} + \cos \frac{3\theta_c}{2} \right) - 3K_{\mathrm{II}} \left(\sin \frac{\theta_c}{2} + \sin \frac{3\theta_c}{2} \right) = 4K_{\mathrm{Ic}}\,. \tag{7.40}$$

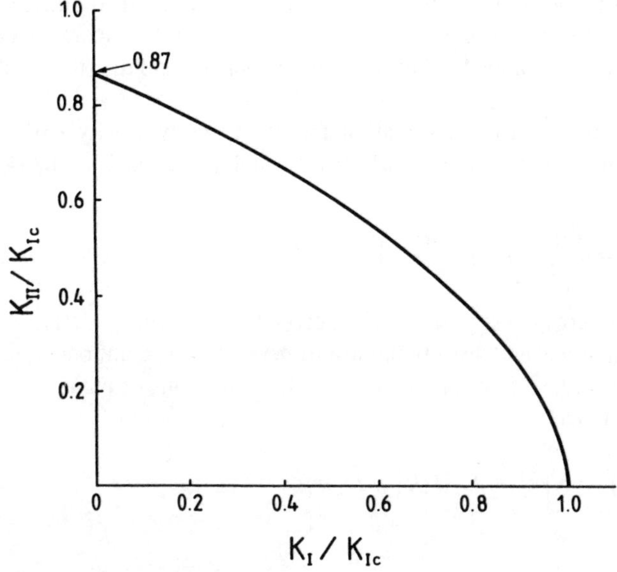

Fig. 7.13. Fracture locus for mixed-mode conditions according to the maximum circumferential stress criterion.

For opening-mode loading ($K_{\mathrm{I}} \neq 0$, $K_{\mathrm{II}} = 0$), Equations (7.38) and (7.40) yield $\theta_c = 0$, $K_{\mathrm{I}} = K_{\mathrm{I}c}$, while for sliding-mode loading they give

$$\theta_c = -\arccos\frac{1}{3} = -70.6^\circ\,,\quad K_{\mathrm{II}} = K_{\mathrm{II}c} = \sqrt{\frac{3}{4}}\,K_{\mathrm{I}c}\,. \tag{7.41}$$

Eliminating θ_c in Equations (7.38) and (7.40) gives the fracture locus in $K_{\mathrm{I}} - K_{\mathrm{II}}$ coordinates shown in Figure 7.13. Figure 7.6 shows the variation of the crack extension angle $-\theta_c$ versus the crack inclination angle for an inclined crack in a plate subjected to a uniaxial stress field. The figure also shows results based on the SED criterion. Note that the results of both criteria agree for large values of β, while for small values of β the stress criterion predicts smaller angles than the SED criterion.

Examples

Example 7.1.

According to the maximum dilatational strain energy density criterion, the crack extends along the direction of maximum dilatational strain energy density which is calculated around the circumference of a circular core area surrounding the crack tip. Derive an equation to determine the crack extension angle in a two-dimensional mixed-mode stress field governed by the K_{I} and K_{II} stress intensity factors under conditions of plane strain. Apply this equation to an inclined crack in an infinite plate subjected to a uniform uniaxial stress at infinity, and compare the crack growth angles with those obtained by the strain energy density and the stress criteria.

Solution: For linear elastic deformation, the strain energy density, $(\mathrm{d}W/\mathrm{d}V)$, can be decomposed into its dilatational, $(\mathrm{d}W/\mathrm{d}V)_\nu$, and distortional, $(\mathrm{d}W/\mathrm{d}V)_d$, components as a sum

$$\left(\frac{\mathrm{d}W}{\mathrm{d}V}\right) = \left(\frac{\mathrm{d}W}{\mathrm{d}V}\right)_\nu + \left(\frac{\mathrm{d}W}{\mathrm{d}V}\right)_d . \tag{1}$$

The strain energy density $(\mathrm{d}W/\mathrm{d}V)$ is given by Equations (7.9), (7.14) and (7.15), while the dilatational and distortional strain energy density components are computed from Equations (7.5), (7.6) and (7.13). Under conditions of plane strain we obtain the dilatational part

$$r\left(\frac{\mathrm{d}W}{\mathrm{d}V}\right) = b_{11}k_{\mathrm{I}}^2 + 2b_{12}k_{\mathrm{I}}k_{\mathrm{II}} + b_{22}k_{\mathrm{II}}^2 \tag{2}$$

such that

$$\begin{aligned} 12\mu b_{11} &= (1-2\nu)\,(1+\nu)\,(1+\cos\,\theta) \\ 12\mu b_{12} &= -(1-2\nu)\,(1+\nu)\,\sin\,\theta \\ 12\mu b_{22} &= (1-2\nu)\,(1+\nu)\,(1-\cos\,\theta)\,, \end{aligned} \tag{3}$$

and for the distortional part

$$r\left(\frac{\mathrm{d}W}{\mathrm{d}V}\right)_d = c_{11}k_{\mathrm{I}}^2 + 2c_{12}k_{\mathrm{I}}k_{\mathrm{II}} + c_{22}k_{\mathrm{II}}^2 \tag{4}$$

such that

$$\begin{aligned} 16\mu c_{11} &= (1+\cos\,\theta)\left[\frac{2}{3}\,(1-2\nu)^2 + 1 - \cos\,\theta\right] \\ 16\mu c_{12} &= 2\,\sin\,\theta\left[\cos\,\theta - \frac{1}{3}\,(1-2\nu)^2\right] \\ 16\mu c_{22} &= \frac{2}{3}\,(1-2\nu)^2\,(1-\cos\,\theta) + 4 - 3\,\sin^2\,\theta \end{aligned} \tag{5}$$

where μ is the shear modulus and r the radial distance from the crack tip.

According to the maximum dilatational strain energy density assumption, the angle θ_c of initial crack extension can be obtained as

$$\frac{\partial}{\partial\theta}\left[\left(\frac{\mathrm{d}W}{\mathrm{d}V}\right)_\nu\right] = 0\,, \quad \frac{\partial^2}{\partial\theta^2}\left[\left(\frac{\mathrm{d}W}{\mathrm{d}V}\right)_\nu\right] < 0\,. \tag{6}$$

Equation (6) gives

$$\sin\theta\, k_{\mathrm{I}}^2 + 2\cos\theta\, k_{\mathrm{I}} k_{\mathrm{II}} - \sin\theta\, k_{\mathrm{II}}^2 = 0\,. \tag{7}$$

Equation (7) determines the angle θ_c when the values of stress intensity factors k_{I} and k_{II} are known.

For an inclined crack in an infinite plate the values of K_{I} and K_{II} are given by Equation (7.17). Substituting these values into Equation (7), we find the crack extension angle θ_c to be

$$\theta_c = 2\beta - \pi\,. \tag{8}$$

Equation (8) for $0 < \beta < \pi/2$ predicts angles θ_c in the range $-\pi < \theta_c < 0$, while according to the strain energy density and stress criteria which are verified by experimental results the angle θ_c varies in the interval $-\pi/2 < \theta_c < 0$. Thus the maximum dilatational strain energy density criterion leads to predictions which are unrealistic and far beyond any experimental observation.

Example 7.2.

An infinite plate contains a circular crack of radius R and angle 2β and is subjected to a uniform uniaxial tensile stress σ at infinity perpendicular to the chord of the crack. The stress intensity factors k_{I} and k_{II} at the crack tip are given by

$$k_{\mathrm{I}} = \frac{\sigma}{2}(R\sin\beta)^{1/2}\left[\frac{\left(1-\sin^2\frac{\beta}{2}\cos^2\frac{\beta}{2}\right)\cos\frac{\beta}{2}}{1+\sin^2\frac{\beta}{2}} + \cos\frac{3\beta}{2}\right] \tag{1a}$$

$$k_{\mathrm{II}} = \frac{\sigma}{2}(R\sin\beta)^{1/2}\left[\frac{\left(1-\sin^2\frac{\beta}{2}\cos^2\frac{\beta}{2}\right)\sin\frac{\beta}{2}}{1+\sin^2\frac{\beta}{2}} + \sin\frac{3\beta}{2}\right]. \tag{1b}$$

Plot the variation of the angle of crack extension θ_c, and the critical stress σ_c for crack growth, versus the angle of the crack 2β, for various values of $\kappa(\kappa = 3 - 4\nu$ for plane strain and $\kappa = (3-\nu)\,/\,(1+\nu)$ for plane stress).

Solution: Substituting the values of stress intensity factors k_{I} and k_{II} from Equation (1) into Equation (7.16a) of the strain energy density criterion we obtain an equation containing the quantities β, κ and θ. The roots of this equation which satisfy inequality (7.16b) give the values of the crack extension angle θ_c. The critical stress σ_c for initiation of crack extension is then determined from equations (7.12), (7.14) and (7.15), where S_c is a material parameter.

Figure 7.14 presents the variation of $-\theta_c$ versus the half angle of the circular crack β for the extreme values of κ equal to 1.0 and 3.0. The figure also shows the

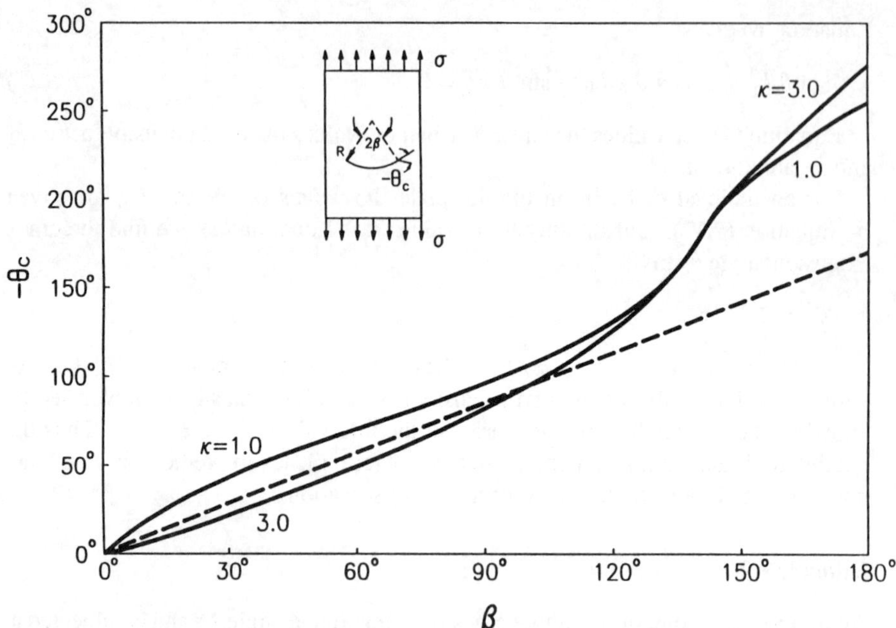

Fig. 7.14. Crack extension angle $(-\theta_c)$ versus half angle β of a circular crack whose chord is perpendicular to the applied tensile stress for $\kappa = 1.0$ and 3.0. The straight dotted line corresponds to the extension of the crack at right angle to the direction of the applied load.

straight line $-\theta_1 = \beta$ corresponding to extension of the crack at a right angle to the direction of the applied load. We see that the angle $-\theta_1$ increases monotonically with β, and that, in the interval $0° < \beta < 137.5°$, the angles $-\theta_1$ for $\kappa = 1.0$ are always greater than those for $\kappa = 3.0$. This rule is reversed in the interval $137.5° < \beta < 180°$. Furthermore, Figure 7.14 illustrates that initial crack extension takes place in a direction almost normal to the applied load for all values of β in the interval $0 < \beta < 120°$. When the angle β is greater than 120° the direction of crack extension deviates from that normal to the load, becoming parallel to the applied load for $\beta = 180°$. The values of the quantity $\sigma_c(R/16\mu S_c)^{1/2}$ for $\kappa = 1.0$, 1.4 and 3.0 are presented in Figure 7.15. It can be seen that all curves for the values of the angle β equal to $\beta = 0°$, 180° and 137.5° tend to infinity. We conclude that a plate weakened by a circular crack requires an infinitely large stress for crack extension, not only for the trivial case of zero angle circular crack, but also for the values of $2\beta = 275°$ and 360°. The infinite value of stress for crack extension is given by the linear theory of fracture. Its physical meaning is that failure of the cracked plate takes place at the same critical load as failure of the uncracked plate. We further observe that the critical stress for crack extension decreases as κ increases, or as the Poisson's ratio ν decreases. We also see that, for each value of κ, there is a specific value of the angle of crack 2β at which the required stress for crack extension reaches

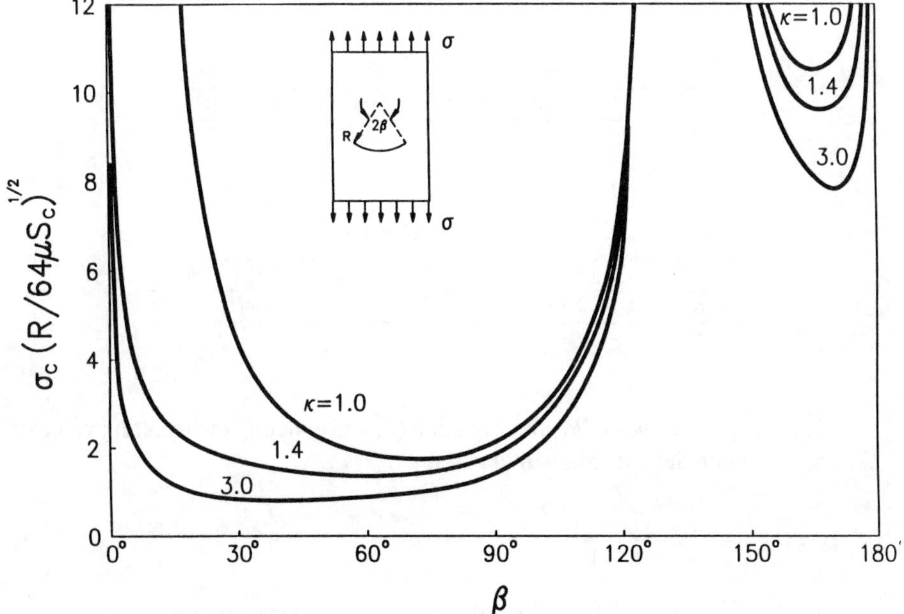

Fig. 7.15. Normalized critical stress for crack extension versus half angle β of a circular crack whose chord is perpendicular to the applied tensile stress for $\kappa = 1.0$, 1.4 and 3.0.

a minimum. This critical value of the angle 2β is equal to 85°, 115° and 140° for $\kappa = 3.0$, 1.4 and 1.0 respectively.

Example 7.3.

For the cylindrical pressure vessel of Example 2.5 determine the angle of initial crack extension θ_c and the critical internal pressure p_c for initiation of crack growth for various values of the crack angle β. Take $\nu = 0.25$.

Solution: Substituting the values of stress intensity factors $K_{\rm I}$ and $K_{\rm II}$ obtained in Example 2.5 into Equation (7.14) we obtain the strain energy density factor

$$S = \left(\frac{pR}{2t}\right)^2 aF(\beta,\theta) \tag{1}$$

where

$$F(\beta,\theta) = a_{11}(1+\sin^2\beta)^2 + a_{12}(1+\sin^2\beta)\sin 2\beta + a_{22}\sin^2\beta\cos^2\beta\,. \tag{2}$$

The coefficients a_{ij} are given by Equation (7.15).

According to the SED criterion, the angle of initial crack extension θ_c is determined from Equation (7.11). Values of $-\theta_c$ for various crack inclination angles β are shown in Table 1.

TABLE 1. Angles of crack extension

β	0°	10°	30°	50°	70°	90°
$-\theta_c$	0°	17.20°	29.36°	27.38°	17.60°	0°

TABLE 2. Critical pressures

β	0°	10°	30°	50°	70°	90°
$p_c R\sqrt{a}/2t$	22.50	21.60	16.95	13.54	11.80	11.30

Substituting the values of θ_c into Equation (1) and equating the resulting value of S to S_c we obtain the critical internal pressure p_c as

$$p_c = \frac{2t}{R\sqrt{a}}\left(\frac{S_c}{F(\beta,\theta_c)}\right)^{1/2} . \tag{3}$$

Values of the dimensionless quantity $p_c R\sqrt{a}/2t$ for different angles β are shown in Table 2.

Observe from this table that a longitudinal crack ($\beta = 90°$) is more dangerous than a circumferential crack ($\beta = 0$) of the same length.

Example 7.4.

A cylindrical bar of radius b contains a circular crack of radius a and is subjected to a force P along the axis of the bar and a torque T (Figure 7.16). The opening-mode and tearing-mode stress intensity factors $k_{\rm I}$ and $k_{\rm III}$ along the crack front created by the force P and torque T, respectively, are given by

$$k_{\rm I} = \frac{\sigma}{1-(a/b)^2}\sqrt{\frac{ac}{b}}\, f_1\left(\frac{a}{b}\right) , \quad \sigma = \frac{P}{\pi b^2} \tag{1a}$$

$$k_{\rm III} = \frac{\tau}{1-(a/b)^4}\sqrt{\frac{ac}{b}}\, f_3\left(\frac{a}{b}\right) , \quad \tau = \frac{2Ta}{\pi b^4} \tag{1b}$$

where

$$f_1\left(\frac{a}{b}\right) = \frac{2}{\pi}\left(1+\frac{1}{2}\frac{a}{b}-\frac{5}{8}\frac{a^2}{b^2}\right)+0.268\,\frac{a^3}{b^3} \tag{2a}$$

$$f_3\left(\frac{a}{b}\right) = \frac{4}{3\pi}\left(1+\frac{1}{2}\frac{a}{b}+\frac{3}{8}\frac{a^2}{b^2}+\frac{5}{16}\frac{a^3}{b^3}-\frac{93}{128}\frac{a^4}{b^4}+0.038\,\frac{a^5}{b^5}\right) . \tag{2b}$$

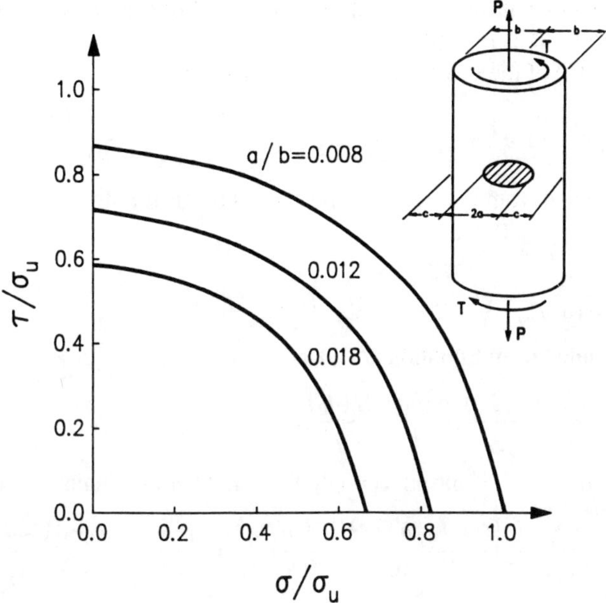

Fig. 7.16. Fracture loci of a cylindrical bar with an internal crack subjected to tension and torsion.

For a bar with $a_0 = 0.016$ cm, $b = 2.0$ cm and $\nu = 1/3$ subjected to the force P only the critical stress for fracture is σ_u. Determine the fracture loci when the bar is subjected to both force P and torque T for $a/b = 0.008$, 0.012 and 0.018.

Solution: The strain energy density function $\mathrm{d}W/\mathrm{d}V$ for tearing-mode deformation is obtained by substituting the values of stresses τ_{xz} and τ_{yz} from Equation (2.56) into Equation (7.2). We obtain

$$\frac{\mathrm{d}W}{\mathrm{d}V} = \frac{K_{\mathrm{III}}^2}{4\pi\mu r} . \tag{3}$$

The strain energy density factor S is computed from Equations (3) and (7.9) as

$$S = \frac{k_{\mathrm{III}}^2}{4\mu} , \quad k_{\mathrm{III}} = \frac{K_{\mathrm{III}}}{\sqrt{\pi}} . \tag{4}$$

The strain energy density factor S for a combination of opening-mode and tearing-mode is therefore given by

$$S = a_{11}k_{\mathrm{I}}^2 + a_{33}k_{\mathrm{III}}^2 , \quad a_{33} = \frac{1}{4\mu} \tag{5}$$

where the coefficient a_{11} is given by Equation (7.15) with $\kappa = 3 - 4\nu$.

The angle of crack growth θ_c is determined from Equation (7.11) as $\theta_c = 0$, that is the crack grows in its own plane.

Equation (7.12) of the strain energy density criterion gives

$$(1-2\nu)\, f_{\mathrm{I}}^2\left(\frac{a}{b}\right)\sigma^2 + \frac{f_3^2(a/b)}{[1+(a/b)^2]^2}\,\tau^2 = \frac{2Eb[1-(a/b)^2]^2\, S_c}{ac(1+\nu)} \,. \tag{6}$$

K_{Ic} is calculated from the fracture stress σ_u of a bar of radius $b = 2.0$ cm with a crack of radius $a_0 = 0.016$ cm as

$$K_{\mathrm{Ic}} = \frac{\sigma_u}{1-(a_0/b)^2}\sqrt{\frac{\pi a_0 c_0}{b}}\, f_1\left(\frac{a_0}{b}\right) \tag{7}$$

and S_c is computed from Equation (7.26) as

$$S_c = \frac{(1+\nu)\,(1-2\nu)}{2E}\,\frac{\sigma_u^2 a_0 c_0 f_{\mathrm{I}}^2(a_0/b)}{b[1-(a_0/b)^2]} \,. \tag{8}$$

When the value of S_c is introduced into Equation (6) we obtain

$$\frac{f_{\mathrm{I}}^2(a/b)}{f_{\mathrm{I}}^2(a_0/b)}\left(\frac{\sigma}{\sigma_u}\right)^2 + \frac{[1+(a/b)^2]^{-2}}{1-2\nu}\,\frac{f_3^2(a/b)}{f_3^2(a_0/b)}\left(\frac{\tau}{\sigma_u}\right)^2 = \frac{a_0 c_0[1-(a/b)^2]^2}{ac[1-(a_0/b)^2]^2} \,. \tag{9}$$

Equation (9) presents the required relation between σ and τ for fracture of the bar (fracture locus). Figure 7.16 shows the fracture loci for $a/b = 0.008$, 0.012 and 0.018 when $a_0 = 0.016$ cm, $b = 2.0$ cm, $\nu = 1/3$. For all combinations of σ and τ that lie outside the curves, fracture of the bar takes place by unstable growth of the circular crack, while for the remaining values of σ and τ the crack does not propagate.

Example 7.5.

An infinite elastic plate is perforated by a circular hole of radius R and a system of n symmetrically located small radial cracks of length l (Figure 7.17). The plate is subjected to a uniform uniaxial tensile stress σ at infinity forming an angle α with the x-axis. The stress intensity factors $k_{\mathrm{I}}^{(j)}$ and $k_{\mathrm{II}}^{(j)}$ at the tip of the j crack are given by

$$k_{\mathrm{I}}^{(j)} = \frac{\sigma\sqrt{R}}{\sqrt{2}}\sqrt{\frac{1-\Delta^{2n}}{n\Delta}}\,[1/2(1+\Delta^n)]^{(2/n)-1}\,[1-4\,\cos 2(\theta_j-\alpha)-$$

$$-2\varepsilon c_{\mathrm{I}}^n\,\cos(\theta_j-2\alpha)+(A_n-\varepsilon c_2^n)\,\cos 2\alpha] \tag{1a}$$

$$k_{\mathrm{II}}^{(j)} = \frac{\sigma\sqrt{R}}{\sqrt{2}}\sqrt{\frac{1-\Delta^{2n}}{n\Delta}}\,[1/2(1+\Delta^n)]^{(2/n)-1}\,[2\,\sin 2(\theta_j-\alpha)+$$

$$+2\varepsilon c_{\mathrm{I}}^n\,\sin(\theta_j-2\alpha)+(A_n-\varepsilon c_2^n)\,\sin 2\alpha] \tag{1b}$$

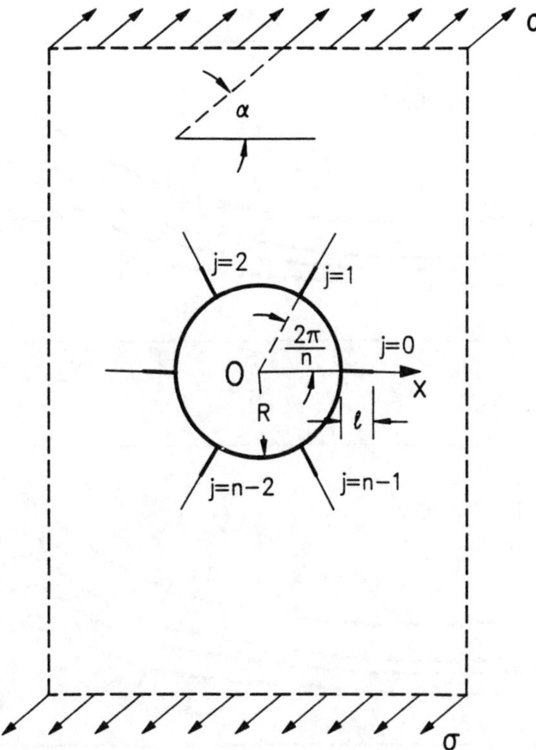

Fig. 7.17. An infinite plate perforated by a circular hole with an array of n small radial cracks subjected to an inclined tension σ.

where

$$\varepsilon c_1^1 = \frac{2\delta^2}{(1+\Delta^{-1})^2}\,, \quad \varepsilon c_2^1 = \frac{\delta^4 + 8\delta^3 + 8\delta^2}{(1+\Delta^{-1})^4}$$

$$\Delta = (1+\delta)^{-1}\,, \quad A_n = 2\left[\frac{1}{2}(1+\Delta^n)\right]^{-2/n}\,, \quad \delta = \frac{l}{R} \tag{2}$$

$$\varepsilon c_1^2 = 0\,, \quad \varepsilon c_2^2 = \left(\frac{1-\Delta^2}{1+\Delta^2}\right)^2 .$$

$$\varepsilon c_1^n = \varepsilon c_2^n = 0 \quad (n \geq 3)\,.$$

Determine the critical fracture stress σ_c of the plate that triggers unstable growth of one of the radial cracks.

Solution: The critical value $\sigma_c^{(j)}$ of the applied stress σ for extension of the j-crack

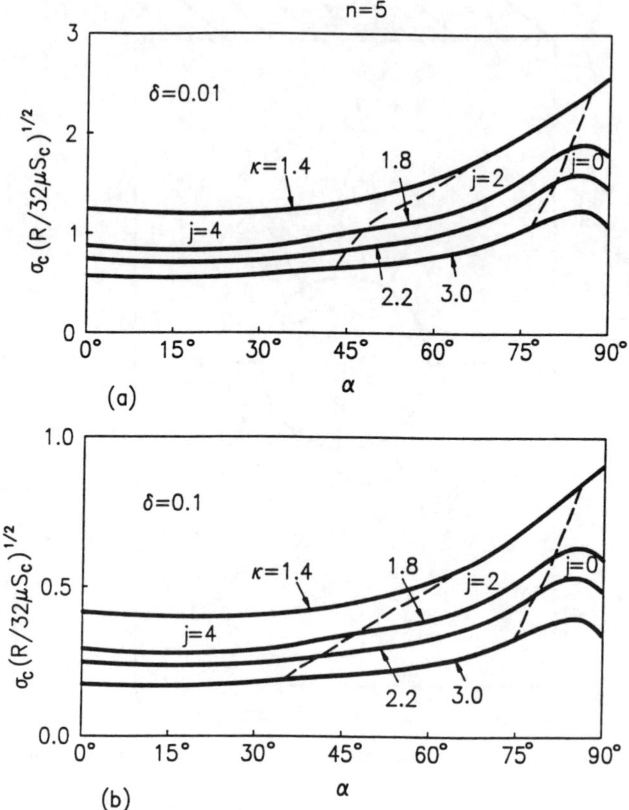

Fig. 7.18. Normalized critical stress for crack extension versus angle α of inclination of the applied stress with respect to the x-axis of Figure 7.17 with $n = 5$ and (a) $\delta = 0.01$ and (b) $\delta = 0.1$, for $\kappa = 1.4$, 1.8, 2.2 and 3.0.

is obtained by using Equations (7.16) and (7.12) of the strain energy density criterion for the j-crack, with the values of the stress intensity factors $k_{\mathrm{I}}^{(j)}$ and $k_{\mathrm{II}}^{(j)}$ given by Equations (1) and (2). It is evident that brittle failure of the plate will take place from the extension of the crack which requires the lowest critical stress $\sigma_c^{(j)}$. Therefore, the critical value σ_c of the applied stress for failure of the plate is given by

$$\sigma_c = \min\left(\sigma_c^0, \sigma_c^{(1)}, \ldots, \sigma_c^{(n-1)}\right).$$

Figure 7.18 presents the variation of the dimensionless critical stress $\sigma_c(R/32\mu S_c)^{-1/2}$ versus the angle α of inclination of the applied stress with respect to the Ox-axis for a plate containing five $(n = 5)$ radial cracks. The ratio δ of crack length l to hole radius R takes the values $\delta = 0.01$ and 0.1. The material constant κ is equal to $\kappa = 1.4, 1.8, 2.2$ and 3.0. The dashed lines in the figure separate the regions in which fracture of the plate starts from unstable growth of the more vulnerable crack $j = 0$, 2 and 4. Values of the quantity $\sigma_c(R/32\mu S_c)^{-1/2}$ for $\alpha = 60°$, $\delta = 0.01$,

TABLE 7.1. Values of $\sigma_c(R/32\mu S_c)^{-1/2}$ for $\alpha = 60°$.

	$\delta = 0.01$			$\delta = 0.04$			$\delta = 0.1$		
n	$\kappa = 1.4$	$\kappa = 2.2$	$\kappa = 3.0$	$\kappa = 1.4$	$\kappa = 2.2$	$\kappa = 3.0$	$\kappa = 1.4$	$\kappa = 2.2$	$\kappa = 3.0$
1	4.0029	2.3111	1.7902	2.0735	1.1972	0.9273	1.3991	0.8079	0.6258
2	4.0021	2.3106	1.7898	2.0731	1.1969	0.9271	1.3974	0.8069	0.6250
3	1.5433	1.1135	0.9305	0.7802	0.5650	0.4728	0.5036	0.3671	0.3080
4	1.9645	1.7359	1.6026	0.9895	0.8744	0.8072	0.6351	0.5612	0.5181
5	1.5800	1.0163	0.8004	0.8524	0.5189	0.4091	0.5172	0.3413	0.2696
6	1.5434	1.1136	0.9305	0.7810	0.5655	0.4732	0.5066	0.3689	0.3094
7	1.5618	1.0454	0.8480	0.7920	0.5323	0.4334	0.5163	0.3494	0.2851
8	1.5947	1.0172	0.8138	0.8105	0.5191	0.4158	0.5314	0.3425	0.2750
9	1.5436	1.0107	0.8008	0.7827	0.5166	0.4098	0.5127	0.3427	0.2723
10	1.5853	1.0165	0.8005	0.8037	0.5205	0.4103	0.5279	0.3469	0.2739
20	1.5551	1.0175	0.8013	0.8018	0.5282	0.4164	0.5621	0.3730	0.2943

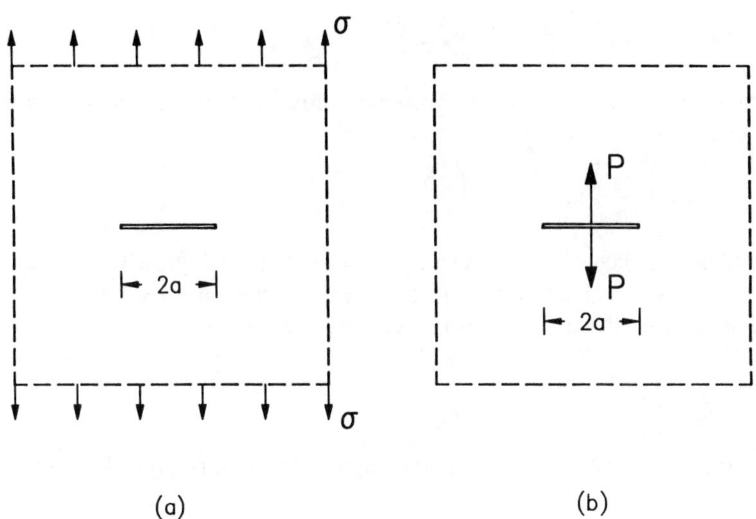

Fig. 7.19. Cracked specimens for (a) unstable crack growth and (b) stable crack growth.

0.04 and 0.1, $\kappa = 1.4$, 2.2 and 3.0 and various radial cracks $(n = 1, 2, \ldots, 20)$ are presented in Table 1.

Example 7.6.

(a) A crack of length $2a$ in a large plate grows in a stable manner under a constant uniform uniaxial stress σ_0 normal to the crack plane (Figure 7.19a). Determine the crack growth increment r_j $(j = 1, 2, \ldots, n)$ during stable crack growth.

(b) As in the previous case for a crack of length $2a$ in an infinite plate subjected to wedge forces P at the middle of the crack (Figure 2.19b).

Solution: (a) Stable crack growth is dictated by Equation (7.10) of the SED criterion. In our case S is calculated from Equation (7.24) as

$$S = \frac{(\kappa - 1)\,\sigma_0^2 a}{8\mu}\,. \tag{1}$$

Equation (7.10) becomes

$$\frac{(\kappa-1)\,\sigma_0^2}{8\mu}\,\frac{a}{r_1} = \frac{(\kappa-1)\,\sigma_0^2}{8\mu}\,\frac{a+r_1}{r_2} = \frac{(\kappa-1)\,\sigma_0^2}{8\mu}\,\frac{a+r_1+r_2}{r_3} = \ldots = \frac{S_c}{r_c} \tag{2}$$

or

$$\frac{a}{r_1} = \frac{a+r_1}{r_2} = \frac{a+r_1+r_2}{r_3} = \ldots \frac{S_c}{r_c}\,. \tag{3}$$

If crack initiation starts when $r_1 = r_0$ the following recursion relation for incremental crack growth is obtained

$$r_{n+1} = \left(1 + \frac{r_0}{a}\right) r_n\,, \quad n \geq 1\,. \tag{4}$$

Equation (4) indicates that each consecutive step of crack growth increases. Unstable crack growth occurs when r_j reaches the critical ligament length r_c.

(b) For this case the stress intensity factor K_{I} is given by (Example 2.2)

$$K_{\mathrm{I}} = \frac{P}{(\pi a)^{1/2}} \tag{5}$$

and the strain energy density factor S is computed from Equation (7.14) with $\theta = 0$, as

$$S = \frac{\kappa - 1}{8\mu}\,\frac{\pi^2}{\pi a}\,. \tag{6}$$

When stable growth of the crack occurs under constant force P, Equation (7.10) becomes

$$\frac{1}{ar_1} = \frac{1}{(a+r_1)\,r_2} = \frac{1}{(a+r_1+r_2)\,r_3} = \ldots = \frac{S_0}{r_0} \tag{7}$$

which for $r_1 = r_0$ gives

$$r_{n+1} = \frac{r_n}{1 + \left(\frac{r_n}{r_0}\right)\left(\frac{r_n}{a}\right)}\,, \quad n \geq 1\,. \tag{8}$$

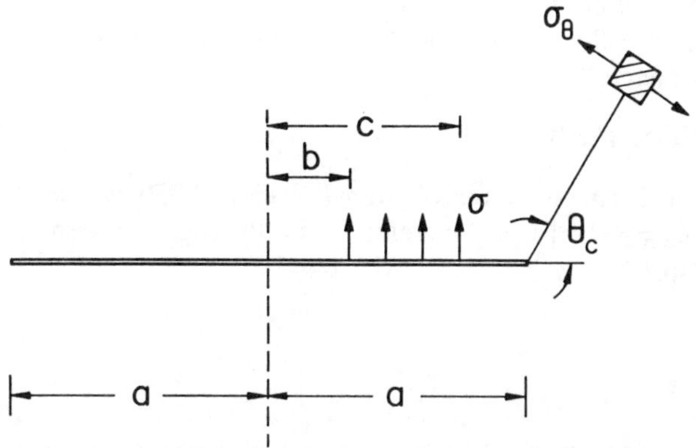

Fig. 7.20. A crack in an infinite plate subjected to a uniform load on part of the upper crack surface.

Equation (8) indicates that each consecutive step of crack growth decreases. Crack arrest occurs when r_j reaches the critical ligament length r_c.

Example 7.7.

The stress intensity factors K_{I} and K_{II} at the right tip of the crack of Figure 7.20 are given by

$$K_{\mathrm{I}} = \frac{\sigma\sqrt{a}}{2\sqrt{\pi}}\left[\sin^{-1}\frac{c}{a} - \sin^{-1}\frac{b}{a} - \sqrt{1-\frac{c^2}{a^2}} + \sqrt{1-\frac{b^2}{a^2}}\right] \tag{1a}$$

$$K_{\mathrm{II}} = \frac{\sigma(c-b)}{2\sqrt{\pi a}}\left(\frac{\kappa-1}{\kappa+1}\right). \tag{1b}$$

Calculate the angle θ_c of extension of the crack from its right tip according to the maximum stress criterion. Take $b/a = 0.5$, $c/a = 0.8$, $\nu = 0.3$ and assume conditions of plane strain.

Solution: The angle θ_c is calculated from Equation (7.38) of the maximum stress criterion. This equation for $\theta_c \neq \pi$, $K_{\mathrm{II}} = 0$ becomes

$$\tan\frac{\theta_c}{2} = \frac{K_{\mathrm{I}} \pm \sqrt{K_{\mathrm{I}}^2 + 8K_{\mathrm{II}}^2}}{4K_{\mathrm{II}}}. \tag{2}$$

We have

$$K_{\mathrm{I}} = \frac{\sigma}{2}\sqrt{\frac{a}{\pi}}\,[\sin^{-1}0.8 - \sin^{-1}0.5 - \sqrt{1-0.8^2} - \sqrt{1-0.5^2}] = \frac{0.67\sigma}{2}\sqrt{\frac{a}{\pi}} \tag{3}$$

$$K_{\mathrm{II}} = \frac{\sigma}{2}\sqrt{\frac{a}{\pi}}\,(0.8-0.5)\left(\frac{1.8-1.0}{1.8+1.0}\right) = \frac{0.086}{2}\sqrt{\frac{a}{\pi}} \tag{4}$$

where for plane strain $\kappa = 3 - 4\nu = 1.8$.

Introducing the values of $K_{\rm I}$ and $K_{\rm II}$ from Equation (3) into Equation (2) we obtain the crack angle

$$\theta_c = -14.14°,\ 152.0° \ . \tag{5}$$

Calculating σ_θ for the two angles θ_c from Equation (7.35b) we find that it becomes maximum for $\theta_c = -14.14°$ and therefore this is the angle of extension of the crack from its right tip.

Problems

7.1. The stress field in the neighborhood of a sharp elliptical notch in a mixed-mode stress field under conditions of plane strain is given by

$$\sigma_x = \frac{1}{\sqrt{2r}} \left[k_1 \cos\frac{\theta}{2}\left(1 - \sin\frac{\theta}{2}\sin\frac{3\theta}{2}\right) - k_1\left(\frac{\rho}{2r}\right)\cos\frac{3\theta}{2} \right.$$

$$\left. -k_2 \sin\frac{\theta}{2}\left(2 + \cos\frac{\theta}{2}\cos\frac{3\theta}{2}\right) + k_2\left(\frac{\rho}{2r}\right)\sin\frac{3\theta}{2} \right]$$

$$\sigma_y = \frac{1}{\sqrt{2r}} \left[k_1 \cos\frac{\theta}{2}\left(1 + \sin\frac{\theta}{2}\sin\frac{3\theta}{2}\right) + k_1\left(\frac{\rho}{2r}\right)\cos\frac{3\theta}{2} \right.$$

$$\left. +k_2 \sin\frac{\theta}{2}\cos\frac{\theta}{2}\cos\frac{3\theta}{2} - k_2\left(\frac{\rho}{2r}\right)\sin\frac{3\theta}{2} \right]$$

$$\tau_{xy} = \frac{1}{\sqrt{2r}} \left[k_1 \sin\frac{\theta}{2}\cos\frac{\theta}{2}\cos\frac{3\theta}{2} - k_1\left(\frac{\rho}{2r}\right)\sin\frac{3\theta}{2} \right.$$

$$\left. +k_2 \cos\frac{\theta}{2}\left(1 - \sin\frac{\theta}{2}\sin\frac{3\theta}{2}\right) - k_2\left(\frac{\rho}{2r}\right)\cos\frac{3\theta}{2} \right]$$

$$\sigma_z = \nu(\sigma_{11} + \sigma_{22})\ , \quad \tau_{yz} = \tau_{zx} = 0$$

where ρ is the notch radius. The radial distance r is measured from the focal point of the notch and the angle θ is measured from a line extended from the major axis of the notch.

Show that the strain energy density factor S is given by

$$S = r\,\frac{\mathrm{d}W}{\mathrm{d}V} = S_1 + \frac{S_2}{r} + \frac{S_3}{r^2}$$

where

$$S_1 = a_{11}k_1^2 + 2a_{12}k_1k_2 + a_{22}k_2^2$$

$$S_2 = b_{11}k_{\mathrm{I}}^2 + 2b_{12}k_1k_2 + b_{22}k_2^2$$

$$S_3 = c_{11}k_{\mathrm{I}}^2 + 2c_{12}k_1k_2 + c_{22}k_2^2 \,.$$

The coefficients a_{ij} are given by Equation (7.15), while b_{ij} and c_{ij} are given by

$$b_{11} = 0\,, \quad b_{12} = -\frac{\rho}{8\mu}\sin\theta\,, \quad b_{22} = -\frac{\rho}{4\mu}\cos\theta$$

$$c_{11} = \frac{\rho^2}{16\mu}\,, \quad c_{12} = 0\,, \quad c_{22} = \frac{\rho^2}{16\mu}\,.$$

7.2. Show that the critical in-plane shear stress τ_c and the critical tensile stress σ_c to fracture an infinite plate with a crack of length $2a$ are related under conditions of plane strain by

$$\frac{\sigma_c}{\tau_c} = \left[\frac{2(1-\nu) - \nu^2}{3(1-2\nu)}\right]^{1/2}\,.$$

7.3. Show that the critical in-plane shear stress τ_c and the critical out-of-plane shear stress s_c to fracture an infinite plate with a crack of length $2a$ are related under conditions of plane strain by

$$\frac{\tau_c}{s_c} = \left[\frac{3}{2(1-\nu) - \nu^2}\right]^{1/2}\,.$$

7.4. Consider a crack of length $2a$ that makes an angle β with the y direction in an infinite plate subjected to stresses σ and $k\sigma$ along the y and x directions, respectively, at infinity (Figure 2.9a). Plot the variation of crack extension angle θ_c and the critical stress σ_c for crack growth versus the crack inclination angle β for $\kappa = 0.2, 0.6, -0.2$ and -1.0 and $\kappa = 1.0, 1.4, 1.8, 2.2, 2.6$ and 3.0 when the stress σ is tensile.

7.5. As in Problem 7.4 when the stress σ is compressive.

7.6. An infinite plate contains a circular crack of radius R and angle 2β and is subjected to a uniform uniaxial tensile stress σ at infinity parallel to the chord of the crack (Figure 7.21). The stress intensity factors k_{I} and k_{II} at the crack tip are given by

$$k_{\mathrm{I}} = \frac{\sigma}{2}(R\sin\beta)^{1/2}\left[\frac{\left(1+\sin^2\frac{\beta}{2}\cos^2\frac{\beta}{2}\right)\cos\frac{\beta}{2}}{1+\sin^2\frac{\beta}{2}} - \cos\frac{3\beta}{2}\right]$$

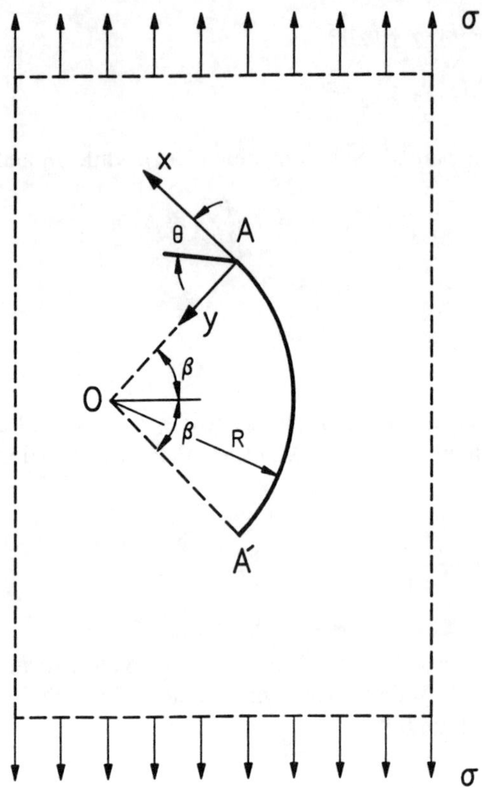

Fig. 7.21. An infinite plate containing a circular crack with its chord parallel to the direction of applied uniaxial stress.

$$k_{\mathrm{II}} = \frac{\sigma}{2}\,(R\,\sin\,\beta)^{1/2}\left[\frac{\left(1+\sin^2\dfrac{\beta}{2}\,\cos^2\dfrac{\beta}{2}\right)\,\sin\,\dfrac{\beta}{2}}{1+\sin^2\dfrac{\beta}{2}} - \sin\,\frac{3\beta}{2}\right].$$

Plot the variation of the angle of crack extension θ_c and the critical stress σ_c for crack growth versus the angle of the crack 2β for various values of κ ($\kappa = 3 - 4\nu$ for plane strain and $\kappa = (3-\nu)\,/\,(1+\nu)$ for plane stress).

7.7. As in Problem 7.6 for compressive applied stress σ.

7.8. As in Example 7.2 for compressive applied stress σ.

7.9. A crack of length $2a$ in an infinite plate is subjected to a concentrated force P applied at the point $x = b$ (Figure 2.16a). The stress intensity factors at the crack tip B are given in Problem 2.5, while the stress intensity factors at the other crack tip A

can be obtained from symmetry considerations as

$$K_{IA}(-a,b) = K_{IB}(a,-b)$$

$$K_{IA}(-a,b) = -K_{IIB}(a,-b)\,.$$

Show that the crack grows from its tip B, and plot the variation of crack extension angle θ_c and critical stress of fracture σ_c versus b/a $(0 < b/a < 1)$ for $\kappa = 1.4$, 1.8, 2.2 and 3.0, when the load P is tensile or compressive.

7.10. A crack of length $2a$ in an infinite plate is subjected to a concentrated force Q applied at the point $x = b$ (Figure 2.16a). The stress intensity factors at the crack tip B are given in Problem 2.5, while the stress intensity factors at the other crack tip A can be obtained from symmetry considerations as

$$K_{IA}(-a,b) = -K_{IB}(a,-b)$$

$$K_{IIA}(-a,b) = K_{II}(a,-b)\,.$$

Plot the variation of the crack extension angle θ_c and the critical stress of fracture σ_c versus b/a $(0 < b/a < 1)$ for $\kappa = 1.4$, 1.8, 2.2 and 3.0. Indicate the regions where extension of the crack starts from its tip A or B.

7.11. Show that the crack of Problem 2.34 propagates along its own plane which is titled at an angle ω to the plane of loading. Show that the critical stress σ_c for crack growth is given by

$$\sigma_c = \frac{2\sqrt{\mu S_c}}{\sqrt{a}\,\sin\omega\,\sqrt{1-2\nu\,\sin^2\omega}}\,.$$

Plot the variation of $\sigma_c\sqrt{a}/2\sqrt{\mu S_c}$ versus ω for $\nu = 0$, 0.1, 0.25, 0.33 and 0.5 and find the angle ω_c at which σ_c becomes minimum for each value of ν.

7.12. A cylindrical vessel of radius R and thickness t contains a through crack of length $2a$ parallel to its axis (Figure 7.22). The edges of the crack are inclined at an angle γ with the surfaces of the vessel wall. The vessel is subjected to a torque T. Determine the crack growth direction and the critical moment T_c for initiation of crack growth. Plot the variation of T_c versus angle γ for various values of Poisson's ratio ν.

7.13. A cylindrical bar of radius b contains a ring-shaped edge crack of depth a in a plane normal to its axis (Figure 7.23). The bar is subjected to a force P along its axis and to a torque T. The opening-mode and tearing-mode stress intensity factors k_I and k_{III} along the crack front created by the force P and torque T, respectively, are given by

$$k_I = \sigma\left(\frac{b}{c}\right)^2\sqrt{\frac{ac}{b}}\,g_1\left(\frac{c}{b}\right)\,,\quad \sigma = \frac{P}{\pi b^2}$$

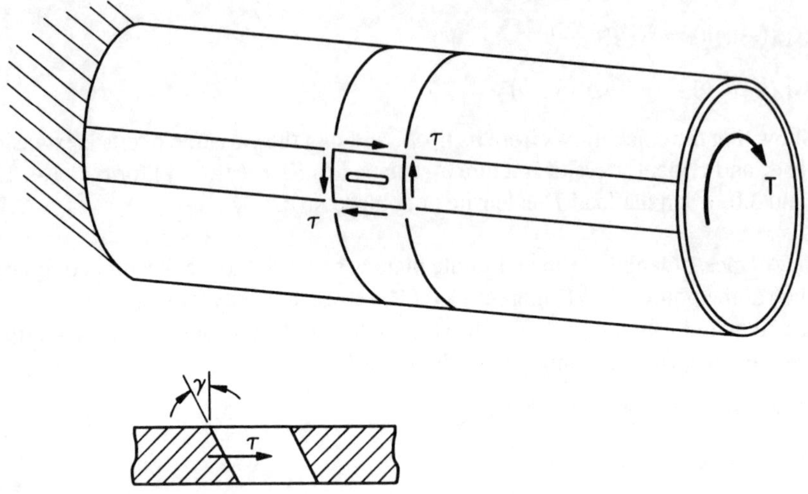

Fig. 7.22. A cylindrical vessel with a through crack parallel to its axis whose edges are inclined with respect to the surface of the vessel wall.

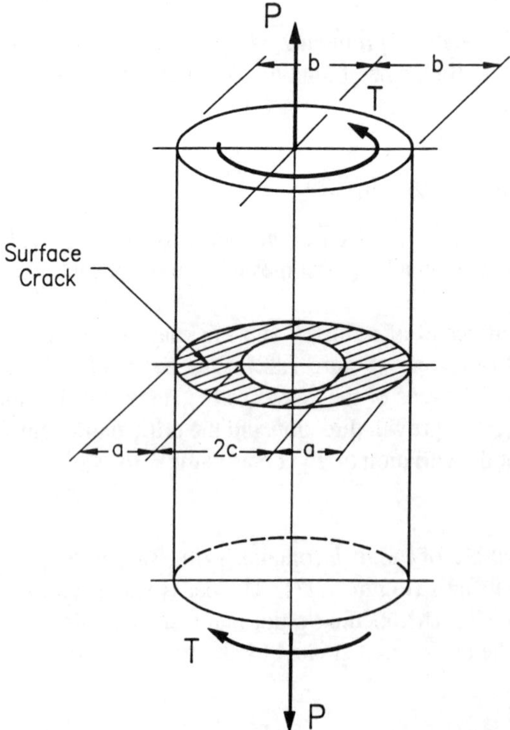

Fig. 7.23. A cylindrical bar with an external crack subjected to tension and torsion.

$$k_{\rm III} = \tau \left(\frac{b}{c}\right)^3 \sqrt{\frac{ac}{b}}\, g_3\left(\frac{c}{b}\right), \quad \tau = \frac{2Ta}{\pi b^4}$$

where

$$g_1\left(\frac{c}{b}\right) = \frac{1}{2}\left[1 + \frac{1}{2}\frac{c}{b} + \frac{3}{8}\frac{c^2}{b^2} - 0.363\frac{c^3}{b^3} + 0.731\frac{c^4}{b^4}\right]$$

$$g_3\left(\frac{c}{b}\right) = \frac{3}{8}\left[1 + \frac{1}{2}\frac{c}{b} + \frac{3}{8}\frac{c^2}{b^2} + \frac{5}{16}\frac{c^3}{b^3} + \frac{25}{128}\frac{c^4}{b^4} + 0.208\frac{c^5}{b^5}\right].$$

For a bar with $a_0 = 0.0052$ cm, $b = 2.0$ cm and $\nu = 1/3$ subjected to the force P only, the critical stress for fracture is σ_u. Determine the fracture loci when the bar is subjected to both force P and torque T for $a/b = 0.003, 0.005$ and 0.011.

7.14. A large thick plate of steel contains a crack of length 5 mm oriented at an angle $\beta = 30°$ with respect to the direction of applied uniaxial tensile stress σ. Calculate the value of the critical stress σ_c for crack growth. $K_{\rm Ic} = 60$ MPa $\sqrt{\rm m}$, $E = 210$ GPa, $\nu = 0.3$.

7.15. As in Problem 7.14 for compressive applied stress σ.

7.16. A cylindrical pressure vessel with closed ends has a radius $R = 1$ m and thickness $t = 40$ mm and it is subjected to internal pressure p. The vessel contains a through crack of length 4 mm oriented at an angle 40° with respect to the circumferential direction. Calculate the maximum pressure p_c the vessel can withstand without failure. $K_{\rm Ic} = 60$ MPa $\sqrt{\rm m}$, $E = 210$ GPa, $\nu = 0.3$.

7.17. A large thick plate containing a crack of length 4 mm oriented at an angle $\beta = 60°$ with respect to the direction of applied uniaxial tensile stress σ fractures at a value $\sigma_c = 1000$ MPa. Calculate $K_{\rm Ic}$ when $E = 210$ GPa, $\nu = 0.3$.

7.18. A large thick plate of steel is subjected to a tensile stress $\sigma = 800$ MPa oriented at an angle 50° with respect to the direction of a through crack. Calculate the maximum permissible crack length the plate can withstand without fracture. $S_c = 1500$ N/m, $E = 210$ GPa, $\nu = 0.3$.

7.19. As in Problem 7.18 when the applied stress σ is compressive.

7.20. A large thick plate contains a crack of length $2a$ oriented at an angle β with respect to the direction of applied uniaxial tensile stress σ. Plot the variation of the quantity $\sigma_c\sqrt{a_c}$ versus β for $S_c = 1500$ N/m, where σ_c and a_c are the critical values of σ and a at crack growth. $E = 210$ GPa, $\nu = 0.3$.

7.21. As in Problem 7.20 when the applied stress is compressive.

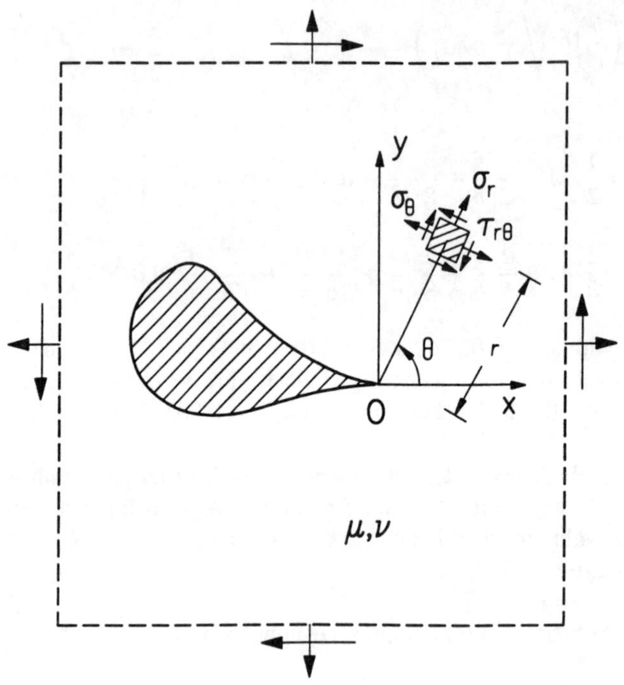

Fig. 7.24. A rigid inclusion with a cuspidal point embedded in a matrix.

7.22. A rigid inclusion with a cuspidal corner O is perfectly bonded to an infinite plate which is subjected to general in-plane loading at infinity (Figure 7.24). The stress field near the point O is given by

$$\sigma_r = \frac{1}{4\sqrt{2r}}\left[k_{\mathrm{I}}\left[5\cos\frac{\theta}{2} + (2\kappa+1)\cos\frac{3\theta}{2}\right] - k_{\mathrm{II}}\left[5\sin\frac{\theta}{2} + (2\kappa-1)\sin\frac{3\theta}{2}\right]\right]$$

$$\sigma_\theta = \frac{1}{4\sqrt{2r}}\left[k_{\mathrm{I}}\left[3\cos\frac{\theta}{2} - (2\kappa+1)\cos\frac{3\theta}{2}\right] - k_{\mathrm{II}}\left[3\sin\frac{\theta}{2} - (2\kappa-1)\sin\frac{3\theta}{2}\right]\right]$$

$$\tau_{r\theta} = \frac{1}{4\sqrt{2r}}\left[k_{\mathrm{I}}\left[\sin\frac{\theta}{2} - (2\kappa+1)\sin\frac{3\theta}{2}\right] + k_{\mathrm{II}}\left[\cos\frac{\theta}{2} - (2\kappa-1)\cos\frac{3\theta}{2}\right]\right]$$

where the coefficients k_{I} and k_{II} are independent of the coordinates r, θ and depend on loading conditions, the plate material and the geometrical shape of the inclusion at the cuspidal point.

Show that the strain energy density factor S is given by

$$S = a_{11}k_1^2 + 2a_{12}k_1k_2 + a_{22}k_2^2$$

where

$$16\mu a_{11} = 2(\kappa - 1)\cos^2\frac{\theta}{2} + \kappa^2 + (2\kappa + 1)\cos^2\theta$$

$$16\mu a_{12} = -[(\kappa - 1) + 2\kappa\cos\theta]\sin\theta$$

$$16\mu a_{22} = 2(\kappa - 1)\sin^2\frac{\theta}{2} + \kappa^2 - (2\kappa - 1)\cos^2\theta\,.$$

7.23. Use Problem 7.22 to show that the angle formed by the fracture path with the tangent of the inclusion at the cuspidal point is given by

$$[(\kappa - 1) + 2(2\kappa + 1)\cos\theta]\sin\theta k_1^2 + 2[(\kappa - 1)\cos\theta + 2\kappa\cos 2\theta]k_1 k_2 +$$
$$+[-(\kappa - 1) - 2(2\kappa - 1)\cos\theta]\sin\theta k_2^2 = 0$$

$$[-(\kappa - 1)\cos\theta - 2(2\kappa + 1)\cos 2\theta]k_1^2 + 2[(\kappa - 1) + 8\kappa\cos\theta]\sin\theta k_1 k_2 +$$
$$+[(\kappa - 1)\cos\theta + 2(2\kappa - 1)\cos 2\theta]k_2^2 > 0\,.$$

7.24. Use Problem 7.23 to show that the angle formed by the fracture path with the tangent of the inclusion at the cuspidal point is calculated according to the maximum stress criterion by

$$(1 + \kappa)k_{\mathrm{I}}\tan^3\frac{\theta}{2} + (3\kappa - 1)k_{\mathrm{II}}\tan^2\frac{\theta}{2} - (1 + 3\kappa)k_{\mathrm{I}}\tan\frac{\theta}{2} - (\kappa - 1)k_{\mathrm{II}} = 0$$

subject to the condition that $\partial^2\sigma_\theta/\partial\theta^2 < 0$ and $\sigma_\theta > 0$.

7.25. A rigid hypocycloidal inclusion is embedded in an infinite plate which is subjected to a uniaxial uniform stress σ at infinity (Figure 7.25). The equation of the inclusion with respect to the frame Oxy is of the form

$$z = \frac{2a}{3}\left(\zeta + \frac{1}{2}\zeta^{-2}\right)$$

with

$$z = re^{i\phi}\,,\quad \zeta = e^{i\theta}$$

and the stress σ subtends an angle β with the x-axis. For this problem the coefficients k_{I} and k_{II} of Problem 7.22 are given by

$$k_{\mathrm{I}}^{(j)} = \frac{\sqrt{2a}}{3\kappa}\,\sigma\left[\frac{\kappa - 1}{2} + \cos\left(\frac{4\pi j}{3} - 2\beta\right)\right]$$

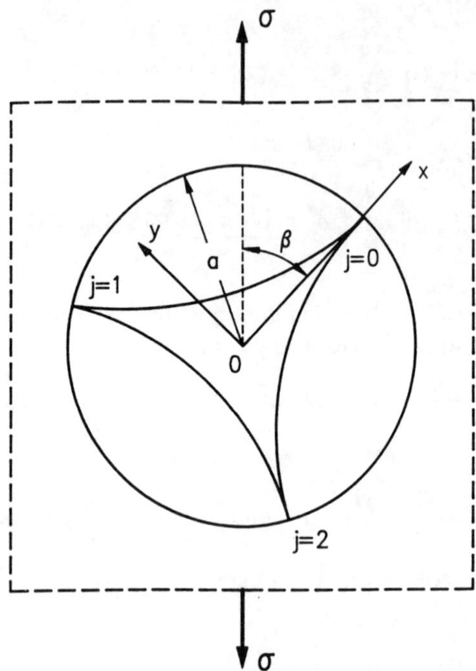

Fig. 7.25. A rigid hypocycloidal inclusion embedded in an infinite plate.

$$k_{\mathrm{II}}^{(j)} = \frac{\sqrt{2a}}{3\kappa}\,\sigma\,\sin\left(\frac{4\pi j}{3} - 2\beta\right)$$

with $j = 0, 1, 2$ for the three cuspidal points of the inclusion.

Plot the variation of the dimensionless quantities $12(r/a)^{1/2}(\sigma_\theta/\sigma)$ and $(72\mu S/\sigma^2 a)$ versus angle θ for $j = 0$ and $\beta = 25°$, $\kappa = 1.8$. Find the angle $(\theta_c)_1$ for which the former quantity becomes maximum, and the angle $(\theta_c)_2$ for which the latter has a local minimum. These angles are the fracture angles according to the maximum stress and the SED criteria.

7.26. For Problem 7.25, plot the variation of the critical stress and the fracture angle of the composite plate for a tensile applied stress σ according to the maximum stress criterion. Indicate the regions in which fracture starts from the more vulnerable corners of the inclusion.

7.27. As in Problem 7.26 according to the SED criterion.

7.28. The stress field at the end points of a rigid rectilinear inclusion embedded in an infinite plate is given by equations of Problem 7.22. For an inclusion of length $2l$ that subtends an angle β with the direction of applied uniaxial stress σ at infinity the

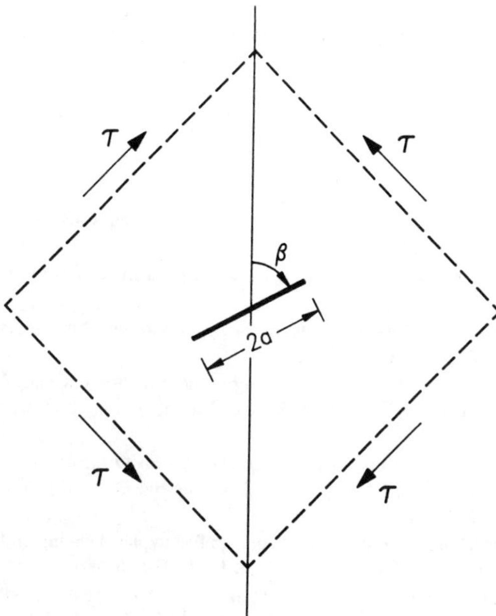

Fig. 7.26. An inclined crack in a pure shear stress field.

coefficients k_{I} and k_{II} are given by

$$k_{\mathrm{I}} = \frac{\sigma\sqrt{l}}{2\kappa}\left(\frac{\kappa-1}{2} + \cos 2\beta\right)$$

$$k_{\mathrm{II}} = -\frac{\sigma\sqrt{l}}{2\kappa}\sin 2\beta\,.$$

Plot the variation of fracture angle and critical stress for fracture of the composite plate versus angle β, for various values of κ $(1 \leq \kappa \leq 3)$ when the applied stress σ is tensile.

7.29. As in Problem 7.28 when the applied stress σ is compressive.

7.30. For a mode-II crack, plot the variation of the circumferential stress σ_θ and the strain energy density factor S versus polar angle θ for $\kappa = 2.0$. Indicate the values of the angle θ for which the former has a local maximum and the latter a local minimum. These values give the crack growth directions according to the maximum stress and the SED criteria.

7.31. A large plate subjected to pure shear stress τ contains a crack of length $2a$ (Figure 7.26). Calculate the angle of crack extension θ_c and the critical shear stress

τ_c for crack growth according to the maximum stress criterion for $\beta = 45°, 60°, 75°$ and $90°$.

References

7.1. Sih, G.C. (1973) 'Some basic problems in fracture mechanics and new concepts', *Engineering Fracture Mechanics* **5**, 365–377.

7.2. Sih, G.C. (1973) 'Energy-density concept in fracture mechanics', *Engineering Fracture Mechanics* **5**, 1037–1040.

7.3. Sih, G.C. (1974) 'Strain-energy-density factor applied to mixed-mode crack problems', *International Journal of Fracture* **10**, 305–321.

7.4. Sih, G.C. (1973) 'A special theory of crack propagation: methods of analysis and solutions of crack problems', in *Mechanics of Fracture*, Vol. 1 (ed. G.C. Sih), Noordhoff Int. Publ., The Netherlands, pp. XXI-XLV.

7.5. Sih, G.C. (1975) 'A three-dimensional strain energy density factor theory of crack propagation', in *Mechanics of Fracture*, Vol. 2 (ed. M.K. Kassir and G.C. Sih), Noordhoff Int. Publ., The Netherlands, pp. XV-LIII.

7.6. Sih, G.C. (1977) 'Strain energy density theory applied to plate bending problems', in *Mechanics of Fracture*, Vol. 3 (ed. G.C. Sih), Noordhoff Int. Publ., The Netherlands, pp. XVII-XLVIII.

7.7. Sih, G.C. (1977) 'Dynamic crack problems-strain energy density fracture theory', in *Mechanics of Fracture*, Vol. 4 (ed. G.C. Sih), Noordhoff Int. Publ., The Netherlands, pp. XVII-XLVII.

7.8. Sih, G.C. (1978) 'Strain energy density and surface layer energy for blunt cracks or notches', in *Mechanics of Fracture*, Vol. 5 (ed. G.C. Sih), Noordhoff Int. Publ., The Netherlands, pp. XIII-CX.

7.9. Sih, G.C. (1981) 'Failure of composites as predicted by the strain energy density theory', in *Mechanics of Fracture*, Vol. 6 (ed. G.C. Sih and E.P. Chen), Martinus Nijhoff Publ., pp. XV-LXXXI.

7.10. Sih, G.C. (1981) 'Experimental fracture mechanics: strain energy density criterion', in *Mechanics of Fracture*, Vol. 7 (ed. G.C. Sih), Martinus Nijhoff Publ., pp. XVII-LVI.

7.11. Gdoutos, E.E. (1984) *Problems of Mixed Mode Crack Propagation*, Martinus Nijhoff Publ.

7.12. Hoek, E. and Bieniawski, Z.T. (1965) 'Fracture propagation mechanics in hard rock', *Technical Report-Rock Mechanics Division*, South African Council for Scientific and Industrial Research.

7.13. Sih, G.C. and Kiefer, B.V. (1979) 'Nonlinear response of solids due to crack growth and plastic deformation', in *Nonlinear and Dynamic Fracture Mechanics* (eds. N. Perrone and S.W. Atluri), The American Society of Mechanical Engineers, AMD, Vol. 35, pp. 136–156.

7.14. Sih, G.C. and Chen, C. (1985) 'Non-self-similar crack growth in elastic-plastic finite thickness plate', *Theoretical and Applied Fracture Mechanics* **3**, 125–139.

7.15. Erdogan, F. and Sih, G.C. (1963) 'On the crack extension in plates under plane loading and transverse shear', *Journal of Basic Engineering, Trans. ASME* **85D**, 519–527.

Chapter 8

Dynamic Fracture

8.1. Introduction

The analysis of crack systems considered so far concerned only quasi-static situations in which the kinetic energy is relatively insignificant compared with the other energy terms and can be omitted. The crack was assumed either to be stationary or to grow in a controlled stable manner, and the applied loads varied quite slowly. The present chapter is devoted entirely to dynamically loaded stationary or growing cracks. In such cases rapid motions are generated in the medium and inertia effects become important.

Elastodynamic analysis of crack problems indicates that stresses and displacements caused by dynamic loading can differ greatly from those associated with the corresponding static loading. At some locations in the body the dynamic stresses are higher than the corresponding static stresses. This result may be explained by the intersection of the elastic waves with the crack faces and other characteristic boundaries of the body. Furthermore, the mechanical properties of the material depend markedly on the time for which the applied loading is maintained in the solid. For example, in most metals, both the yield and ultimate strength increase with the rate of loading. Dynamic loads give rise to high stress levels near cracks and fracture takes place so rapidly that there is insufficient time for yielding to develop. Energy is therefore released within a short time, leading to rapid crack propagation; this explains the experimental observation that dynamic loads generally promote brittle fractures.

Broadly speaking, problems of dynamic fracture mechanics fall into two main categories. The first concerns the situation where a crack reaches a point of instability and moves rapidly under slowly varying applied loading. Motion of the crack leads to a sudden unloading along the crack path. The second category of dynamic problems arises when a body with a stationary crack is subjected to a rapidly varying load – for example, an impact or impulsive load. Problems of interest concern initiation of rapid crack growth, crack speed, crack branching and crack arrest.

A dynamic crack problem may be stated in its most general form as follows:

A solid body with an initial crack is subjected to a time-dependent loading. We are seeking the conditions of crack initiation, growth and arrest. Even in its most general form, the formulation of the problem is not an easy task. Study is usually restricted to plane symmetric or axisymmetric problems in which the crack area can be characterized by a length parameter a, and the crack path is known beforehand. Solution of the problem requires determination of the three displacement components and the crack length, as a function of time. The three equations of motion, coupled with a fracture criterion, provide four equations for the determination of these four unknown quantities.

The present chapter presents the basic concepts and the salient points of dynamic fracture mechanics. We first describe the theory advanced by Mott for the prediction of the speed of a moving crack. The theory, in spite of its limitations, constitutes the first attempt to include the kinetic energy term into the Griffith energy balance equation. We then outline further extensions and improvements of Mott's results, and describe the stress field around a crack moving at constant velocity. We focus attention on the stress field near the crack tip; this has an inverse square root singularity and is expressed in terms of the dynamic stress intensity factor, just as in the static case. We then introduce the concept of the strain energy release rate and relate it to the stress intensity factor. We treat the problems of crack branching and crack arrest. The chapter concludes with a description of the main methods for the experimental study of dynamic crack problems.

8.2. Mott's model

The theory proposed by Mott [8.1] constitutes the first attempt for a quantitative prediction of the speed of a rapidly moving crack. Mott extended Griffith's theory by adding a kinetic energy term to the expression of the total energy of the system and sought the configuration which keeps this total energy constant. The problem considered by Mott was the propagation of a central crack in an infinite plate subjected to a uniform time-independent uniaxial stress σ perpendicular to the plane of the crack. He made the following key assumptions:

(i) The stress and displacement fields for the dynamic problem are the same as those for the static problem, with the same crack length.
(ii) The crack is travelling at a constant speed.
(iii) The crack speed is small compared to the shear wave speed in the body.

The kinetic energy term K entering into the energy balance equation (4.1) is given by

$$K = \int_V \frac{1}{2} \rho \, \dot{u}_k \dot{u}_k \, \mathrm{d}V \tag{8.1}$$

where ρ denotes the mass density, u_k the displacement component and V the body volume. A dot over a letter denotes ordinary differentiation with respect to time.

To obtain an expression for the kinetic energy term K we must know the displacement field in the solid. Mott derived an expression for K on dimensional grounds. The components of the velocity at a given point in the body due to a rapidly propagating crack may be written as

$$\dot{u} = \frac{\mathrm{d}u}{\mathrm{d}t} = \frac{\partial u}{\partial a}\frac{\mathrm{d}a}{\mathrm{d}t} = V\frac{\partial u}{\partial a} \tag{8.2a}$$

$$\dot{v} = \frac{\mathrm{d}v}{\mathrm{d}t} = \frac{\partial v}{\partial a}\frac{\mathrm{d}a}{\mathrm{d}t} = V\frac{\partial v}{\partial a} \tag{8.2b}$$

where $V = \mathrm{d}a/\mathrm{d}t$ is the crack speed.

The kinetic energy takes the form

$$K = \frac{1}{2}\rho V^2 \iint_R \left[\left(\frac{\partial u}{\partial a}\right)^2 + \left(\frac{\partial v}{\partial a}\right)^2\right] \mathrm{d}x\,\mathrm{d}y\,. \tag{8.3}$$

The displacement components u and v behind the crack tip given from Equations (2.30) can be put in the form

$$u = \frac{\sigma}{E}\sqrt{ar}\, f_1(\theta)\,, \quad v = \frac{\sigma}{E}\sqrt{ar}\, f_2(\theta)\,. \tag{8.4}$$

For a fixed element, the distance r from the moving crack tip is proportional to the crack length, and the displacements u and v can be put in the form

$$u = \frac{c_1\sigma a}{E}\,, \quad v = \frac{c_2\sigma a}{E}\,. \tag{8.5}$$

Thus, Equation (8.3) becomes

$$K = \frac{1}{2}\rho V^2 \iint_R (c_1^2 + c_2^2)\,\mathrm{d}x\,\mathrm{d}y\,. \tag{8.6}$$

For an infinite plate the crack length $2a$ is the only characteristic length. Thus, the area integral in Equation (8.6) must be proportional to a, and K becomes

$$K = \frac{1}{2}k^2\rho V^2 a^2 \frac{\sigma^2}{E^2} \tag{8.7}$$

where k is a constant.

Integrating the energy balance equation (Equation (4.1)) with respect to time and omitting the strain energy dissipated to plastic deformation we obtain

$$W - U^e - K - \Gamma = c \tag{8.8}$$

where c is a constant.

For a crack propagating under either fixed-grips or constant stress conditions we introduce the value of $W - U^e$ from Equation (4.12) and obtain, for conditions of generalized plane stress,

$$\frac{\pi\sigma^2 a^2}{E} - \frac{1}{2}k^2\rho V^2 a^2 \frac{\sigma^2}{E^2} - 4\gamma a = c\,. \tag{8.9}$$

Differentiating Equation (8.9) with respect to a and making use of the second assumption above ($\partial a/\partial a = 0$) we find the crack speed

$$V = \sqrt{\frac{2\pi}{k^2}}\sqrt{\frac{E}{\rho}}\sqrt{1-\frac{a_0}{a}} \tag{8.10}$$

where from Griffith's equation (Equation (4.14))

$$\gamma = \frac{\sigma^2\pi a_0}{2E} \tag{8.11}$$

with a_0 being the half crack length at $t = 0$.

For $a \gg a_0$ Equation (8.10) predicts a limiting crack speed

$$V = \sqrt{\frac{2\pi}{k^2}}\,v_s\,,\quad v_s = \sqrt{\frac{E}{\rho}} \tag{8.12}$$

where v_s is the speed of longitudinal waves in the material which is equal to the speed of sound.

Roberts and Wells [8.2] found that $\sqrt{2\pi/k^2} = 0.38$ for a Poisson's ratio $\nu = 0.25$. Thus Equation (8.10) becomes

$$V = 0.38 v_s\sqrt{1-\frac{a_0}{a}}\,. \tag{8.13}$$

Berry [8.3] and Dulaney and Brace [8.4] re-examined Mott's theory by relieving the assumption $\partial V/\partial a = 0$. If the applied stress σ is greater than the critical stress σ_0 for the initiation of crack extension, Equation (8.9) gives

$$\frac{\pi\sigma^2 a^2}{E} - \frac{k^2\rho V^2 a^2\sigma^2}{2E^2} - \frac{2\pi a_0 a\sigma_0^2}{E} = c\,. \tag{8.14}$$

Putting $V = 0$ and $a = a_0$ at $t = 0$, we find that Equation (8.14) takes the form

$$\frac{\pi\sigma^2 a_0^2}{E} - \frac{2\pi a_0^2\sigma_0^2}{E} = c\,. \tag{8.15}$$

By eliminating the constant c from Equations (8.14) and (8.15) we obtain

$$V^2 = \frac{2\pi E}{k^2\rho}\left(1-\frac{a_0}{a}\right)\left[1-(2n^2-1)\frac{a_0}{a}\right] \tag{8.16}$$

where

$$\sigma_0 = n\sigma\,,\quad n \le 1\,. \tag{8.17}$$

For $n = 1$ Equation (8.16) gives

$$V = 0.38 v_s\left(1-\frac{a_0}{a}\right) \tag{8.18}$$

using the result obtained by Roberts and Wells to determine k.

Experimentally obtained crack speeds are below the theoretical values given by Equation (8.18). Roberts and Wells [8.2] reported values of a/v_s in the range 0.20–0.37 for various materials. Kanninen [8.5], based on the Dugdale model, predicted a crack speed $a = 0.1v_s$ in a steel plate. Duffy *et al.* [8.6] measured cleavage fracture speeds in steel pipes of the order of 700 m/s and shear fracture speeds of the order of 200 m/s.

8.3. Stress field around a rapidly propagating crack

The singular stress field around a crack propagating with speed $V = \dot{a}$ is given by [8.7]

$$\sigma_{11} = \frac{K(t)\,B}{\sqrt{2\pi r}}\left[(1+2\beta_1^2-\beta_2^2)\sqrt{\frac{r}{r_1}}\cos\frac{\theta_1}{2} - \frac{4\beta_1\beta_2}{1+\beta_2^2}\sqrt{\frac{r}{r_2}}\cos\frac{\theta_2}{2}\right] \quad (8.19a)$$

$$\sigma_{22} = \frac{K(t)\,B}{\sqrt{2\pi r}}\left[-(1+\beta_2^2)\sqrt{\frac{r}{r_1}}\cos\frac{\theta_1}{2} + \frac{4\beta_1\beta_2}{1+\beta_2^2}\sqrt{\frac{r}{r_2}}\cos\frac{\theta_2}{2}\right] \quad (8.19b)$$

$$\sigma_{12} = \frac{2K(t)\,B\beta_1}{\sqrt{2\pi r}}\left[\sqrt{\frac{r}{r_1}}\sin\frac{\theta_1}{2} - \sqrt{\frac{r}{r_2}}\sin\frac{\theta_2}{2}\right] \quad (8.19c)$$

where

$$B = \frac{1+\beta_2^2}{4\beta_1\beta_2-(1+\beta_2^2)^2}\,, \quad (8.20)$$

$$\beta_1^2 = 1-\frac{V^2}{C_1^2}\,,\quad \beta_2^2 = 1-\frac{V^2}{C_2^2} \quad (8.21)$$

$$x+i\beta_1 y = r_1 e^{i\theta_1}\,,\quad x+i\beta_2 y = r_2 e^{i\theta_2}\,. \quad (8.22)$$

C_1 and C_2 are the dilatational and shear wave speeds given by

$$C_1^2 = \frac{\kappa+1}{\kappa-1}\frac{\mu}{\rho}\,,\quad C_2^2 = \frac{\mu}{\rho}\,. \quad (8.23)$$

The stress intensity factor $K(t)$ is defined as in the static case by

$$K(t) = \lim_{r\to 0}\left[\sqrt{2\pi r}\,\sigma_{22}(r,0,t)\right]. \quad (8.24)$$

Equations (8.19) for $V \to 0$ coincide with Equations (2.28) of the stress field around a static crack.

The particle velocity field in the vicinity of the crack tip is given by

$$\dot{u}_1 = -\frac{K(t)\,BV}{\sqrt{2\pi r}\,\mu}\left[\sqrt{\frac{r}{r_1}}\cos\frac{\theta_1}{2} - \frac{2\beta_1\beta_2}{1+\beta_2^2}\sqrt{\frac{r}{r_2}}\cos\frac{\theta_2}{2}\right] \tag{8.25a}$$

$$\dot{u}_2 = -\frac{K(t)\,\beta_1 V}{\sqrt{2\pi r}\,\mu}\left[\sqrt{\frac{r}{r_1}}\sin\frac{\theta_1}{2} - \frac{2}{1+\beta_2^2}\sqrt{\frac{r}{r_2}}\sin\frac{\theta_2}{2}\right]. \tag{8.25b}$$

Relations for the dynamic stress intensity factor have been developed by various investigators. Broberg [8.8] and Baker [8.9] gave the following equation

$$K(t) = k(V)\,K(0) \tag{8.26}$$

where $K(0)$ is the static stress intensity factor and $k(V)$ is a geometry independent function of crack speed. $K(0)$ can be approximated as the value of stress intensity factor for a static crack of length equal to the length of the moving crack. The quantity $k(V)$ decreases monotonically with crack speed and can be approximated by

$$k(V) = 1 - \frac{V}{C_R} \tag{8.27}$$

where C_R denotes the Rayleigh wave speed. Observe that the dynamic stress intensity factor becomes equal to zero when the crack speed V becomes equal to C_R.

Rose [8.10] gave the following approximation for $k(V)$:

$$k(V) = \left(1 - \frac{V}{C_R}\right)\,(1 - hV)^{-1/2} \tag{8.28}$$

where

$$h = \frac{2}{C_1}\left(\frac{C_2}{C_R}\right)^2\left[1 - \left(\frac{C_2}{C_1}\right)\right]^2 \tag{8.29}$$

and C_1 and C_2 are given by Equation (8.23).

It is of fundamental importance to refer two basic properties of the singular elastodynamic stress field resulting from Equations (8.19). The first is related to the angular variation of the circumferential stress σ_θ and the second to the stress triaxiality ahead of the crack tip.

Figure 8.1 presents the angular variation of the stress σ_θ normalized to its value for $\theta = 0$, for different values of crack speed, to the Rayleigh wave speed ratio $\dot{a}/C_R$. Note that C_R is somewhat smaller than the shear wave speed and observe that σ_θ presents a maximum for an angle θ different from zero when the crack speed becomes large. This result explains the experimentally observed phenomenon of crack branching at large crack speeds.

The ratio of the principal stresses σ_{22} and σ_{11} ahead of the crack ($\theta = 0$) is expressed by

$$\frac{\sigma_{22}}{\sigma_{11}} = \frac{4\beta_1\beta_2 - (1+\beta_2^2)^2}{(1+\beta_2^2)\,(1+2\beta_1^2-\beta_2^2) - 4\beta_1\beta_2}. \tag{8.30}$$

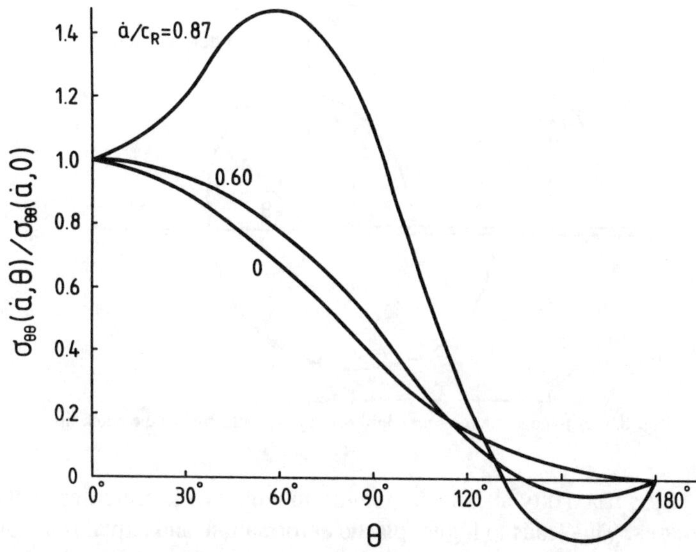

Fig. 8.1. Normalized circumferential stress $\sigma_{\theta\theta}(\dot{a},\theta)/\sigma_{\theta\theta}(\dot{a},0)$ versus polar angle θ for different values of $\dot{a}/c_R$.

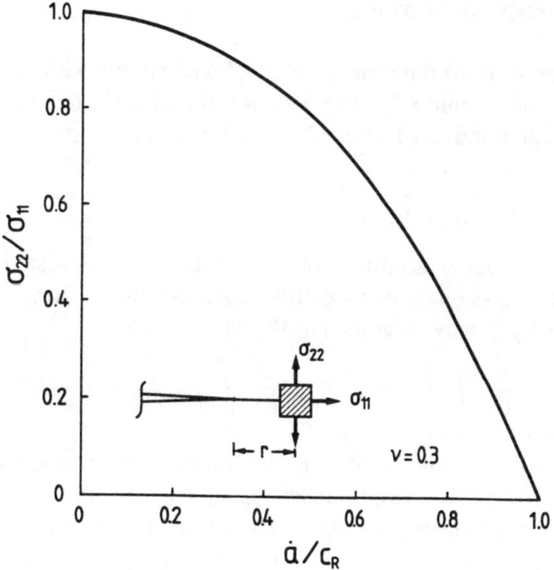

Fig. 8.2. Variation of σ_{22}/σ_{11} versus $\dot{a}/c_R$.

Figure 8.2 shows the variation of σ_{22}/σ_{11} with $\dot{a}/C_R$. Observe that σ_{22}/σ_{11} decreases continuously form 1, at zero crack speed, to 0 at the Rayleigh speed.

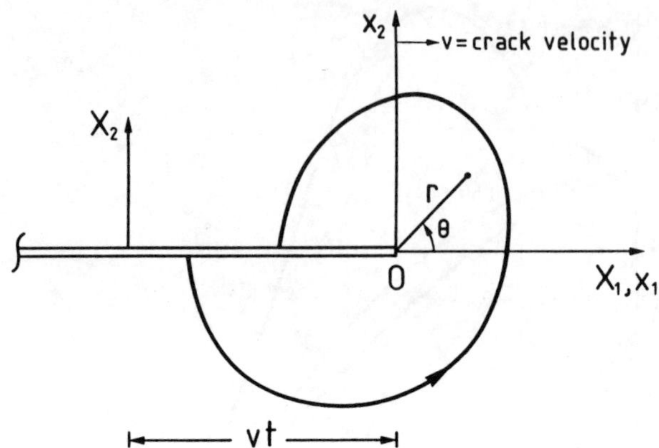

Fig. 8.3. A fixed and a moving coordinate system attached to the crack tip.

Thus, the stress triaxiality ahead of a rapidly moving crack decreases as the crack speed increases. This leads to higher plastic deformation, and explains the observed phenomenon of increasing crack growth resistance at high crack speeds.

8.4. Strain energy release rate

Consider a crack in a two-dimensional elastic body moving with constant velocity V in the X_1 direction (Figure 8.3). The body is referred to the fixed coordinate system X_1X_2, while a set of moving rectangular coordinates $x_1,\ x_2$ are attached to the crack tip. We have

$$x_1 = X_1 - Vt\,,\quad x_2 = X_2\,. \tag{8.31}$$

Let us now consider an arbitrary contour C which encompasses the crack tip and travels at the same speed as the crack. The balance of energy flow across the moving region bounded by C may be written as [8.11]:

$$\int\limits_C T_i\dot{u}_i\,\mathrm{d}s = \frac{\mathrm{d}}{\mathrm{d}t}\iint\limits_R \rho E\,\mathrm{d}A + \frac{1}{2}\frac{\mathrm{d}}{\mathrm{d}t}\iint\limits_R \rho\dot{u}_i\dot{u}_i\,\mathrm{d}A + VG\,. \tag{8.32}$$

The left-hand side of this equation represents the work of tractions across C; the first and second terms on the right-hand side are the rate of increase of internal and kinetic energies stored inside the region R enclosed by the curve C; the third term is the rate at which energy is dissipated by the moving crack. $T_i = \sigma_{ij}\eta_j$ denotes the traction components acting across C, and η_j the components of the unit normal vector $\mathbf{n}$ of curve C. When Equation (8.32) is referred to the moving coordinates $x_1,\ x_2$ defined from Equation (8.31) it takes the form

$$G = \iint\limits_R \frac{\partial}{\partial x_1}(\rho E)\,\mathrm{d}A + \frac{1}{2}V^2\iint\limits_R \frac{\partial}{\partial x_1}\left(\rho\frac{\partial u_i}{\partial x_1}\frac{\partial u_i}{\partial x_1}\right)\mathrm{d}A-$$

$$\int_C T_i \frac{\partial u_i}{\partial x_1} \, \mathrm{d}s \,. \tag{8.33}$$

Using Green's convergence theorem, we find that Equation (8.33) takes the form

$$G = \int_C \left(\rho E + \frac{1}{2} \rho V^2 \frac{\partial u_i}{\partial x_1} \frac{\partial u_i}{\partial x_1} \right) \mathrm{d}x_2 - T_i \frac{\partial \dot{u}_i}{\partial x_1} \, \mathrm{d}s \,. \tag{8.34}$$

Conservation of the mechanical energy of the system leads to

$$\iint_R \rho \dot{E} \, \mathrm{d}A = \iint_R \sigma_{ij} \dot{\varepsilon}_{ij} \, \mathrm{d}A \,. \tag{8.35}$$

Introducing the strain energy density function $\omega = \omega(\varepsilon_{ij})$ from

$$\sigma_{ij} = \frac{\partial \omega}{\partial \varepsilon_{ij}} \tag{8.36}$$

and applying Green's theorem, we obtain

$$\int_C \rho E \, \mathrm{d}x_2 = \iint_R \sigma_{ij} \frac{\partial \varepsilon_{ij}}{\partial x_1} \, \mathrm{d}A = \iint_R \frac{\partial \omega}{\partial x_1} \, \mathrm{d}A = \int_C \omega \, \mathrm{d}x_2 \,. \tag{8.37}$$

Using Equation (8.37), we find that Equation (8.34) becomes

$$G = \int_C \left(\omega + \frac{1}{2} \rho V^2 \frac{\partial u_i}{\partial x_1} \frac{\partial u_i}{\partial x_1} \right) \mathrm{d}x_2 - T_i \frac{\partial u_i}{\partial x_1} \, \mathrm{d}s \,. \tag{8.38}$$

For a static crack ($V = 0$), Equation (8.38) reduces to the J-integral (Equation (6.16)) which is equal to the strain energy release rate G for elastic behavior (Equation (6.30)). Using derivations similar to those in the J-integral (Section 6.2) we can easily prove that G is path independent – that is, it retains its value for an arbitrary choice of the integration path C surrounding the crack tip.

As for the J-integral (Section 6.3), we can easily calculate G by choosing C as a circle of radius R centered at the crack tip, and using the asymptotic expressions of stresses and displacements given from Equation (8.19) and (8.25)

$$\begin{aligned} G = \; & R \int_{-\pi}^{\pi} \left[\omega \, \cos\theta + \frac{1}{2} \rho V^2 \left[\left(\frac{\partial u_1}{\partial x_1} \right)^2 + \left(\frac{\partial u_2}{\partial x_1} \right)^2 \right] \cos\theta \right. \\ & \left. - (\sigma_{11} \, \cos\theta + \sigma_{12} \, \sin\theta) \frac{\partial u_1}{\partial x_1} + (\sigma_{12} \, \cos\theta + \sigma_{22} \, \sin\vartheta) \frac{\partial u_2}{\partial x_1} \right] \mathrm{d}\theta \end{aligned} \tag{8.39}$$

or

$$G = \left[\frac{\beta_1 (1 - \beta_2^2)}{4\beta_1\beta_2 - (1 + \beta_2^2)^2} \, \frac{1}{2\mu} \right] K^2(t) \,. \tag{8.40}$$

Equation (8.40) establishes a relation between the strain energy release rate or crack driving force G and the dynamic stress intensity factor $K(t)$ under plane strain conditions. For the static crack problem ($V = 0$) Equation (8.40) reduces to Equation (4.21).

A fracture criterion for dynamic crack propagation based on Equation (8.40) can be established as in the static case. It is assumed that dynamic crack propagation occurs when the strain energy release rate G becomes equal to a critical value; this is equivalent to a critical stress intensity factor fracture criterion. For small-scale yielding the concept of K-dominance for stationary cracks can be extended to dynamic cracks. In these circumstances the fracture criterion for a propagating crack takes the form

$$K(t) = K_{ID}(V, t) \tag{8.41}$$

where $K_{ID}(V, t)$ represents the resistance of the material to dynamic crack propagation and is assumed to be a material property. The dynamic stress intensity factor $K(t)$ is determined from the solution of the corresponding elastodynamic problem and is a function of loading, crack length and geometrical configuration of the cracked body. The material parameter K_{ID} can be determined experimentally and depends on crack speed and environmental conditions. A brief description of the available experimental methods used for the determination of K and K_{ID} will be presented in Section 8.6.

8.5. Crack branching

When a crack propagates at a high speed it may divide into two branches. In many cases these divide further until a multiple crack branching pattern is obtained.

Sih [8.12] used the model of a finite crack spreading at both ends at constant speed in conjunction with the strain energy density criterion, to predict the crack bifurcation angle. For this case the strain energy density factor is given by

$$\begin{aligned} S = \; & \frac{K_{\mathrm{I}}^2}{8\pi\mu} F^2(\beta_1, \beta_2) \, [2(1+\beta_2^2)^2 \, [2(1-\nu)\,(1+\beta_1^2)^2 - \\ & -2(1-2\nu)\,(2\beta_1^2 + 1 - \beta_2^2)\,(1+\beta_2^2)] \, g^2(\beta_1) + \\ & +32\beta_1^2\beta_2^2 g^2(\beta_2) - 16\beta_1\beta_2(1+\beta_1^2)\,(1+\beta_2^2)\, g(\beta_1)\, g(\beta_2) + \\ & +8\beta_1^2(1+\beta_2^2)\,[h(\beta_1) - h(\beta_2)]^2] \end{aligned} \tag{8.42}$$

where

$$\begin{aligned} & g^2(\beta_j) + h^2(\beta_j) = \sec\theta(1 + \beta_j^2 \tan^2\theta)^{-1/2} \\ & g^2(\beta_j) - h^2(\beta_j) = \sec\theta(1 + \beta_j^2 \tan^2\theta)^{-1} \end{aligned} \tag{8.43}$$

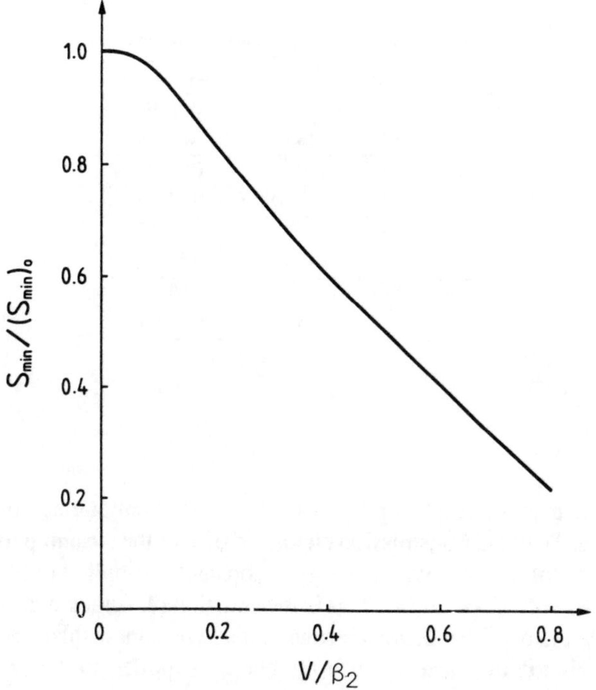

Fig. 8.4. Strain energy density factor versus crack speed [8.12].

and

$$F(\beta_1, \beta_2) = \beta_1 [[(1+\beta_2^2)^2 - 4\beta_1^2\beta_2^2]\, K(\beta_1) - 4\beta_1^2(1-\beta_2^2)\, K(\beta_2) - \\ -[4\beta_1^2 + (1+\beta_2^2)^2]\, E(\beta_1) + 8\beta_1^2 E(\beta_2)]^{-1} \,. \tag{8.44}$$

Here K and E are complete elliptic integrals of the first and second kind, respectively.

According to the strain energy density failure criterion, the crack extends in the direction which makes S a minimum. The critical value $S_{\min}$ is then computed from Equation (8.42) as a function of crack speed. Figure 8.4 shows the variation of $S_{\min}/(S_{\min})_0$ with V/β_2. Here $(S_{\min})_0 = (1-2\nu)\,\sigma\sqrt{a}/4\mu$ and V represents the crack speed. Observe that $S_{\min}$ decreases smoothly from its largest value at $V = 0$ as the crack speed increases. Table 8.1 shows values of the half branch angle θ_0 with the corresponding values of $S_{\min}/(S_{\min})_0$, for different values of crack speed, and Poisson's ratio. Note that as ν is varied from 0.21 to 0.24 θ_0 changes from $\pm 18.84°$ to $\pm 15.52°$ which is very close to experimental observation.

TABLE 8.1. Angles of crack bifurcation.

ν	V/β_2	θ_0	$S_{\min}/(S_{\min})_0$
0.21	0.45	$0°$	0.56209
	0.46	$\pm 18.84°$	0.55199
0.22	0.46	$0°$	0.55196
	0.47	$\pm 17.35°$	0.54204
0.23	0.47	$0°$	0.54185
	0.48	$\pm 16.27°$	0.53210
0.24	0.48	$0°$	0.53177
	0.49	$\pm 15.52°$	0.52217

8.6. Crack arrest

The problem of arrest of a rapidly propagating crack is of major theoretical and practical importance. The load transmission characteristics of the system play a significant role in the arrest of a crack. When energy is constantly supplied to the crack-tip region, the crack continues to move. This is the situation of a crack in a uniform tensile stress field. On the other hand, crack growth under constant displacement conditions eventually leads to crack arrest, since the energy supplied to the crack-tip region progressively decreases with time. When the distance between the energy source and crack tip increases with time, the capability of a system to arrest a crack increases. This occurs, for example, in the splitting of a long cantilever beam specimen.

A crack arrest criterion based on the stress intensity factor can be put in the form

$$K(t) = K_{IA} = \min\left[K_{ID}(V)\right] \tag{8.45}$$

where $K(t)$ is the dynamic stress intensity factor and K_{ID} the material fracture toughness for dynamic crack propagation. Experimental studies indicate that K_{ID} depends on crack speed $\dot{a}$ [8.13–8.15]. Figure 8.5 shows a typical form of the curve $K_{ID} = K_{ID}(V)$ for many metals and polymers. Observe that K_{ID} is nearly speed independent at low crack speeds, and increases as the crack speed increases.

The capability of a multi-member structural system to arrest a crack increases when the load is taken up and transmitted to other structural elements. Usually arrest strips are used, a method that finds application in aircraft structures. For further information on crack arrest procedures used in structural design refer to [8.16].

8.7. Experimental determination of crack velocity and dynamic stress intensity factor

Experimental studies play a key role in improving our understanding of dynamic fracture behavior of materials and structures, and in the measurement of the relevant

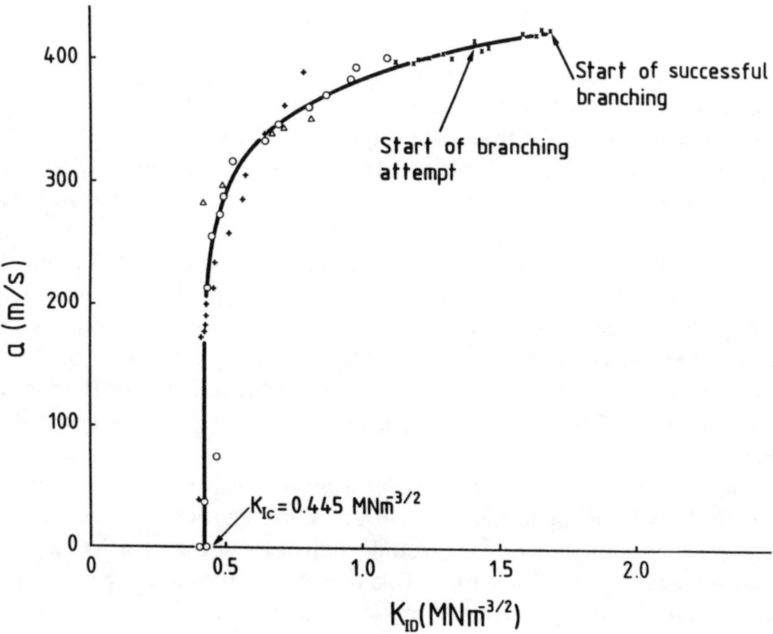

Fig. 8.5. Dynamic fracture toughness versus crack speed for Homalite 100 [8.13].

dynamic fracture material properties. In this section we present the most widely used experimental methods for measuring the crack velocity and the dynamic stress intensity factor.

(a) Crack velocity

Initial measurements of crack velocity were conducted using a series of conducting wires placed at certain intervals along the crack path and perpendicular to the direction of the crack propagation. The wires form one leg of a bridge which is connected to an oscilloscope. As the crack propagates the wires break and the corresponding times are obtained from the trace on the oscilloscope. This technique allows measurement of the average speed over the gage length between the wires.

High-speed photography is perhaps the most widely used method of recording rapid crack propagation. The multiple-spark Cranz-Schardin camera, which is capable of operating at rates of up to 10^6 frames per second, is widely employed. Although best results are obtained for transparent materials, the method can also be used for nontransparent materials by polishing the surface of the specimens.

(b) Dynamic stress intensity factor

The method of dynamic photoelasticity was first used by Wells and Post [8.17] to determine the state of stress and the speed of a rapidly propagating crack. The stress intensity factor was obtained from the analysis of the isochromatic pattern around the crack tip. Further studies of this problem have been performed by Kobayashi and coworkers [8.18, 8.19]. These investigations were based on the static solution of the stress field near the crack tip. Kobayashi and Mall [8.20] estimated that the error introduced when the static stress field was used, is small for crack propagating speeds less than $0.15C_1$. Extensive studies on the dynamic photoelastic investigation of crack problems have been performed at the Photomechanics Laboratory of the University of Maryland [8.13, 8.14, 8.21, 8.22] using the dynamic stress field around a moving crack. A K versus V relationship was established and for Homalite 100 it was found to be independent of the specimen geometry, for crack speeds below 300 m/s.

The optical method of caustics has also been used extensively for the experimental study of crack initiation, rapid crack growth, crack arrest and crack branching [8.23–8.28]. A dynamic correction for the determination of the stress intensity factor from the obtained optical pattern was used. This method has proved to be very efficient and powerful for the study of dynamic crack problems.

For the determination of dynamic fracture toughness, K_{ID}, several types of specimens – including the double cantilever beam specimen, the single edge-notched specimen and the wedge-loaded specimen – have been proposed. The last type of specimen presents a number of advantages over the others and is mainly used in dynamic fracture testing. Duplex specimens, with crack initiation taking place in a hardened starter section welded into the test material, are sometimes preferred in situations where large monolithic specimens are needed. The dynamic fracture toughness K_{ID} is determined as a function of crack speed by measuring the critical stress intensity factor for crack initiation and the crack length at arrest, and using appropriate dynamic analysis for the specific type of specimen. The critical value K_{IA} of the stress intensity factor at crack arrest is determined as the minimum value of K_{ID} taken from a number of measurements at various crack speeds.

Besides the dynamic fracture toughness K_{ID}, which depends on crack speed, the critical value K_{Id} of the stress intensity factor for crack initiation under a rapidly applied load is of interest in practical applications. K_{Id} depends on the loading rate and temperature and is considered to be a material parameter. K_{Id} is determined experimentally by using the three-point bend specimen. The specimen is loaded by a falling weight and K_{Id} is determined by static analysis. K_{Id} decreases with increasing loading rate, below the transition temperature, and increases with load rate above the transition temperature.

Examples

Example 8.1.

The singular elastodynamic stress and displacement fields of a crack subjected to tearing-mode of deformation are expressed by

$$\sigma_{13} = \frac{K_{\mathrm{III}}}{\sqrt{2\pi r}}\, F(\beta_2)\, h(\beta_2)\,, \quad \sigma_{23} = \frac{K_{\mathrm{III}}}{\sqrt{2\pi r}}\, \beta_2\, F(\beta_2)\, g(\beta_2) \tag{1a}$$

$$u_3 = -\frac{K_{\mathrm{III}}}{\mu}\sqrt{\frac{2r}{\pi}}\,(\cos^2\theta + \beta_2^2\sin^2\theta)^{1/2}\, F(\beta_2)\, h(\beta_2) \tag{1b}$$

where the functions $g(\beta_2)$ and $h(\beta_2)$ are given by Equation (8.43).

For an infinite plate with a central crack extending on both ends at constant speed the function $F(\beta_2)$ is given by

$$F(\beta_2) = \frac{2}{\pi}\, K\left(\frac{V}{C_2}\right) \tag{2}$$

where K denotes the complete elliptic integral of the first kind with argument V/C_2.

Calculate the strain energy release rate G_{III} and plot its variation with V/C_2.

Solution: The strain energy release rate G_{III} is calculated from Equation (8.38) by choosing the integration path C as a circle of radius R centered at the crack tip, as

$$G_{\mathrm{III}} = R\int_{-\pi}^{\pi}\left[\omega\cos\theta + \frac{1}{2}\rho V^2\left(\frac{\partial u_3}{\partial x}\right)^2\cos\theta - (\sigma_{13}\cos\theta + \sigma_{23}\sin\theta)\frac{\partial u_3}{\partial x}\right]\mathrm{d}\theta\,. \tag{3}$$

The strain energy density function ω is given by

$$\omega = \frac{1}{2\mu}\,(\sigma_{13}^2 + \sigma_{23}^2)\,. \tag{4}$$

Substituting the values of stresses σ_{13} and σ_{23} and displacement u_3 from Equation (1) into Equations (3) and (4) and performing the integration in Equation (3) we obtain

$$\frac{G_{\mathrm{III}}}{G_{\mathrm{III}}^*} = \beta_2\, F^2(\beta_2) \tag{5}$$

where G_{III}^* is the static value of G and can be obtained by putting $V = 0$ as (Equation (4.27))

$$G_{\mathrm{III}}^* = G_{\mathrm{III},V=0} = \frac{K_{\mathrm{III}}^2}{2\mu}\,. \tag{6}$$

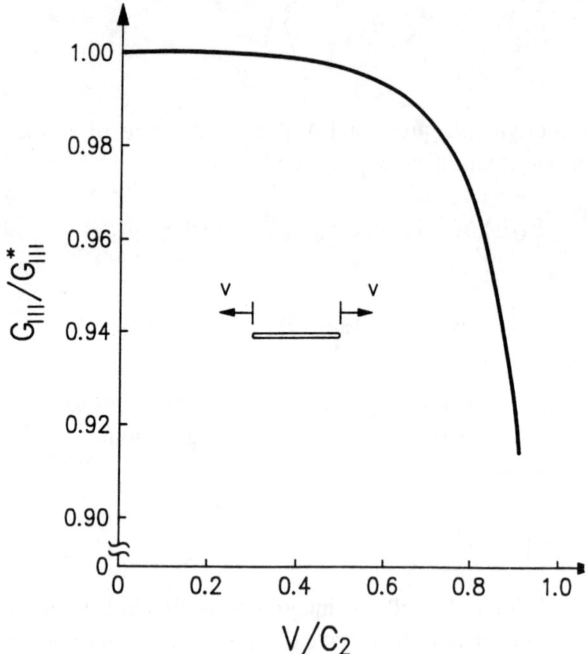

Fig. 8.6. Normalized strain energy release rate versus crack speed of an expanding crack under anti-plane shear.

The variation of $G_{\text{III}}/G^*_{\text{III}}$ versus V/C_2 is shown in Figure 8.6. Note that the influence of crack speed on G_{III} is negligible for small values of V/C_2, while at very high values of V/C_2, G_{III} diminishes rapidly and becomes zero at $V = C_2$.

Example 8.2.

A double cantilever beam (DCB) of height $2h$ with a crack of length a (Figure 4.14) is made of a nonlinear material whose stress-strain relation is described by

$$\sigma = \alpha\varepsilon(\varepsilon)^{(\beta-1)/2}, \quad 0 < \beta \leq 1 \tag{1}$$

where α measures the stiffness of the material and is equal to the modulus of elasticity E for $\beta = 1$ (linear material). Equation (1) is shown in Figure 8.7. The DCB is subjected to an end load P that remains constant during rapid crack propagation. Let a_0 denote the initial crack length and P_c the load at crack propagation.

Calculate the speed V and acceleration a_c of the crack and plot their variation versus a_0/a for various values of β [8.11].

Solution: The energy balance equation during crack growth (Equation (4.1)) takes the form

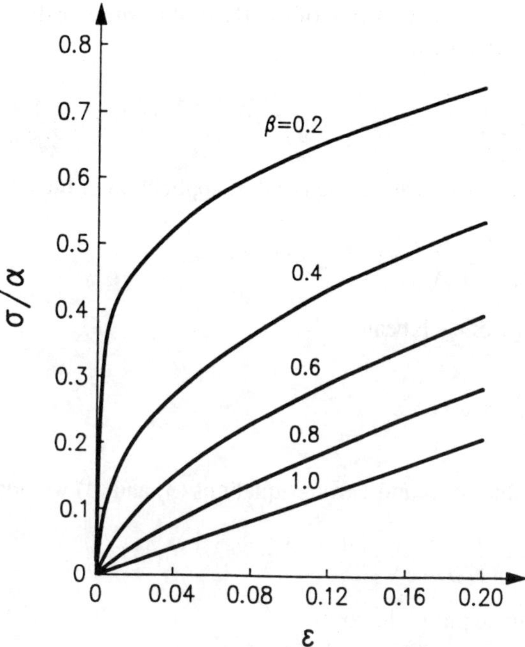

Fig. 8.7. Nonlinear stress-strain curves.

$$P_c(u - u_c) = U(a) - U(a_0) + K + \gamma(a - a_0) \; . \tag{2}$$

The left-hand side of Equation (2) represents the work supplied to the system during growth of the crack from its initial length a_0 to the length a. u_c and u represent the load-point displacements at crack lengths a_0 and a, respectively. The right-hand side of Equation (2) is composed of the term $U(a) - U(a_0)$ that represents the change of strain energy, the term K that represents the kinetic energy and the term $\gamma(a - a_0)$ that represents the change of the surface energy.

From beam theory we have for the stress σ and strain ε at position x of the DCB

$$\sigma = \frac{Pxy}{I}\,(y^2)^{(\beta-1)/2} \tag{3}$$

$$\varepsilon = y\left(\frac{aI}{Px}\right)^{-1/\beta} \tag{4}$$

where

$$I = 2\int_0^{h/2} y^{\beta+1}\,\mathrm{d}y \; . \tag{5}$$

The deflection $y(x)$ of each beam of the DCB at position x during crack growth is calculated from beam theory as

$$y(x) = \frac{\beta}{\beta+1}\left(\frac{P_c}{aI}\right)^{1/\beta}\left[\frac{\beta}{2\beta+1}\,x^{(2\beta+1)/\beta} - a^{(\beta+1)/\beta}x + \frac{\beta+1}{2\beta+1}\,a^{(2\beta+1)/\beta}\right] \quad (6)$$

and the deflection of each beam at the point of application of the load is

$$u = y(0) = \frac{\beta}{2\beta+1}\left(\frac{P_c}{aI}\right)^{1/\beta} a^{(2\beta+1)/\beta}\,. \quad (7)$$

The strain energy of each beam is

$$U(a) = \int_V \Big(\int_0^{\varepsilon} \sigma\,\mathrm{d}\varepsilon\Big)\,\mathrm{d}V\,. \quad (8)$$

Substituting the values of σ and ε from Equations (3) and (4) we obtain

$$U(a) = \frac{\beta}{(\beta+1)\,(2\beta+1)}\,P_c^{(\beta+1)/\beta}\,a^{(2\beta+1)/\beta}(aI)^{-1/\beta}\,. \quad (9)$$

Equation (9) can be put in the form

$$U(a) = \frac{P_c u}{\beta+1}\,. \quad (10)$$

The kinetic energy K due to the motion of the beams along the y-direction is

$$K = \frac{1}{2}\int_0^a \rho h\left(\frac{\mathrm{d}y}{\mathrm{d}t}\right)^2 \mathrm{d}x\,. \quad (11)$$

Substituting the value of $y = y(x)$ from Equation (6) into Equation (11) we obtain

$$K = \frac{1}{6}\,\rho h\left(\frac{P_c}{aI}\right)^{2/\beta} a^{(2+3\beta)/\beta}\,V^2 \quad (12)$$

where $V = \mathrm{d}a/\mathrm{d}t$ is the crack velocity.

Substituting the values of u, $u_c = u(a = a_0)$, $U(a)$, $U(a_0)$ and K from Equations (7), (9) and (12) into the energy balance Equation (2) we obtain for the crack speed during crack growth

$$V^2 = \frac{6\beta^2}{(2\beta+1)\,(\beta+1)}\,\frac{(aI)^{1/\beta}}{\rho h}\,P_c^{(\beta-1)/\beta}\,a^{-(\beta+1)/\beta}$$

$$\left[1 - \left(\frac{a_0}{a}\right)^{(2\beta+1)/\beta} - n\left(1 - \frac{a_0}{a}\right)\left(\frac{a_0}{a}\right)^{(\beta+1)/\beta}\right]. \quad (13)$$

Here

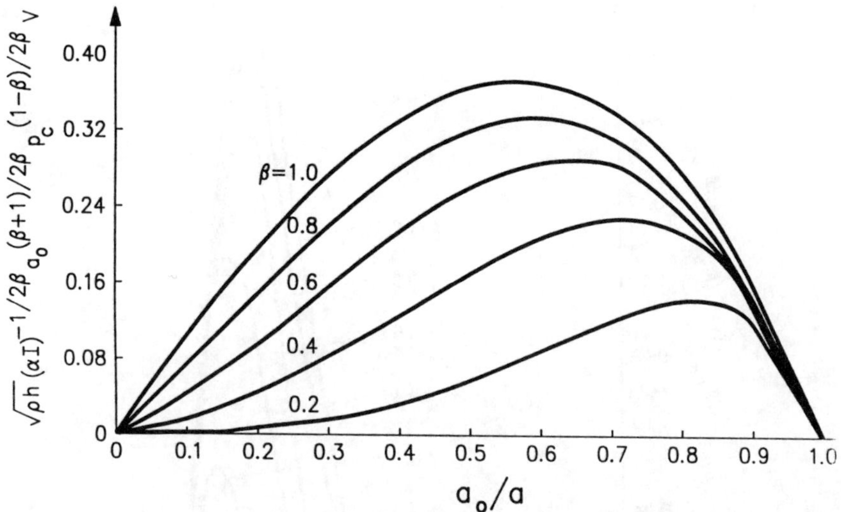

Fig. 8.8. Normalized crack velocity versus crack length at constant force.

$$na_0^{(\beta+1)/\beta} = \frac{(\beta+1)\,(2\beta+1)}{\beta^2}\,\gamma(aI)^{1/\beta}\,(P_c)^{-(1+\beta)/\beta}\,. \tag{14}$$

By differentiation from Equation (13) with respect to time we obtain the crack acceleration $a_c = \mathrm{d}V/\mathrm{d}t$

$$a_c = \frac{3\beta^2}{(2\beta+1)\,(\beta+1)}\,\frac{(aI)^{1/\beta}}{\rho h}\,P_c^{(\beta-1)/\beta}\,a^{-(2\beta+1)/\beta}$$

$$\left[(1-n)\left(\frac{3\beta+2}{\beta}\right)\left(\frac{a_0}{a}\right)^{(2\beta+1)/\beta} + \frac{\beta+1}{\beta}\left[2n\left(\frac{a_0}{a}\right)^{(\beta+1)/\beta} - 1\right]\right]. \tag{15}$$

As the initial acceleration of the crack $a_c = a_c(a = a_0)$ must be positive we obtain from Equation (15)

$$(1-n)\left(\frac{3\beta+2}{\beta}\right) + (2n-1)\left(\frac{\beta+1}{\beta}\right) > 0 \tag{16}$$

so that

$$n < \frac{2\beta+1}{\beta}\,. \tag{17}$$

Numerical results were obtained for $n = 0.99[(2\beta+1)/\beta]$ and various values of β. Figure 8.8 presents the variation of normalized crack speed V versus a_0/a for $\beta = 0.2$, 0.4, 0.6, 0.8 and 1.0. The value $\beta = 1.0$ corresponds to a linear material. Note that crack speed increases during crack growth, from zero value at $a_0/a = 1$ reaches a maximum, and then decreases and becomes zero at $a_0/a \to 0$.

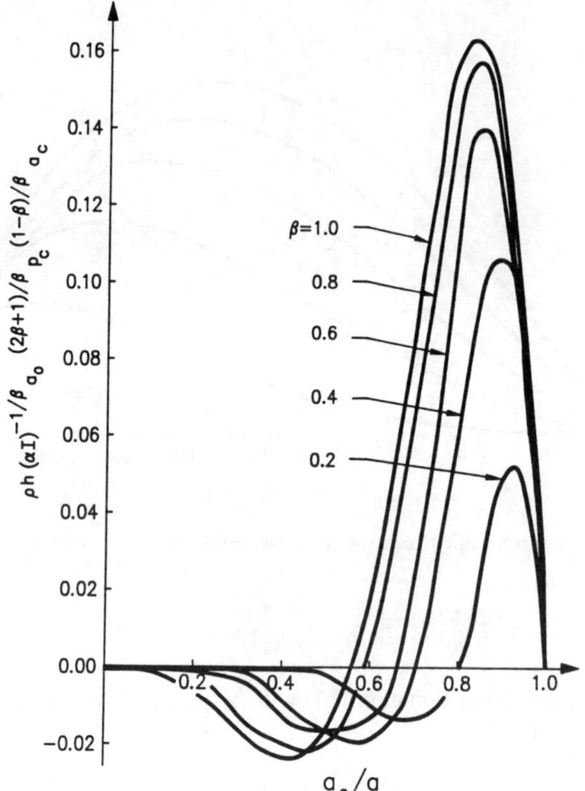

Fig. 8.9. Normalized crack acceleration versus crack length at constant force.

Figure 8.9 shows the variation of normalized crack acceleration a_c versus a_0/a for various values of β. The crack first accelerates, and then decelerates before coming to a complete stop at $a_0/a \to 0$. From Figures 8.8 and 8.9 we observe that the crack travels more slowly as β decreases. This should be expected as the material becomes stiffer with decreasing β.

Example 8.3.

An infinite strip of height $2h$ with a semi-infinite crack is rigidly clamped along its upper and lower faces at $y = \pm h$ (Figure 6.16). The upper and lower faces are moved in the positive and negative y-direction over distances u_0, respectively. Determine the dynamic stress intensity factor $K(t)$ during steady state crack propagation.

Solution: The dynamic stress intensity factor $K(t)$ is computed from Equation (8.40) as

$$K(t) = \left[\frac{EG}{1+\nu} \frac{4\beta_1\beta_2 - (1+\beta_2^2)^2}{\beta_1(1-\beta_2^2)} \right]^{1/2} . \tag{1}$$

The strain energy release rate G is computed from Equation (8.38) by taking the same integration path as in Example 6.1. Observing that $\partial u_i/\partial x = 0$ along the line BC we find that G for the dynamic problem is equal to its value for the static problem. Substituting the value of $G = J$ from Equation (9) of Example 6.1 into Equation (1) we obtain for $K(t)$

$$K(t) = \left[\frac{(1-\nu)[4\beta_1\beta_2 - (1+\beta_2^2)^2]}{(1+\nu)^2\,(1-2\nu)\,\beta_1(1-\beta_2^2)} \right]^{1/2} \frac{Eu_0}{h^{1/2}} \tag{2}$$

under conditions of plane strain, and

$$K(t) = \left[\frac{[4\beta_1\beta_2 - (1+\beta_2^2)^2]}{(1+\nu)^2\,(1-\nu)\,\beta_1(1-\beta_2^2)} \right]^{1/2} \frac{Eu_0}{h^{1/2}} \tag{3}$$

under conditions of generalized plane stress.

Example 8.4.

A crack of length 20 mm propagates in a large steel plate under a constant stress of 400 MPa. The dynamic toughness of the material K_{ID} can be expressed by the following empirical equation

$$K_{ID} = \frac{K_{IA}}{1 - \left(\dfrac{V}{V_l}\right)^m} \tag{1}$$

where K_{IA} is the arrest toughness of the material, V_l is the limiting crack speed and m is an empirical parameter. Using Rose's approximation for the dynamic stress intensity factor determine the speed of crack during propagation. Take: $C_1 =$ 5940 m/sec, $C_2 =$ 3220 m/sec, $C_R =$ 2980 m/sec; $K_{IA} =$ 100 MPa $\sqrt{\text{m}}$, $m = 2$, $V_l =$ 1600 m/sec.

Solution: Crack propagation is governed by the following equation (Equation (8.41))

$$K_I(t) = K_{ID} \tag{2}$$

of the material.

Using Rose's approximation for $K_I(t)$ we have (Equation (8.26))

$$K_I(t) = \sigma\sqrt{\pi a}\,k(V) \tag{3}$$

or

$$K_I(t) = 800\ \text{MPa}\sqrt{\pi \times (10\times 10^{-3})\,\text{m}}\;k(V) = 141.8\,k(V)\ \text{MPa}\sqrt{\text{m}} . \tag{4}$$

$K(V)$ is computed from Equation (8.28), where h is given by Equation (8.29). We have

$$h = \frac{2}{5940}\left(\frac{3220}{2980}\right)^2\left(1-\frac{3220}{5940}\right)^2 = 0.0824\times10^{-3}\ \mathrm{s/m} \tag{5}$$

and

$$k(V) = \left(1-\frac{V}{2980}\right)(1-0.0824\times10^{-3}V)^{-1/2}\,. \tag{6}$$

K_{ID} is computed as

$$K_{ID} = \frac{100\ \mathrm{MPa}\sqrt{\mathrm{m}}}{1-\left(\dfrac{V}{1600}\right)^2}\,. \tag{7}$$

Substituting the values of $K_I(t)$ and K_{ID} from Equations (3) and (7) into Equation (2) we obtain an equation involving the unknown crack speed. From a numerical solution of this equation we obtain

$$V = 571.3\ \mathrm{m/s}\,. \tag{8}$$

Problems

8.1. Show that the strain energy release rate G defined by Equation (8.38) is path independent.

8.2. The double cantilever beam of Example 8.2 is subjected to an end displacement u_c which remains constant during rapid crack propagation. Calculate the speed and acceleration of the crack and plot their variation versus a_0/a for various values of β. Take $n = 0.6(2\beta+1)/\beta$.

8.3. The double cantilever beam of Example 8.2 is made of a linear material. Calculate the speed of the crack and plot its variation versus a/a_0 when the applied load P is kept constant during crack propagation. Take various values of the ratio $n = P/P_0$, where P_0 is the load at crack initiation.

8.4. The double cantilever beam of Example 8.2 is made of a linear material. Calculate the speed of the crack and plot its variation versus a/a_0 when the applied displacement u is kept constant during crack propagation. Take various values of the ratio $n = u/u_0$, where u_0 is the displacement at crack initiation.

8.5. The Yoffé crack model considers a crack of fixed length propagating with constant speed in an infinite plate under a uniform tensile stress normal to the crack line. It is assumed that the crack retains its original length during propagation by resealing itself at the trailing end. For this problem the strain energy release rate G is given by

$$\frac{G}{G^*} = \frac{4}{\kappa+1}\,\beta_1(1-\beta_2^2)\,[4\beta_1\beta_2-(1+\beta_2^2)^2]\,F^2(\beta_1,\beta_2)$$

where G^* is the strain energy release rate at zero crack speed. G^* and F are given by

$$G^* = \frac{\kappa+1}{8\mu}\,K^2$$

$$F = \left[4\beta_1\beta_2-(1+\beta_2^2)^2\right]^{-1}.$$

Plot the variation of G/G^* versus crack speed V/C_2 under conditions of plane strain for various values of Poisson's ratio ν.

8.6. For Problem 8.5 show that G becomes infinite at a crack speed V computed from the following equation

$$(\kappa+1)\left(\frac{V}{C_2}\right)^4\left[\left(\frac{V}{C_2}\right)^2-8\right]+8(\kappa+5)\left(\frac{V}{C_2}\right)^2-32=0\,.$$

This equation gives the Rayleigh wave speed $C_R(C_R < C_2 < C_1)$.

8.7. Plot the variation of normalized Rayleigh wave speed C_R/C_2 versus Poisson's ratio ν under conditions of plane stress and plane strain according to Equation of Problem 8.6. Also plot the variation of C_1/C_2 versus ν.

8.8. Calculate the dilatational, C_1, the shear, C_2, and the Rayleigh, C_R, wave speeds for (a) steel with $E = 210$ GPa, $\rho = 7800$ kg/m^3, $\nu = 0.3$ and (b) copper with $E = 130$ GPa, $\rho = 8900$ kg/m^3, $\nu = 0.34$.

8.9. Broberg considered a crack in an infinite plate extending on both ends with constant speed. For this case the strain energy release rate G is computed from Equation of Problem 8.5 where the function $F(\beta_1,\beta_2)$ is given by

$$F(\beta_1,\beta_2) = \beta_1[[(1+\beta_2^2)^2-4\beta_1^2\beta_2^2]\,K(\beta_1)-4\beta_1^2(1-\beta_2^2)\,K(\beta_2)$$

$$-[4\beta_1^2+(1+\beta_2^2)^2]\,E(\beta_1)+8\beta_1^2E(\beta_2)]^{-1}$$

and K and E are complete elliptic integrals of the first and second kind respectively. They are given by

$$K(k) = \frac{\pi}{2}\left[1+\left(\frac{1}{2}\right)^2k^2+\left(\frac{3}{2.4}\right)^2k^4+\left(\frac{3.5}{2.4.6}\right)^2k^6+\ldots\right]$$

$$E(k) = \frac{\pi}{2}\left[1 - \left(\frac{1}{2^2}\right)^2 k^2 - \left(\frac{3^2}{2^2.4^2}\right)^2 \frac{k^4}{3} - \left(\frac{3^2.5^2}{2^2.4^2.6^2}\right)^2 \frac{k^6}{5} - \dots\right].$$

Plot the variation of G/G^* versus crack speed V/C_2 under conditions of plane strain for various values of Poisson's ratio ν. Note that G becomes zero as ν tends to the Rayleigh wave speed C_R.

8.10. An infinite strip of height $2h$ with a semi-infinite crack is rigidly clamped along its upper and lower faces at $y = \pm h$ (Figure 6.16). The upper and lower faces are moved in the positive and negative z-direction over distances w_0, respectively. Determine the dynamic stress intensity factor $K_{\text{III}}(t)$ during steady state crack propagation. Note that $K_{\text{III}}(t)$ is related to the strain energy release rate G_{III} by

$$G_{\text{III}} = \frac{K_{\text{III}}^2(t)}{2\mu\beta_2}.$$

8.11. Show that the strain energy release rate G_{III} for mode-I dynamic crack propagation can be determined by the following equation

$$G_{\text{I}} = \lim_{\delta\to 0} \frac{1}{\delta} \int_0^{\delta} \sigma_{22}(x_1, 0)\, u_2(\delta - x_1, \pi)\, dx_1$$

where δ is a segment of crack extension along the x_1-axis. Note that this expression of G_{I} is analogous to Equation (4.19) of the static crack.

8.12. Establish a relationship between the dynamic, G, and static, $G(0)$, energy release rates using Rose's approximation to $k(V)$. Plot the variation of $G/G(0)$ versus V/C_R for a thick steel plate with: $C_1 = 5940$ m/sec, $C_2 = 3220$ m/sec and $C_R = 2980$ m/sec.

8.13. A crack of length $2a$ propagates in a large steel plate under a constant stress of 300 MPa. The dynamic toughness of the material K_{ID} can be expressed by the following empirical equation

$$K_{ID} = \frac{K_{IA}}{1 - \left(\frac{V}{V_l}\right)^m}$$

where K_{IA} is the arrest toughness of the material, V_l is the limiting crack speed and m is an empirical parameter. Plot the variation of crack speed V versus initial crack length for 10 mm $< 2a <$ 100 mm, using Rose's approximation for the dynamic stress intensity factor. Take $C_1 = 5940$ m/sec, $C_2 = 3220$ m/sec, $C_R = 2980$ m/sec; $K_{IA} = 100$ MPa $\sqrt{\text{m}}$, $m = 2$, $V_l = 1600$ m/sec.

References

8.1. Mott, N.F. (1948) 'Fracture of metals: Theoretical considerations', *Engineering* **165**, 16–18.

8.2. Roberts, D.K. and Wells, A.A. (1954) 'The velocity of brittle fracture', *Engineering* **178**, 820–821.

8.3. Berry, J.P. (1960) 'Some kinetic considerations of the Griffith criterion for fracture', *Journal of the Mechanics and Physics of Solids* **8**, 194–216.

8.4. Dulaney, E.N. and Brace, W.F. (1960) 'Velocity behavior of a growing crack', *Journal of Applied Physics* **31**, 2233–2236.

8.5. Kanninen, M.F. (1968) 'An estimate of the limiting speed of a propagating ductile crack', *Journal of the Mechanics and Physics of Solids* **16**, 215–228.

8.6. Duffy, A.R., McClure, G.M., Eiber, R.J., and Maxey, W.A. (1969) 'Fracture design practices for pressure piping', in *Fracture – An Advanced Treatise*, Vol. V (ed. H. Liebowitz), Academic Press, pp. 159–232.

8.7. Gdoutos, E.E. (1990) *Fracture Mechanics Criteria and Applications*, Kluwer Academic Publishers, pp. 234–237.

8.8. Broberg, K.B. (1960) 'The propagation of a brittle crack', *Arkiv for Fysik* **18**, 159–192.

8.9. Baker, B.R. (1962) 'Dynamic stresses created by a moving crack', *Journal of Applied Mechanics, Trans. ASME* **29**, 449–458.

8.10. Rose, L.R.F. (1976) 'An approximate (Wiener-Hopf) kernel for dynamic crack problems in linear elasticity and viscoelasticity', *Proceedings of the Royal Society of London* **A349**, 497–521.

8.11. Sih, G.C. (1970) 'Dynamic aspects of crack propagation', in *Inelastic Behavior of Solids* (eds. M.F. Kanninen, W.F. Adler, A.R. Rosenfield and R.I. Jaffee), McGraw-Hill, pp. 607–639.

8.12. Sih, G.C. (ed.) (1977) 'Mechanics of Fracture, Vol. 4, Elastodynamic Crack Problems', Noordhoff Int. Publ., The Netherlands, pp. XXXIX-XLIV.

8.13. Irwin, G.R., Dally, J.W., Kobayashi, T., Fourney, W.F., Etheridge, M.J., and Rossmanith, H.P. (1979) 'On the determination of the $\dot{a} - K$ relationship for birefringent polymers', *Experimental Mechanics* **19**, 121–128.

8.14. Dally, J.W. (1979) 'Dynamic photoelastic studies of fracture', *Experimental Mechanics* **19**, 349–361.

8.15. Rosakis, A.J. and Zehnder, A.T. (1985) 'On the dynamic fracture of structural metals', *International Journal of Fracture* **27**, 169–186.

8.16. Bluhm, J.I. (1969) 'Fracture arrest', in *Fracture – An Advanced Treatise* (ed. H. Liebowitz), Academic Press, Vol. 5, pp. 1–63.

8.17. Wells, A. and Post, D. (1958) 'The dynamic stress distribution surrounding a running crack – A photoelastic analysis', *Proceedings of the Society for Experimental Stress Analysis* **16**, 69–92.

8.18. Bradley, W.B. and Kobayashi, A.S. (1970) 'An investigation of propagating cracks by dynamic photoelasticity', *Experimental Mechanics* **10**, 106–113.

8.19. Kobayashi, A.S. and Chan, C.F. (1976) 'A dynamic photoelastic analysis of dynamic-tear-test specimen', *Experimental Mechanics* **16**, 176–181.

8.20. Kobayashi, A.S. and Mall, S. (1978) 'Dynamic fracture toughness of Homalite 100', *Experimental Mechanics* **18**, 11–18.

8.21. Rossmanith, H.P. and Irwin, G.R. (1979) 'Analysis of dynamic isochromatic crack-tip stress patterns', *University of Maryland Report.*

8.22. Dally, J.W. and Kobayashi, T. (1979) 'Crack arrest in duplex specimens', *International Journal of Solids and Structures* **14**, 121–129.

8.23. Katsamanis, F., Raftopoulos, D., and Theocaris, P.S. (1977) 'Static and dynamic stress intensity factors by the method of transmitted caustics', *Journal of Engineering Materials and Technology* **99**, 105–109.

8.24. Theocaris, P.S. and Katsamanis, F. (1978) 'Response of cracks to impact by caustics', *Engineering Fracture Mechanics* **10**, 197–210.

8.25. Theocaris, P.S. and Papadopoulos, G. (1980) 'Elastodynamic forms of caustics for running cracks under constant velocity', *Engineering Fracture Mechanics* **13**, 683–698.

8.26. Rosakis, A.J. (1982) 'Experimental determination of the fracture initiation and dynamic crack propagation resistance of structural steels by the optical method of caustics', Ph.D. Thesis, Brown University.

8.27. Rosakis, A.J. and Freund, L.B. (1981) 'The effect of crack-tip plasticity on the determination of dynamic stress-intensity factors by the optical method of caustics', *Journal of Applied Mechanics, Trans. ASME* **48**, 302–308.

8.28. Beinert, J. and Kalthoff, J.F. (1981) 'Experimental determination of dynamic stress intensity factors by shadow patterns', in *Mechanics of Fracture*, Vol. 7 (ed. G.C. Sih), Martinus Nijhoff Publ., pp. 281–330.

Chapter 9

Fatigue and Environment-Assisted Fracture

9.1. Introduction

It was first realized in the middle of the nineteenth century that engineering components and structures often fail when subjected to repeated fluctuating loads whose magnitude is well below the critical load under monotonic loading. Early investigations were primarily concerned with axle and bridge failures which occurred at cyclic load levels less than half their corresponding monotonic load magnitudes. Failure due to repeated loading was called "fatigue failure".

Early studies in fatigue did not account for the details of the failure mode nor for the existence and growth of initial imperfections in the material, but tried to determine the fatigue life in terms of global measurable quantities, like stress, strain, mean stress, etc. The results of tests performed on small laboratory specimens subjected to repeated sinusoidally fluctuating loads were interpreted in diagrams expressing the stress amplitude versus the number of cycles to failure, known as $S-N$ curves. The fatigue life was found to increase with decreasing stress level, and below a certain stress level, known as the fatigue limit, failure did not occur for any number of loading cycles. The mean stress level, defined as the average of the maximum and minimum stress on the cyclic loading, plays an important role on fatigue life. It was found that the cyclic life decreased with increasing mean stress for a given maximum applied stress level. A number of empirical relationships for fatigue life, derived from the curve fitting of test data, have been proposed in the literature. The $S-N$ curve method and other available procedures based on gross specimen quantities lead to an inaccurate prediction of the fatigue life of engineering components due to the large scatter of experimental results as influenced by specimen size and geometry, material and the nature of the fluctuating load. Furthermore, the physical phenomena and mechanisms governing the fatigue process are completely ignored.

A better understanding of the fatigue phenomenon can be obtained by modelling the fatigue crack initiation and propagation processes. Crack initiation is analyzed at the microscopic level, while the continuum mechanics approach is used for crack propagation. The necessity for addressing these two processes separately arises from the inability of the current theory to bridge the gap between material damage that occurs at microscopic and macroscopic levels. It is generally accepted that, when a structure is subjected to repeated external load, energy is accumulated in the neighborhood of voids and microscopic defects which grow and coalesce, forming microscopic cracks. Eventually larger macroscopic cracks are formed. A macrocrack may be defined as one that is large enough to permit the application of the principles of homogeneous continuum mechanics. A macrocrack is usually referred to as a fatigue crack. The number of cycles required to initiate a fatigue crack is the fatigue crack initiation life N_i.

Following the initiation of a fatigue crack, slow stable crack propagation begins, until the crack reaches a critical size corresponding with the onset of global instability leading to catastrophic failure. Thus, the fatigue life of an engineering component may be considered to be composed of three stages: the initiation or stage I; the propagation or stage II; and the fracture or stage III, in which the crack growth rate increases rapidly as global instability is approached. The number of cycles required to propagate a fatigue crack until it reaches its critical size is the fatigue crack propagation life N_p. Depending on the material, the amplitude of the fluctuating load and environmental conditions, the fatigue crack initiation life may be a small or a substantial part of the total fatigue life.

It has long been recognized that the strength of solids depends greatly on the environment in which they are located. Under the influence of the environment, a body may behave in a brittle or a ductile manner, and its strength may increase or decrease. The failure of engineering components subjected to an aggressive environment may occur under applied stresses well below the strength of the material. Environmental conditions greatly influence the processes of local failure at the tip of a crack and cause subcritical crack growth and gradual failure of structural components. Failure under such conditions involves an interaction of complex chemical, mechanical and metallurgical processes. The basic subcritical crack growth mechanisms include stress corrosion cracking, hydrogen embrittlement and liquid embrittlement. An aggressive environment has a deleterious effect on the fatigue life of an engineering component. The corrosion-fatigue behavior of a structural system subjected to a fluctuating load in the presence of an environment is extremely complicated.

Because of their practical importance, the problems of fatigue and environment-assisted crack growth have been given much attention in the literature. A vast number of data are available, and a number of different theories have been proposed, based mainly on experimental data correlations. It is the purpose of the present chapter to present the phenomena of fatigue and environment-assisted crack growth within the framework of the macroscopic scale. The basic fatigue crack growth laws based on the stress intensity factor are described. A phenomenological analysis of the problem of stress corrosion cracking is also presented.

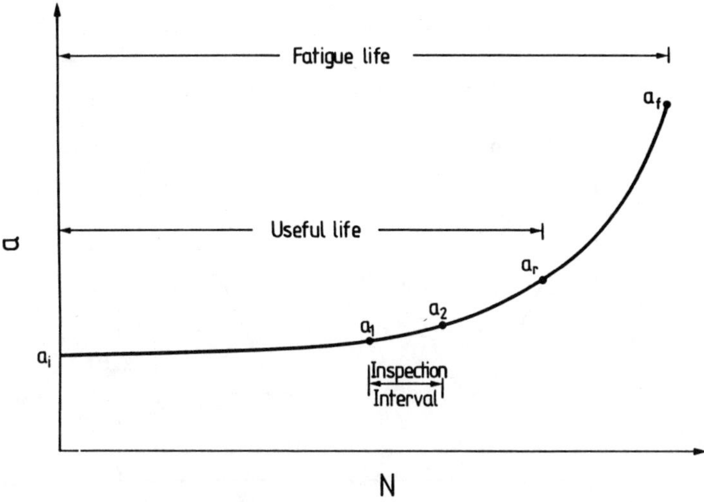

Fig. 9.1. Typical form of crack size versus number of cycles curve for constant amplitude loading.

9.2. Fatigue crack propagation laws

(a) General considerations

Fatigue crack propagation, referred to as stage II, represents a large portion of the fatigue life of many materials and engineering structures. Accurate prediction of the fatigue crack propagation stage is of utmost importance for determining the fatigue life. The main question of fatigue crack propagation may be stated in this form: *Determine the number of cycles N_c required for a crack to grow from a certain initial crack size a_0 to the maximum permissible crack size a_c, and the form of this increase $a = a(N)$, where the crack length a corresponds to N loading cycles.* Figure 9.1 presents a plot of a versus N which is required for predicting, say, the life of a particular engineering component. a_i represents the crack length that is big enough for fracture mechanics to apply, but too small for detection, while a_1 is the non-destructive inspection detection limit. The crack first grows slowly until the useful life of the component is reached. The crack then begins to propagate very rapidly, reaching a length a_f at which catastrophic failure begins.

Fatigue crack propagation data are obtained from precracked specimens subjected to fluctuating loads, and the change in crack length is recorded as a function of loading cycles. The crack length is plotted against the number of loading cycles for different load amplitudes. The stress intensity factor is used as a correlation parameter in analyzing the fatigue crack propagation results. The experimental results are usually plotted in a log (ΔK) versus log $(\mathrm{d}a/\mathrm{d}N)$ diagram, where ΔK is the amplitude of the stress intensity factor and $\mathrm{d}a/\mathrm{d}N$ is the crack propagation rate. The load is usually sinusoidal with constant amplitude and frequency (Figure 9.2). Two of the

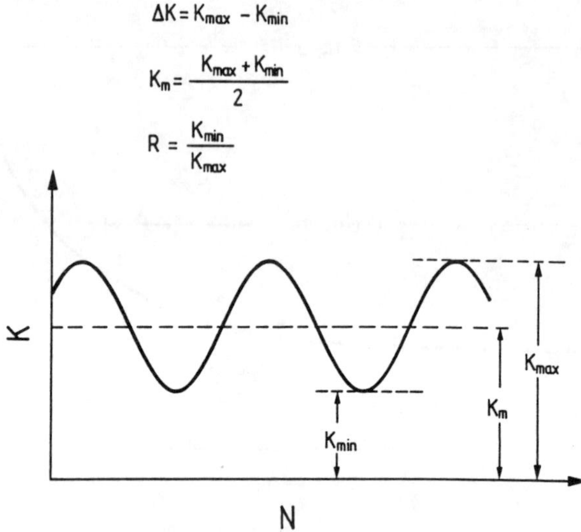

Fig. 9.2. A sinusoidal load with constant amplitude and frequency.

four parameters $K_{\max}$, $K_{\min}$, $\Delta K = K_{\max} - K_{\min}$ or $R = K_{\min}/K_{\max}$ are needed to define the stress intensity factor variation during a loading cycle.

A typical plot of the characteristic sigmoidal shape of a $\log(\Delta K) - \log(\mathrm{d}a/\mathrm{d}N)$ fatigue crack growth rate curve is shown in Figure 9.3. Three regions can be distinguished. In region I, $\mathrm{d}a/\mathrm{d}N$ diminishes rapidly to a vanishingly small level, and for some materials there is a threshold value of the stress intensity factor amplitude ΔK_{th} meaning that for $\Delta K < \Delta K_{\mathrm{th}}$ no crack propagation takes place. In region II there is a linear $\log(\Delta K) - \log(\mathrm{d}a/\mathrm{d}N)$ relation. Finally, in region III the crack growth rate curve rises and the maximum stress intensity factor $K_{\max}$ in the fatigue cycle becomes equal to the critical stress intensity factor K_c, leading to catastrophic failure. Experimental results indicate that the fatigue crack growth rate curve depends on the ratio R, and is shifted toward higher $\mathrm{d}a/\mathrm{d}N$ values as R increases.

(b) Crack propagation laws

A number of different quantitative continuum mechanics models of fatigue crack propagation have been proposed in the literature. All these models lead to relations based mainly on experimental data correlations. They relate $\mathrm{d}a/\mathrm{d}N$ to such variables as the external load, the crack length, the geometry and the material properties. Representative examples of such relations will be analyzed in this section.

One of the earlier mathematical models of fatigue crack propagation was proposed by Head [9.1]. He considered an infinite plate with a central crack of length $2a$ subjected to a sinusoidally applied stress $\pm\sigma$. Modelling the material elements ahead of the crack tip as rigid-plastic work-hardening tensile bars and the remaining

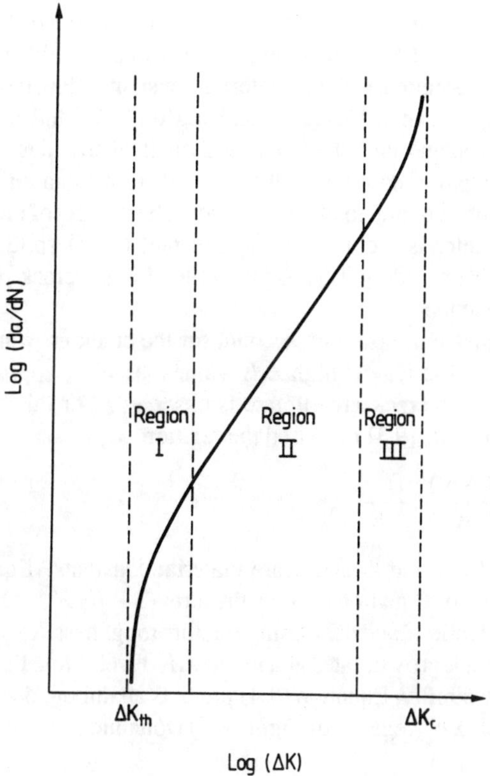

Fig. 9.3. Typical form of the fatigue crack growth rate curve.

elements as elastic bars, he arrived at the relation

$$\frac{\mathrm{d}a}{\mathrm{d}N} = C_1 \sigma^3 a^{3/2} \tag{9.1}$$

where C_1 is a constant which depends on the mechanical properties of the material and has to be determined experimentally. Equation (9.1) can be written in terms of the stress intensity factor as

$$\frac{\mathrm{d}a}{\mathrm{d}N} = C K_{\mathrm{I}}^3 \,. \tag{9.2}$$

One of the most widely used fatigue crack propagation laws is that proposed by Paris and Erdogan [9.2] and is usually referred to in the literature as the "Paris law". It has the form

$$\frac{\mathrm{d}a}{\mathrm{d}N} = C(\Delta K)^m \tag{9.3}$$

where $\Delta K = K_{\max} - K_{\min}$, with $K_{\max}$ and $K_{\min}$ referring to the maximum and minimum values of the stress intensity factor in the load cycle. The constants C

and m are determined empirically from a $\log(\Delta K) - \log(\mathrm{d}a/\mathrm{d}N)$ plot. The value of m is usually put equal to 4, resulting in the so-called "4th power law" while the coefficient C is assumed to be a material constant. Equation (9.3) represents a linear relationship between $\log(\Delta K)$ and $\log(\mathrm{d}a/\mathrm{d}N)$ and is used to describe the fatigue crack propagation behavior in region II of the diagram of Figure 9.3. Fatigue crack propagation data are well predicted from Equation (9.3) for specific geometrical configurations and loading conditions. The effect of mean stress, loading and specimen geometry is included in the constant C. Despite these drawbacks, Equation (9.3) has been widely used to predict the fatigue crack propagation life of engineering components.

Equation (9.3) does not, however, account for the crack growth characteristics at low and high levels of ΔK. At high ΔK values, as K_{max} approaches the critical level K_c, an increase in crack growth rate is observed. For this case (region III of Figure 9.3) Forman *et al.* [9.3] proposed the relation

$$\frac{\mathrm{d}a}{\mathrm{d}N} = \frac{C(\Delta K)^n}{(1-R)\,K_c - \Delta K} \tag{9.4}$$

where $R = K_{\mathrm{min}}/K_{\mathrm{max}}$ and C and n are material constants. Equation (9.4) arose from the modification of Equation (9.3) by the term $(1-R)\,K_c - \Delta K$; this decreases with increasing load ratio R and decreasing fracture toughness K_c, both of which give rise to increasing crack growth rates at a given ΔK level. Note that for $K_{\mathrm{max}} = K_c$, corresponding to instability, Equation (9.4) predicts an unbounded value of $\mathrm{d}a/\mathrm{d}N$.

For low values of ΔK (region I of Figure 8.3) Donahue *et al.* [9.4] have suggested the relation

$$\frac{\mathrm{d}a}{\mathrm{d}N} = K(\Delta K - \Delta K_{\mathrm{th}})^m \tag{9.5}$$

where ΔK_{th} denotes the threshold value of ΔK. According to Klesnil and Lucas [9.5], ΔK_{th} is given by

$$\Delta K_{\mathrm{th}} = (1-R)^\gamma\, \Delta K_{th(0)} \tag{9.6}$$

where $\Delta K_{th(0)}$ is the threshold value at $R = 0$ and γ is a material parameter.

A generalized fatigue crack propagation law which can describe the sigmoidal response exhibited by the data of Figure 9.3 has been suggested by Erdogan and Ratwani [9.6], and has the form

$$\frac{\mathrm{d}a}{\mathrm{d}N} = \frac{C(1+\beta)^m\,(\Delta K - \Delta K_{\mathrm{th}})^n}{K_c - (1+\beta)\,\Delta K} \tag{9.7}$$

where $C,\ m,\ n$ are empirical material constants, and

$$\beta = \frac{K_{\mathrm{max}} + K_{\mathrm{min}}}{K_{\mathrm{max}} - K_{\mathrm{min}}}\,. \tag{9.8}$$

The factor $(1+\beta)^m$ has been introduced to account for the effect of the mean stress level on fatigue crack propagation, while the factor $[K_c - (1+\beta)\,\Delta K]$ takes care of the experimental data at high stress levels. Finally, the factor $(\Delta K - \Delta K_{\text{th}})^n$ accounts for the experimental data at low stress levels and the existence of a threshold value ΔK_{th} of ΔK at which no crack propagation occurs. By proper choice of constants, Equation (9.8) can be made to fit the experimental data over a range from 10^{-8} to 10^{-2} in/cycle.

Attempts have been made to apply the J-integral concept to elastic-plastic fatigue crack propagation. A relation of the form

$$\frac{\mathrm{d}a}{\mathrm{d}N} = C(\Delta J)^m \tag{9.9}$$

in complete analogy to the Paris law has been suggested [9.7, 9.8].

9.3. Fatigue life calculations

When a structural component is subjected to fatigue loading, a dominant crack reaches a critical size under the peak load during the last cycle leading to catastrophic failure. The basic objective of the fatigue crack propagation analysis is the determination of the crack size, a, as a function of the number of cycles, N (Figure 9.1). Thus, the fatigue crack propagation life N_p is obtained. When the type of applied load and the expression of the stress intensity factor are known, application of one of the foregoing fatigue laws enables a realistic calculation of the fatigue crack propagation life of the component.

As an example, consider a plane fatigue crack of length $2a_0$ in a plate subjected to a uniform stress σ perpendicular to the plane of the crack. The stress intensity factor K is given by

$$K = f(a)\,\sigma\,\sqrt{\pi a} \tag{9.10}$$

where $f(a)$ is a geometry dependent function.

Integrating the fatigue crack propagation law expressed by Equation (9.3), gives

$$N - N_0 = \int_{a_0}^{a} \frac{\mathrm{d}a}{C(\Delta K)^m} \tag{9.11}$$

where N_0 is the number of load cycles corresponding to the half crack length a_0. Introducing the stress intensity factor range ΔK, where K is given from Equation (9.10), into Equation (9.11) we obtain

$$N - N_0 = \int_{a_0}^{a} \frac{\mathrm{d}a}{C[f(a)\,\Delta\sigma\,\sqrt{\pi a}]^m}\,. \tag{9.12}$$

Assuming that the function $f(a)$ is equal to its initial value $f(a_0)$ so that

$$\Delta K = \Delta K_0\sqrt{\frac{a}{a_0}}\,, \quad \Delta K_0 = f(a_0)\,\Delta\sigma\,\sqrt{\pi a_0}\,. \tag{9.13}$$

Equation (9.12) gives

$$N - N_0 = \frac{2a_0}{(m-2)\,C(\Delta K_0)^m}\left[1 - \left(\frac{a_0}{a}\right)^{m/2-1}\right] \quad \text{for} \quad m \neq 2\,. \tag{9.14}$$

Unstable crack propagation occurs when

$$K_{\max} = f(a)\,\sigma_{\max}\,\sqrt{\pi a} \tag{9.15}$$

from which the critical crack length a_c is obtained. Then Equation (9.14) for $a = a_c$ gives the fatigue crack propagation life $N_p = N_c - N_0$.

Usually, however, $f(a)$ varies with the crack length a and the integration of Equation (9.12) cannot be performed directly, but only through the use of numerical methods.

9.4. Variable amplitude loading

The fatigue crack propagation results discussed so far have been concerned with constant amplitude load fluctuation. Although this type of loading occurs frequently in practice, the majority of engineering structures are subjected to complex fluctuating loading. Unlike the case of constant cyclic load where ΔK increases gradually with increasing crack length, abrupt changes take place in ΔK due to changes in applied load. Thus, there occur load interaction effects which greatly influence the fatigue crack propagation behavior.

It was first recognized empirically in the early 1960s that the application of a tensile overload in a constant amplitude cyclic load leads to crack retardation following the overload; that is, the crack growth rate is smaller than it would have been under constant amplitude loading. This effect is shown schematically in Figure 9.4. The amount of crack retardation is dramatically decreased when a tensile-compressive overload follows a constant amplitude cyclic load.

An explanation of the crack retardation phenomenon may be obtained by examining the behavior of the plastic zone ahead of the crack tip. The overload has left a large plastic zone behind. The elastic material surrounding this plastic zone after unloading acts like a clamp on this zone causing compressive residual stresses. As the crack propagates into the plastic zone, the residual compressive stresses tend to close the crack. Hence the crack will propagate at a decreasing rate into the zone of residual stresses. When these stresses are overcome and the crack is opened again, subsequent fluctuating loading causes crack growth.

Due to the crack retardation phenomenon, the determination of the fatigue life under a variable amplitude loading by simply summing the fatigue lives of the various constant amplitude loads in the loading history leads to conservative predictions. Of the various methods proposed for this reason, we will briefly present the root-mean-square model and the models based on crack retardation and crack closure.

The root-mean-square model proposed by Barsom [9.9] applies to variable amplitude narrow-band random loading spectra. It is assumed that the average fatigue

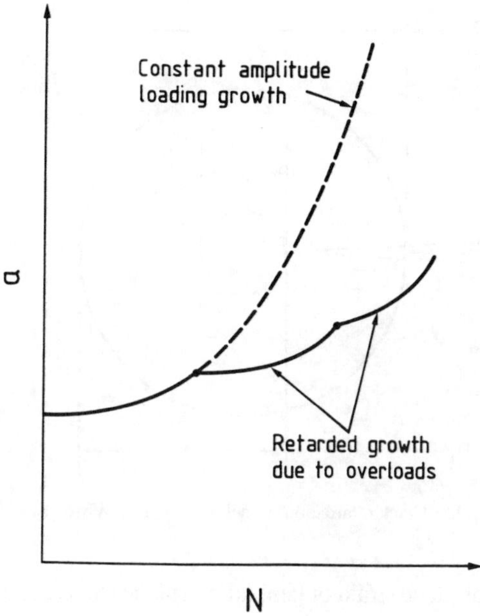

Fig. 9.4. Typical form of crack length versus number of cycles curve for constant amplitude loading and constant amplitude plus overloading.

crack growth rate under a variable amplitude random loading fluctuation is approximately equal to the rate of fatigue crack growth under constant amplitude cyclic load; this is equal to the root-mean-square of the variable amplitude loading. Thus, the fatigue crack propagation laws presented in Section 9.2 can be equally applied for a variable amplitude random loading when ΔK is replaced by the root-mean-square value of the stress intensity factor ΔK_{rms} given by

$$\Delta K_{\mathrm{rms}} = \sqrt{\frac{\Sigma (\Delta K_i)^2 \, n_i}{\Sigma \, n_i}} \tag{9.16}$$

where n_i is the number of loading amplitudes with a stress intensity factor range of ΔK_i.

The model based on crack retardation proposed by Wheeler [9.10] assumes that, after a peak load, there is a load interaction effect when the crack-tip plastic zones for the subsequent loads are smaller than the plastic zone due to the peak load. Consider that at a crack length a_0 an overload stress σ_0 creates a crack-tip plastic zone of length c_{po} which according to Equation (3.6) or (3.7) is given by

$$c_{po} = \frac{1}{A} \frac{\sigma_0^2 a_0}{\sigma_Y^2} \tag{9.17}$$

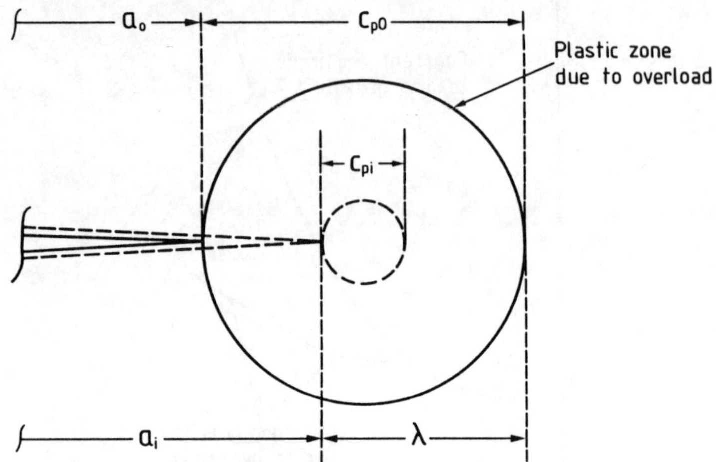

Fig. 9.5. Crack retardation model proposed by Wheeler [9.10].

where $A = 1$ or 3 for plane strain or generalized plane stress conditions, respectively (Figure 9.5). When the crack has propagated to a length a_i a stress σ_i will produce a plastic zone of length c_{pi} given by

$$c_{pi} = \frac{1}{A}\frac{\sigma_i^2 a_i}{\sigma_Y^2}\,. \tag{9.18}$$

The plastic zone due to the stress σ_i is included inside the plastic zone due to the overload. Then a retardation factor ϕ is introduced given by

$$\phi = \left(\frac{c_{pi}}{\lambda}\right)^m, \quad \text{for} \quad a_i + c_{pi} < a_0 + c_{po} \tag{9.19}$$

where $\lambda = a_0 + c_{po} - a_i$ and m is an empirical parameter. Then, the crack growth increment for $a_i + c_{pi} < a_0 + c_{po}$ is given by

$$\left(\frac{\mathrm{d}a}{\mathrm{d}N}\right)_R = \phi\left(\frac{\mathrm{d}a}{\mathrm{d}N}\right) \tag{9.20}$$

where $\mathrm{d}a/\mathrm{d}N$ is the constant amplitude crack growth rate corresponding to the stress intensity factor range ΔK_i of load cycle i. When $a_i + c_{pi} > a_0 + c_{po}$ the crack has propagated through the overload plastic zone and the retardation factor is $\phi = 1$.

Elber [9.11, 9.12] introduced a model based on crack closure. It is based on the observation that the faces of fatigue cracks subjected to zero-tension loading close during unloading, and compressive residual stresses act on the crack faces at zero load. Thus, the crack closes at a tensile rather than zero or compressive load. An effective stress intensity factor range is defined by

$$(\Delta K)_{\text{eff}} = K_{\max} - K_{\text{op}} \tag{9.21}$$

where K_{op} corresponds to the point at which the crack is fully open. Then, one of the crack propagation laws of Section 9.2 can be employed. Using, for example, the Paris law (Equation (9.3)) we can state this in the form

$$\frac{da}{dN} = C(U\Delta K)^m \tag{9.22}$$

where

$$U = \frac{K_{max} - K_{op}}{K_{max} - K_{min}}, \quad \Delta K = K_{max} - K_{min}. \tag{9.23}$$

A number of empirical relations have been proposed for the determination of U. Elber suggested that U can be given in the form

$$U = 0.5 + 0.4R, \tag{9.24}$$

where

$$R = \frac{K_{min}}{K_{max}} \quad \text{for} \quad -0.1 \le R \le 0.7. \tag{9.25}$$

Schijve [9.13] proposed the relation

$$U = 0.55 + 0.33R + 0.12R^2 \tag{9.26}$$

which extends the previous equation to negative R ratios in the range $-1.0 < R < 0.54$. Relations (9.25) and (9.26) were obtained for a 2024–T3 aluminum. De Koning [9.14] suggested a method of determining K_{op}.

9.5. Environment-assisted fracture

It has long been recognized that failure of engineering components subjected to an aggressive environment may occur under applied stresses well below the strength of the material. Failure under such conditions involves an interaction of complex chemical, mechanical and metallurgical processes. The basic subcritical crack growth mechanisms include stress corrosion cracking, hydrogen embrittlement and liquid metal embrittlement. Although a vast number of experiments have been performed and a number of theories have been proposed, a general mechanism for environment-assisted cracking is still lacking. It is not the intent of this section to cover the material in depth, but rather to present a brief account of the phenomenological aspect of the problem of stress corrosion cracking.

The experimental methods for the evaluation of the stress corrosion susceptibility of a material under given environmental conditions fall into two categories: the time-to-failure tests and the growth rate tests. Both kinds of tests are performed on fatigue precracked specimens. The most widely used specimens are the cantilever beam specimen subjected to constant load, and the wedge-loaded specimen subjected to

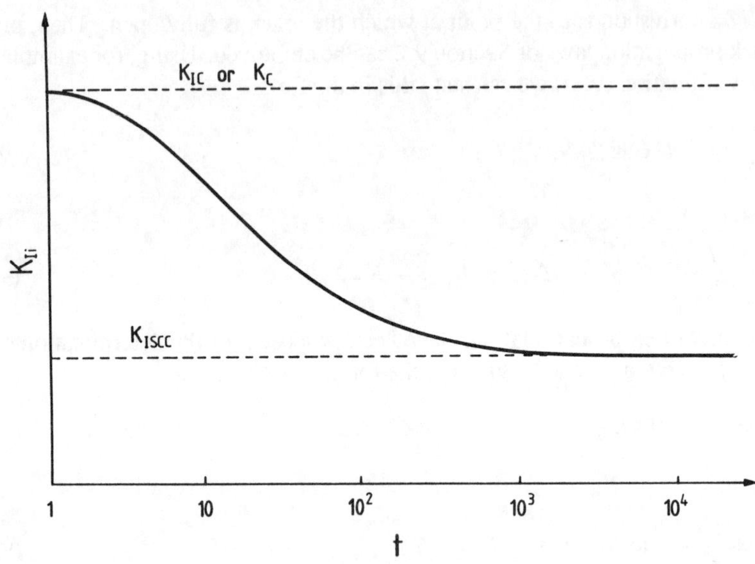

Fig. 9.6. Initial stress intensity factor, $K_{\mathrm{I}i}$, versus time to failure, t, under environment assisted fracture.

constant displacement. The stress intensity factors for the specimens are calculated by appropriate calibration formulas.

In the time-to-failure tests the specimens are loaded to various initial stress intensity factor levels $K_{\mathrm{I}i}$ and the time required to failure is recorded. The test results are represented in a $K_{\mathrm{I}i}$, versus time t diagram, a representative form of which is shown in Figure 9.6. Observe that, as $K_{\mathrm{I}i}$ decreases, the time to failure increases. The maximum value of $K_{\mathrm{I}i}$ is equal to K_{Ic} or K_c, where K_{Ic} is the plane strain fracture toughness and K_c is the fracture toughness at thicknesses smaller than the critical thickness for which plane strain conditions apply. A threshold stress intensity factor K_{ISCC} is obtained, below which there is no crack growth. It is generally accepted that K_{ISCC} is a unique property of the material-environment system. The time required for failure can be divided into the incubation time (the time interval during which the initial crack does not grow) and the time of subcritical crack growth. The incubation time depends on the material, environment and $K_{\mathrm{I}i}$, while the time of subcritical crack growth depends on the type of load, the specimen geometry and the kinetics of crack growth caused by the interaction of material and environment.

In the crack growth rate method for the study of stress corrosion cracking, the rate of crack growth per unit time, $\mathrm{d}a/\mathrm{d}t$, is measured as a function of the instantaneous stress intensity factor, K_{I}. Figure 9.7 shows a typical form of the curve $\log(\mathrm{d}a/\mathrm{d}t) - K_{\mathrm{I}}$. This can be divided into three regions. In regions I and III the rate of crack growth, $\mathrm{d}a/\mathrm{d}t$, depends strongly on the stress intensity factor, K_{I}, while in region II $\mathrm{d}a/\mathrm{d}t$ is almost independent of K_{I}. This behavior in region II indicates that crack growth is not of a mechanical nature, but it is caused by chemical, metallurgical and other

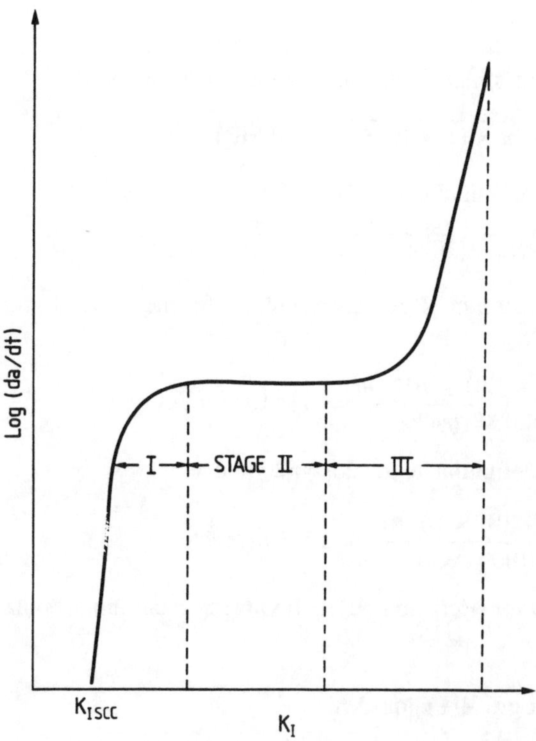

Fig. 9.7. Logarithm of subcritical crack growth rate, $\log(da/dt)$, versus stress intensity factor, K_{I}, under environment assisted cracking.

processes occurring at the crack tip. Note that in region I the threshold stress intensity factor corresponds to K_{ISCC}.

Examples

Example 9.1.

A large plate contains a crack of length $2a_0$ and is subjected to a constant-amplitude tensile cyclic stress normal to the crack which varies between 100 MPa and 200 MPa. The following data were obtained: for $2a_0 = 2$ mm it was found that $N = 20,000$ cycles were required to grow the crack to $2a_f = 2.2$ mm, while for $2a = 20$ mm it was found that $N = 1000$ cycles were required to grow the crack to $2a_f = 22$ mm. The critical stress intensity factor is $K_c = 60$ MPa $\sqrt{m}$. Determine the constants in the Paris (Equation (9.3)) and Formam (Equation (9.4)) equations.

Solution: The stress intensity factor range ΔK is calculated as

$$\Delta K = \Delta\sigma\,\sqrt{\pi a_0}\,. \tag{1}$$

For the crack of initial length $2a_0 = 2$ mm we have

$$\Delta K = 100\,\sqrt{\pi \times (1 \times 10^{-3})} = 5.60\ \text{MPa}\,\sqrt{\text{m}} \tag{2}$$

and for the crack of initial length $2a_0 = 20$ mm we have

$$\Delta K = 100\,\sqrt{\pi \times (10 \times 10^{-3})} = 17.72\ \text{MPa}\,\sqrt{\text{m}}\,. \tag{3}$$

The crack growth rate $\mathrm{d}a/\mathrm{d}N$ is calculated for the crack of initial length $2a_0 =$ 2 mm as

$$\frac{\mathrm{d}a}{\mathrm{d}N} = \frac{(1.1 - 1.0) \times 10^{-3}\ \text{m}}{20000\ \text{cycle}} = 5 \times 10^{-9}\ m/\text{cycle} \tag{4}$$

and for the crack of initial length $2a_0 = 20$ mm as

$$\frac{\mathrm{d}a}{\mathrm{d}N} = \frac{(11 - 10) \times 10^{-3}\ \text{m}}{1000\ \text{cycle}} = 10^{-6}\ m/\text{cycle}\,. \tag{5}$$

(a) *Paris equation* (Equation (9.3)). Taking the logarithm of both sides of Equation (9.3) we obtain

$$\log\frac{\mathrm{d}a}{\mathrm{d}N} = \log C + m\ \log \Delta K\,. \tag{6}$$

Introducing into Equation (6) the data of our problem we obtain the following two equations for the determination of the two unknowns C and m

$$\log(5 \times 10^{-9}) = \log C + m\ \log 5.6$$

$$\log(10^{-6}) = \log C + m\ \log 17.72 \tag{7}$$

or

$$-8.30 = \log C + 0.748\,m$$

$$-6.00 = \log C + 1.248\,m\,. \tag{8}$$

From Equation (7) and (8) we obtain

$$C = 1.82 \times 10^{-12}\,MN^{-4.6}\,m^{7.9}/\text{cycle}\,,\quad m = 4.6\,. \tag{9}$$

(b) *Forman equation* (Equation (9.4)). Taking the logarithm of both sides of Equation (9.4) we obtain

$$\log\left[[(1 - R)\,K_c - \Delta K]\,\frac{\mathrm{d}a}{\mathrm{d}N}\right] = \log C + n\ \log \Delta K\,. \tag{10}$$

Introducing into Equation (10) the data of our problem with $R = K_{min}/K_{max} = \sigma_{min}/\sigma_{max} = 100/200 = 0.5$ we obtain the following two equations for the determination of the two unknowns C and n

$$\log[(0.5 \times 60 - 5.6) \times 5 \times 10^{-9}] = \log C + n \log 5.6$$

$$\log[(0.5 \times 60 - 17.72) \times 10^{-6}] = \log C + n \log 17.72 \tag{11}$$

or

$$-6.914 = \log C + 0.748\, n$$

$$-4.911 = \log C + 1.248\, n\,. \tag{12}$$

From Equation (12) we obtain

$$C = 1.229\ MN^{-3.3}\, m^{5.95}/\text{cycle}\,, \quad n = 4.006\,. \tag{13}$$

Example 9.2.

A large thick plate contains a crack of length $2a_0 = 10$ mm and is subjected to a constant-amplitude tensile cyclic stress normal to the crack which varies between $\sigma_{min} = 100$ MPa and $\sigma_{max} = 200$ MPa. The critical stress intensity factor is $K_{Ic} = 60$ MPa $\sqrt{m}$ and fatigue crack growth is governed by the equation

$$\frac{\mathrm{d}a}{\mathrm{d}N} = 0.42 \times 10^{-11} (\Delta K)^3 \tag{1}$$

where $\mathrm{d}a/\mathrm{d}N$ is expressed in m/cycle and ΔK in MPa $\sqrt{m}$. Plot a curve of crack growth, a, versus number of cycles, N, up to the point of crack instability.

If a lifetime of 10^6 cycles is required for the plate, discuss the options the designer has for an improved lifetime.

Solution: The critical crack length a_c at instability is calculated from equation

$$K_I = K_{IC} \tag{2}$$

which becomes

$$200\sqrt{\pi a_c} = 60 \tag{3}$$

and gives

$$a_c = 28.65 \text{ mm}\,. \tag{4}$$

TABLE 1. Fatigue crack growth calculations

a_0(mm)	a(mm)	ΔK_0(MPa $\sqrt{m}$)	ΔN(cycles)	N(cycles)
5	7	12.53	187,412	187,412
7	9	14.83	120,696	308,108
9	11	16.81	86,056	394,164
11	13	18.59	65,339	459,503
13	15	20.21	51,791	511,294
15	17	21.71	42,358	553,652
17	19	23.11	35,480	589,132
19	21	24.43	30,282	619,414
21	23	25.69	26,241	645,655
23	25	26.88	23,027	668,682
25	27	28.02	20,417	689,099
27	28.65	29.12	15,209	704,308

The curve of crack growth, a, versus number of cycles N_c up to the crack length a_c is calculated by Equation (9.3) with $m = 3$, $C = 0.42 \times 10^{-11}\ MN^{-3}\ m^{5.5}$/cycle and $\Delta K_0 = 100\sqrt{\pi \times (5 \times 10^{-3})} = 12.53$ MPa $\sqrt{\text{m}}$. Thus, for the crack to grow from its initial length $a_0 = 5$ mm to a length $a = 7$ mm the number of cycles N required is given by

$$N = \frac{2 \times (5 \times 10^{-3})}{1 \times (0.42 \times 10^{-11})\,(12.53)^3} \left[1 - \left(\frac{5}{7}\right)^{0.5}\right] = 187{,}412 \text{ cycles}\,.$$

In a similar manner the number of cycles required for a crack of length $a_0 = 7$ mm to grow to a length $a = 9$ mm is given by

$$N = \frac{2 \times (7 \times 10^{-3})}{1 \times (0.42 \times 10^{-11})\,[100\sqrt{\pi \times (7 \times 10^{-3})}]^3} \left[1 - \left(\frac{7}{9}\right)^{0.5}\right] = 120{,}696 \text{ cycles}\,.$$

The above procedure is repeated for crack growth of steps 2 mm. Results are shown in Table 1 and are plotted in Figure 9.8.

The total number of cycles required to propagate a crack from 5 mm to 28.65 mm is calculated as

$$N_c = \frac{2 \times (5 \times 10^{-3})}{1 \times (0.42 \times 10^{-11}) \times (12.53)^3} \left[1 - \left(\frac{5}{28.65}\right)^{0.5}\right] = 704{,}697 \text{ cycles}$$

which is close to the value 704,308 obtained previously (Table 1).

If a lifetime of 10^6 cycles is required for the plate the designer may make the following changes:

(a) Employ a different metal with higher K_{Ic}, so as to increase the critical crack length a_c at instability.

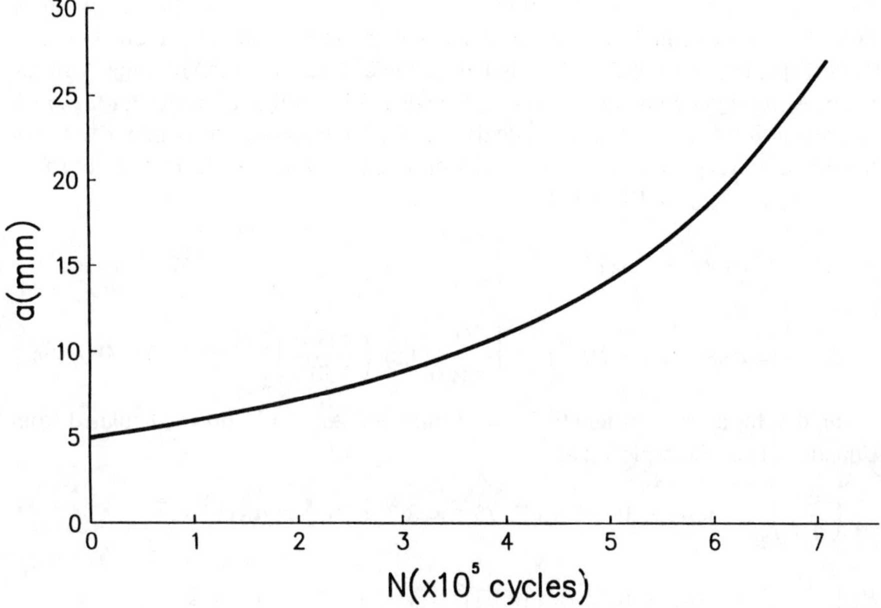

Fig. 9.8. Crack length versus number of cycles.

(b) Reduce the maximum value of the applied stress σ_{max}.

(c) Reduce the stress range $\Delta\sigma$.

(d) Improve the inspection so as to reduce the assumed initial crack length. If for example the initial crack length was reduced from $2a_0 = 10$ mm to $2a_0 = 6$ mm, the lifetime of the plate would be increased to 1,056,097 cycles, which is more than the required number of 10^6 cycles.

Example 9.3.

A plate of width $2b = 50$ mm contains a central crack of length $2a_0 = 10$ mm, and is subjected to a constant-amplitude tensile cyclic stress normal to the crack which varies between $\sigma_{min} = 100$ MPa and $\sigma_{max} = 200$ MPa. The crack growth is dictated by Equation (1) of Example 9.2. Calculate the number of cycles required for the crack to propagate to a length $2a = 20$ mm. The stress intensity factor K_I of the plate is given by

$$K_I = \sigma(\pi a)^{1/2} \left[\frac{2b}{\pi a} \tan\left(\frac{\pi a}{2b}\right)\right]^{1/2} . \tag{1}$$

Solution: In this case the function $f(a) = [(2b/\pi a)\ \tan(\pi a/2b)]^{1/2}$ of Equation (9.10) is not equal to its initial value $f(a_0)$, but depends on the crack length. Thus, Equation (9.14) cannot be applied for the calculation of the lifetime N of the plate. A step-by-step procedure is adapted where the number of cycles for the crack to grow at intervals of 2 mm is calculated. We have for the stress intensity factor fluctuations ΔK_{10} and ΔK_{12} for cracks of length 10 mm and 12 mm under stress $\Delta\sigma = \sigma_{\max} - \sigma_{\min} = 100$ MPa:

$$\Delta K_{10} = 100(\pi \times 5 \times 10^{-3})^{1/2} \left[\frac{50}{\pi \times 5} \tan\left(\frac{\pi \times 5}{50}\right)\right]^{1/2} = 12.75 \text{ MPa} \sqrt{m}$$

$$\Delta K_{12} = 100(\pi \times 6 \times 10^{-3})^{1/2} \left[\frac{50}{\pi \times 6} \tan\left(\frac{\pi \times 6}{50}\right)\right]^{1/2} = 14.07 \text{ MPa} \sqrt{m}\,.$$

$\mathrm{d}a/\mathrm{d}N$ for a crack of length $2a = 10$ mm and $2a = 12$ mm is calculated from Equation (1) of Example 9.2 as

$$\left(\frac{\mathrm{d}a}{\mathrm{d}N}\right)_{10} = 0.42 \times 10^{-11} \times (12.75)^3 = 8.7 \times 10^{-9}\ m/\text{cycle}$$

$$\left(\frac{\mathrm{d}a}{\mathrm{d}N}\right)_{12} = 0.42 \times 10^{-11} \times (14.07)^3 = 11.70 \times 10^{-9}\ m/\text{cycle}\,.$$

The mean value of $\mathrm{d}a/\mathrm{d}N$ is then calculated as

$$\frac{\mathrm{d}a}{\mathrm{d}N} = \frac{8.7 \times 10^{-9} + 11.70 \times 10^{-9}}{2} = 10.2 \times 10^{-9}\ m/\text{cycle}$$

and the number of cycles $\mathrm{d}N$ to propagate a crack from length $2a_0 = 10$ mm to $2a = 12$ mm is calculated as

$$\mathrm{d}N = \frac{1 \times 10^{-3}}{10.2 \times 10^{-9}} = 98{,}039 \text{ cycles}\,.$$

Crack growth calculations are arranged in Table 1.

Thus, the number of cycles required to propagate the crack of initial length $2a_0 =$ 10 mm to a length of $2a_c = 20$ mm is $N_c = 315{,}351$ cycles.

If the variation of $f(a)$ during crack growth were ignored, N_c would be calculated from Equation (9.14) as

$$N_c = \frac{2 \times (5 \times 10^{-3})}{1 \times (0.42 \times 10^{-11}) \times (12.75)^3} \left[1 - \left(\frac{5}{10}\right)^{0.5}\right] = 336.457 \text{ cycles}\,.$$

Example 9.4.

A double cantilever beam (DCB) (Figure 4.14) with height $2h = 60$ mm, thickness $B = 20$ mm and an initial crack of length $a_0 = 200$ mm is subjected to a tensile load which varies between $P_{\min} = 5$ kN and $P_{\max} = 10$ kN. The critical stress intensity factor of the beam is $K_{IC} = 100$ MPa $\sqrt{m}$ and crack growth is dictated by equation

TABLE 1. Fatigue crack growth calculations.

a(mm)	da(mm)	ΔK(MPa $\sqrt{m}$)	da/dN (m/cycle)	da/dN(m/cycle) (mean value)	dN cycles
5		12.75	8.7×10^{-9}		
	1			10.2×10^{-9}	98,039
6		14.07	11.70×10^{-9}		
	1			13.43×10^{-9}	74,460
7		15.34	15.16×10^{-9}		
	1			17.15×10^{-9}	58,309
8		16.58	19.14×10^{-9}		
	1			21.43×10^{-9}	46,664
9		17.81	23.73×10^{-9}		
	1			26.40×10^{-9}	37,879
10		19.06	29.08×10^{-9}		
					315,351

$$\frac{\mathrm{d}a}{\mathrm{d}N} = 5 \times 10^{-15}(\Delta K)^4 \tag{1}$$

where $\mathrm{d}a/\mathrm{d}N$ is expressed in m/cycle and ΔK in MPa $\sqrt{m}$. Calculate the critical crack length a_c for instability and the fatigue lifetime of the beam.

Solution: The critical crack length a_c at instability is calculated from equation

$$K_\mathrm{I} = K_\mathrm{Ic} \tag{2}$$

where the stress intensity factor K_I for the DCB is given by Equation (7) of Example 4.2.

We have for $P_{\max} = 10$ kN

$$K_\mathrm{I} = \sqrt{\frac{12}{(30 \times 10^{-3}\ m)^3}}\ \frac{(10 \times 10^{-3}\ MN)\, a_c}{20 \times 10^{-3}\ m} = 333 a_c\ \mathrm{MPa}/\sqrt{m}\,. \tag{3}$$

From Equations (2) and (3) with $K_\mathrm{Ic} = 100$ MPa $\sqrt{m}$ we obtain

$$a_c = 300\ \mathrm{mm}. \tag{4}$$

The fatigue life N_c of the beam is calculated from Equation (9.3) where

$$\Delta K = \sqrt{\frac{12}{h^3}}\ \frac{(\Delta P)\, a}{B}\,. \tag{5}$$

We have for $\Delta P = 5$ kN

$$\Delta K = \sqrt{\frac{12}{(30 \times 10^{-3})^3}}\ \frac{(5 \times 10^{-3})\, a}{20 \times 10^{-3}} = 166.7\, a\,. \tag{6}$$

Equation (9.3) with $C = 5 \times 10^{-15}$ and $m = 4$ becomes

$$\mathrm{d}N = \frac{\mathrm{d}a}{(5 \times 10^{-15})\,(166.7\,a)^4} \tag{7}$$

or

$$\mathrm{d}N = \frac{2.59 \times 10^5\,\mathrm{d}a}{a^4}\,. \tag{8}$$

Equation (8) gives

$$N_c = \frac{2.59 \times 10^5}{3}\left(\frac{1}{a_0^3} - \frac{1}{a_c^3}\right) \tag{9}$$

or

$$N_c = \frac{2.59 \times 10^5}{3}\left(\frac{1}{0.2^3} - \frac{1}{0.3^3}\right) = 76 \times 10^6 \text{ cycles}\,. \tag{10}$$

Example 9.5.

A large plate with an initial crack of length $2a_0$ is subjected to a series of cyclic stress amplitudes $\Delta\sigma_i$ $(i = 1, 2, \ldots, n)$ normal to the crack. Assume that the final crack length at instability $2a_f$ is the same for all stress amplitudes $\Delta\sigma_i$. If fatigue crack growth is governed by equation

$$\frac{\mathrm{d}a}{\mathrm{d}N} = C(\Delta K)^2 \tag{1}$$

show that

$$\sum_{i=1}^{n} \frac{N_i}{(N_i)_f} = 1 \tag{2}$$

where N_i is the number of cycles required to grow the crack from $2a_{i-1}$ to $2a_i$ and $(N_i)_f$ is the total number of cycles required to grow the crack from its initial length $2a_0$ to its final length $2a_f$ at instability.

Equation (2) is known as the Miner rule of fatigue crack growth under variable load amplitudes.

Solution: From Equation (1) we obtain for the number of cycles N_1 required to grow the crack from its initial length $2a_0$ to a length $2a_1$

$$\mathrm{d}N_1 = \frac{\mathrm{d}a}{C(\Delta\sigma_1\sqrt{\pi a})^2} \tag{3}$$

and by integration we have

$$N_1 = \frac{1}{\pi C(\Delta\sigma_1)^2} \int_{a_0}^{a_1} \frac{da}{a} = \frac{1}{\pi C(\Delta\sigma_1)^2} \ln\left(\frac{a_1}{a_0}\right) . \tag{4}$$

In a similar manner we obtain for the number of cycles $(N_1)_f$ required to grow the crack from its initial length $2a_0$ to the final length $2a_f$ at instability

$$(N_1)_f = \frac{1}{\pi C(\Delta\sigma_1)^2} \ln\left(\frac{a_f}{a_0}\right) . \tag{5}$$

From Equations (4) and (5) we obtain

$$\frac{N_1}{(N_1)_f} = \ln\left(\frac{a_1}{a_0}\right) / \ln\left(\frac{a_f}{a_0}\right) . \tag{6}$$

In a similar manner we obtain

$$\frac{N_2}{(N_2)_f} = \ln\left(\frac{a_2}{a_1}\right) / \ln\left(\frac{a_f}{a_0}\right) \tag{7}$$

and

$$\frac{N_i}{(N_i)_f} = \ln\left(\frac{a_i}{a_{i-1}}\right) / \ln\left(\frac{a_f}{a_0}\right) \tag{8}$$

and so on.

From Equations (6) to (8) we obtain

$$\begin{aligned} &\frac{N_1}{(N_1)_f} + \frac{N_2}{(N_2)_f} + \ldots + \frac{N_i}{(N_i)_f} + \ldots \frac{N_n}{(N_n)_f} = \\ &\left[\ln\left(\frac{a_1}{a_0}\right) + \ln\left(\frac{a_2}{a_1}\right) + \ldots + \ln\left(\frac{a_i}{a_{i-1}}\right) + \ldots + \ln\left(\frac{a_n}{a_{n-1}}\right)\right] / \ln\left(\frac{a_f}{a_0}\right) \end{aligned} \tag{9}$$

or

$$\sum_{i=1}^{n} \frac{N_i}{(N_i)_f} = \ln\left[\left(\frac{a_1}{a_0}\right)\left(\frac{a_2}{a_1}\right)\ldots\left(\frac{a_n}{a_{n-1}}\right)\ldots\left(\frac{a_n}{a_{n-1}}\right)\right] / \ln\left(\frac{a_f}{a_0}\right) \tag{10}$$

or for $a_n = a_f$

$$\sum_{i=1}^{n} \frac{N_i}{(N_i)_f} = 1 \tag{11}$$

which is Equation (2).

Example 9.6.

A large thick plate contains a crack of length $2a_0 = 20$ mm and is subjected to a series of triangular stress sequences normal to the crack as shown in Figure 9.9. The stress varies between $\sigma_{\min} = 0$ and $\sigma_{\max} = 200$ MPa and it takes 1000 cycles in the triangle to reach $\sigma_{\max}$. The critical stress intensity factor is $K_{Ic} = 100$ MPa $\sqrt{m}$ and fatigue crack growth is governed by equation

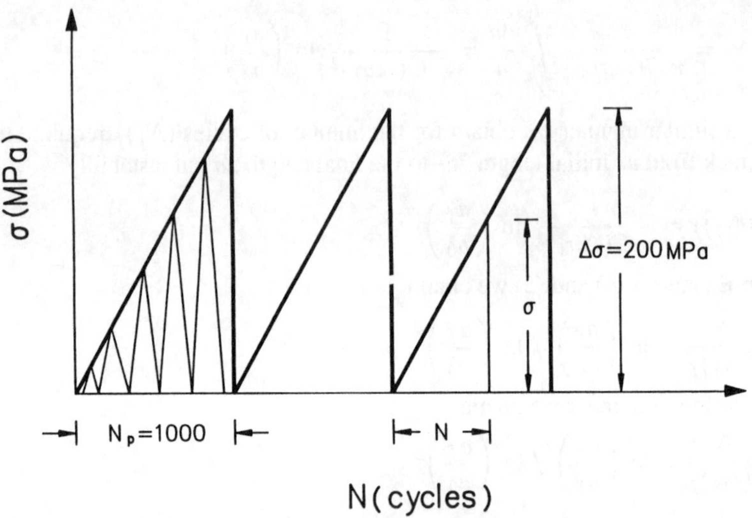

Fig. 9.9. Triangular cyclic stress profile.

$$\frac{\mathrm{d}a}{\mathrm{d}N} = 10^{-12}(\Delta K)^3 \tag{1}$$

where $\mathrm{d}a/\mathrm{d}N$ is expressed in m/cycle and ΔK in MPa $\sqrt{m}$.

Calculate the number of triangular stress sequences $(N_c)_1$ required to grow the crack to instability. Compare $(N_c)_1$ to the number of cycles $(N_c)_2$ required to grow the crack to instability when the plate is subjected to a constant amplitude stress cycle between the stresses $\sigma_{\mathrm{min}} = 0$ and $\sigma_{\mathrm{max}} = 200$ MPa.

Solution: The root-mean-square value of the stress intensity factor range ΔK_{rms} is calculated from Equation (9.16). We have for the stress σ at N cycles in the triangular stress sequence

$$\sigma = \Delta\sigma \, \frac{N}{N_p} \tag{2}$$

so that

$$\Delta K = \Delta\sigma \, \sqrt{\pi a} \, \frac{N}{N_p} \, . \tag{3}$$

Since we have a large number of cycles ($N_p = 1000$) in each triangular stress sequence ΔK_{rms} can be calculated by integration as

$$(\Delta K_{\mathrm{rms}})^2 = \frac{1}{N_p} \int_0^{N_p} (\Delta K)^2 \, \mathrm{d}N \tag{4}$$

and substituting the value of ΔK from Equation (3) we obtain

$$(\Delta K_{\rm rms})^2 = \frac{(\Delta\sigma\sqrt{\pi a})^2}{N_p}\int_0^{N_p}\frac{N^2}{N_p^2}\,{\rm d}N \tag{5}$$

or

$$(\Delta K_{\rm rms})^2 = \frac{(\Delta\sigma)^2\,\pi a}{3}\,. \tag{6}$$

The critical crack length a_c at instability is calculated from equation

$$K_{\rm I} = K_{\rm Ic} \tag{7}$$

which becomes

$$200\sqrt{\pi a_c} = 100 \tag{8}$$

and gives

$$a_c = 79.6\ {\rm mm}\,. \tag{9}$$

Equation (1) with $\Delta K = \Delta K_{\rm rms}$ gives

$$(N_c)_1 = \int_{0.01}^{0.0796}\frac{{\rm d}a}{10^{-12}(200)^3\,(\pi a)^{1.5}\,(1/3)^{1.5}} \tag{10}$$

or

$$(N_c)_1 = 116,645\times 2(0.01^{-0.5} - 0.0796^{-0.5}) = 1.5\times 10^6\ \text{cycles}\,. \tag{11}$$

The number of cycles $(N_c)_2$ required to grow the crack to instability when the plate is subjected to a constant amplitude stress cycle between the stress $\sigma_{\rm min} = 0$ and $\sigma_{\rm max} = 200$ MPa is calculated from Equation (1) as

$$(N_c)_2 = \int_{0.01}^{0.0796}\frac{{\rm d}a}{10^{-12}(200)^3\,(\pi a)^{1.5}} \tag{12}$$

or

$$(N_c)_2 = 0.29\times 10^6\ \text{cycles} \tag{13}$$

which is equal to $(N_c)_1$ divided by $3^{1.5}$.

Problems

9.1. A large thick plate contains a crack of length $2a_0$ and is subjected to a constant-amplitude cyclic stress normal to the crack which varies between $\sigma_{\rm max}$ and $\sigma_{\rm min}$. Fatigue crack growth is governed by the equation

$$\frac{{\rm d}a}{{\rm d}N} = C[(\Delta K)^2 - (\Delta K_{\rm th})^2]$$

where C is a material parameter and ΔK_{th} is the threshold value of ΔK.

Show that the number of cycles N_c required to grow the crack to instability is given by

$$N_c = \frac{1}{C\pi(\Delta\sigma)^2} \ln \frac{(K_{IC}/\sigma_{\max})^2 - (\Delta K_{\text{th}}/\Delta\sigma)^2}{\pi a_0 - (\Delta K_{\text{th}}/\Delta\sigma)^2}$$

where K_{Ic} is the critical stress intensity factor and $\Delta\sigma = \sigma_{\max} - \sigma_{\min}$.

9.2. Show that, if $m = 4$ in Equation (9.3), the time it takes for a crack in a large plate to quadruple its length is equal to 1.5 times the time it takes to double its length.

9.3. A crack grows at a rate $(\mathrm{d}a/\mathrm{d}N)_1 = 8.84 \times 10^{-7}$ *m*/cycle when the stress intensity factor amplitude is $(\Delta K)_1 = 50$ MPa $\sqrt{m}$ and at a rate $(\mathrm{d}a/\mathrm{d}N)_2 = 4.13 \times 10^{-5}$ when $(\Delta K)_2 = 150$ MPa $\sqrt{\mathrm{m}}$. Determine the parameters C and m in Paris equation (Equation (9.3)).

9.4. A plate of width $2b = 40$ mm contains a center crack of length $2a_0$ and is subjected to a constant-amplitude tensile cyclic stress which varies between 100 MPa and 200 MPa. The following data were obtained: For $2a_0 = 2$ mm $N = 20,000$ cycles were required to grow the crack to $2a_f = 2.2$ mm, while for $2a_0 = 20$ mm $N = 1000$ cycles were needed to grow the crack to $2a_f = 22$ mm. The critical stress intensity factor is $K_{Ic} = 60$ MPa $\sqrt{m}$. Determine the constants in the Paris (Equation (9.3)) and Forman (Equation (9.4)) equations.

9.5. A large thick plate contains a crack and is subjected to a cyclic stress normal to the crack which varies between 0 and 400 MPa. Determine the maximum allowable crack length the plate can withstand when $K_{Ic} = 90$ MPa $\sqrt{m}$. If the initial crack had a length 6 mm calculate the number of loading cycles the plate can withstand, when $C = 10^{-20}\, MN^{-3}\, m^{5.5}$/cycle, $m = 3$.

9.6. A large thick plate contains a crack of length 8 mm and is subjected to a constant-amplitude tensile cyclic stress normal to the crack which varies between 200 MPa and 400 MPa. Plot a curve of crack growth versus number of cycles up to the point of crack instability. The material parameters are as follows: $K_{Ic} = 90$ MPa $\sqrt{\mathrm{m}}$, $C = 10^{-12}\, MN^{-4}\, m^7$/cycle, $m = 4$.

9.7. A large thick plate contains an edge crack of length 1 mm and is subjected to a constant-amplitude tensile cyclic stress normal to the crack which varies between 100 MPa and 300 MPa. Which of the following three materials gives the longest lifetime for the plate.

Material	K_{Ic}(MPa $\sqrt{m}$)	C	m
A	55	$5 \times 10^{-14}\, MN^{-4}\, m^{7}$/cycle	4.0
B	70	$2 \times 10^{-12}\, MN^{-3.5}\, m^{6.25}$/cycle	3.5
C	90	$3 \times 10^{-11}\, MN^{-3}\, m^{5.5}$/cycle	3.0

9.8. A large thick plate contains a crack of length $2a_0$ and is subjected to a constant-amplitude tensile cyclic stress normal to the crack with maximum stress $\sigma_{max} =$ 200 MPa and stress range $\Delta\sigma = \sigma_{max} - \sigma_{min}$. The fatigue crack growth is governed by equation

$$\frac{da}{dN} = 3.9 \times 10^{-14} (\Delta K)^{3.7}$$

where da/dN is expressed in m/cycle and ΔK in MPa $\sqrt{m}$. Determine the fatigue lifetime of the plate for the following conditions:

(a) $2a_0 = 2$ mm, $K_{Ic} = 24$ MPa $\sqrt{m}$, $\Delta\sigma = 50$ MPa

(b) $2a_0 = 2$ mm, $K_{Ic} = 24$ MPa $\sqrt{m}$, $\Delta\sigma = 100$ MPa

(c) $2a_0 = 2$ mm, $K_{Ic} = 44$ MPa $\sqrt{m}$, $\Delta\sigma = 50$ MPa

(d) $2a_0 = 1$ mm, $K_{Ic} = 24$ MPa $\sqrt{m}$, $\Delta\sigma = 50$ MPa

(e) $2a_0 = 1$ mm, $K_{Ic} = 44$ MPa $\sqrt{m}$, $\Delta\sigma = 100$ MPa.

Discuss the influence on fatigue lifetime of the initial crack length $2a_0$, the fracture toughness K_{Ic} and the stress range $\Delta\sigma$.

9.9. A large thick plate contains a crack of length $2a_0 = 2$ mm and is subjected to a constant-amplitude tensile cyclic stress normal to the crack which varies between 0 and 100 MPa. The fatigue crack growth is governed by equation of Problem 9.6. The material of the plate has been heat treated to the following conditions with the corresponding values of K_{Ic}.

Condition	A	B	C
K_{Ic}(MPa $\sqrt{m}$)	24	30	44

For each of the three material conditions, determine the fatigue lifetime N_f of the plate for crack lengths varying between $2a_0$ and $2a_f$, where the latter length corresponds to crack instability. Plot the variation of N_c versus $a(a_0 < a < a_f)$ for the three material conditions.

9.10. A large thick plate has two equal cracks of length $a = 1$ mm emanating from both sides of a hole with radius $R = 10$ mm. The plate is subjected to a constant-amplitude tensile cyclic stress normal to the crack which varies between 100 MPa and

200 MPa and the cracks grow at equal rates. Calculate the number of loading cycles the plate can withstand when $K_{Ic} = 90$ MPa $\sqrt{m}$, and crack growth is governed by equation

$$\frac{da}{dN} = 0.42 \times 10^{-14} (\Delta K)^3$$

where da/dN is expressed in m/cycle and ΔK in MPa $\sqrt{m}$. The stress intensity factor at the crack tip is given by

$$K_I = k\sigma \sqrt{\pi a}$$

where k depends on the a/R according to the following table.

a/R	0	0.1	0.2	0.3	0.4	0.5	0.6	0.8	1.0	1.5	2.0	3.0	∞
k	3.39	2.73	2.41	2.15	1.96	1.83	1.71	1.58	1.45	1.29	1.21	1.14	1.00

9.11. A thick plate of width $b = 50$ mm contains an edge crack of length 10 mm and is subjected to a constant-amplitude tensile cyclic stress normal to the crack which varies between 100 MPa and 200 MPa. Fatigue crack growth is dictated by equation of Problem 9.10 and $K_{Ic} = 55$ MPa $\sqrt{m}$. Calculate the number of cycles the plate can withstand.

9.12. A cylindrical pressure vessel has a radius $R = 1$ m and thickness $t = 40$ mm and contains a long axial surface crack of depth $a = 2$ mm. The vessel is subjected to internal pressure p which varies between 0 and 200 MPa. Calculate the number of loading cycles the vessel can withstand. $K_{Ic} = 55$ MPa $\sqrt{m}$ and fatigue crack growth is dictated by equation of Problem 9.10.

9.13. An ASTM three-point bend specimen of span $S = 30$ cm, width $W = 8$ cm, thickness $B = 4$ cm with a crack of length $a_0 = 3.5$ cm is subjected to a constant-amplitude cyclic load which varies between 30 kN and 50 kN. Calculate the number of cycles required to grow the crack to a length of $a_f = 4.0$ cm and plot a curve of crack length versus number of cycles.

9.14. An ASTM compact tension specimen of width $W = 12$ cm, thickness $B = 6$ cm with a crack of length $a_0 = 5$ cm is subjected to a constant-amplitude cyclic load which varies between 20 kN and 40 kN. Calculate the number of cycles required to grow the crack to a length $a_f = 5$ cm and plot a curve of crack length versus number of cycles.

9.15. A thick plate contains a semicircular surface crack and is subjected to a constant-amplitude tensile cyclic stress normal to the crack which varies between 200 MPa and 400 MPa. The fatigue crack growth is governed by Equation of Problem 9.8. Assume that during fatigue the crack grows in a self-similar manner. Calculate the number of cycles required to grow the crack from an initial radius a_0 to a final radius a_f for the following cases (a) $a_0 = 2$ mm, $a_f = 20$ mm, (b) $a_0 = 2$ mm, $a_f = 60$ mm, (c) $a_0 = 4$ mm, $a_f = 20$ mm and (d) $a_0 = 4$ mm, $a_f = 60$ mm. The stress intensity factor along the crack is given by

$$K_{\rm I} = \frac{2.24}{\pi} \sigma \sqrt{\pi a} .$$

9.16. A large thick plate contains a crack of length 2 mm and is subjected to a cyclic stress normal to the crack which varies between 0 and σ. Crack growth is governed by Paris' equation with $C = 5 \times 10^{-14}\ MN^{-4}\ m^7$/cycle and $m = 4$. The critical stress intensity factor is $K_{\rm Ic} = 100$ MPa $\sqrt{m}$. Calculate the value of σ if it is required that the plate withstand 10^8 cycles.

9.17. A large plate contains a central crack of length 2 mm and is subjected to a cyclic stress normal to the crack which varies between 100 MPa and 300 MPa. Crack growth is governed by Paris' equation with $C = 5 \times 10^{-15}\ MN^{-4}\ m^7$/cycle and $m = 4$. Calculate the crack length after 10^6 cycles.

9.18. A large plate contains a central crack of length 2 mm and is subjected to a cyclic wedge load which varies between 1 MN and 3 MN (Figure 2.8, Example 2.2). Crack growth is governed by Paris' equation with $C = 5 \times 10^{-15}\ MN^{-4}\ m^7$/cycle and $m = 4$. Calculate the crack length after 10^3 cycles.

9.19. A double cantilever beam (Figure 4.14) with height $2h = 40$ mm, thickness $B = 20$ mm and a crack of length $a = 100$ mm is subjected to a tensile cyclic load which varies between 0 and 8 kN. Crack growth is governed by Paris' equation with $C = 5 \times 10^{-15}\ MN^{-4}\ m^7$/cycle and $m = 4$. Calculate the crack length after 10^6 cycles.

9.20. A large plate contains an edge crack of length 1 mm and is subjected to a tensile cyclic stress which varies between 0 and σ. Determine the value of σ so that the crack does not grow in fatigue. $\Delta K_{\rm th} = 7$ MPa $\sqrt{m}$.

9.21. In Problem 9.18 the threshold value $\Delta K_{\rm th} = 5$ MPa $\sqrt{m}$. Calculate the number of cycles to crack arrest.

9.22. A large plate with a crack of length 10 mm was tested to stress corrosion in salt water. The time to failure for an applied stress normal to the crack $\sigma = 100, 200$ and 300 MPa was 2000, 50 and 2 hours. Estimate K_{ISSC} and calculate the amount of crack growth for each applied stress. $K_{\rm Ic} = 55$ MPa $\sqrt{m}$.

References

9.1. Head, A.K. (1953) 'The growth of fatigue crack', *Philosophical Magazine* **44**, 924–938.

9.2. Paris, P. and Erdogan, F. (1963) 'A critical analysis of crack propagation laws', *Journal of Basic Engineering, Trans. ASME* **85**, 528–534.

9.3. Forman, R.G., Kearney, V.E., and Engle, R.M. (1967) 'Numerical analysis of crack propagation in cyclic-loaded structures', *Journal of Basic Engineering, Trans. ASME* **89**, 459–464.

9.4. Donahue, R.J., Clark, H.M., Atanmo, P., Kumble, R., and McEvily, A.J. (1972) 'Crack opening displacement and the rate of fatigue crack growth', *International Journal of Fracture Mechanics* **8**, 209–219.

9.5. Klesnil, M. and Lucas, P. (1972) 'Effect of stress cycle asymmetry on fatigue crack growth', *Material Science Engineering* **9**, 231–240.

9.6. Erdogan, F. and Ratwani, M. (1970) 'Fatigue and fracture of cylindrical shells containing a circumferential crack', *International Journal of Fracture Mechanics* **6**, 379–392.

9.7. Dowling, N.E. and Begley, J.A. (1976) 'Fatigue crack growth during gross plasticity and the *J*-integral', in *Mechanics of Crack Growth, ASTM STP* **590**, American Society for Testing and Materials, Philadelphia, pp. 82–103.

9.8. Dowling, N.E. (1977) 'Crack growth during low-cycle fatigue of smooth axial specimens', in *Cyclic Stress-Strain and Plastic Deformation Aspects of Fatigue Crack Growth, ASTM STP* **637**, American Society for Testing and Materials, Philadelphia, pp. 97–121.

9.9. Barsom, J.M. (1973) 'Fatigue-crack growth under variable-amplitude loading in ASTM A514-B steel', in *Progress in Flaw Growth and Fracture Toughness Testing, ASTM STP* **536**, American Society for Testing and Materials, Philadelphia, pp. 147–167.

9.10. Wheeler, O.E. (1972) 'Spectrum loading and crack growth', *Journal of Basic Engineering, Trans. ASME* **94**, 181–186.

9.11. Elber, W. (1970) 'Fatigue crack closure under cyclic tension', *Engineering Fracture Mechanics* **2**, 37–45.

9.12. Elber, W. (1976) 'Equivalent constant-amplitude concept for crack growth under spectrum loading', *Fatigue Crack Growth Under Spectrum Loads, ASTM STP* **595**, American Society for Testing and Materials, Philadelphia, pp. 236–250.

9.13. Schijve, J. (1981) 'Some formulas for the crack opening stress level', *Engineering Fracture Mechanics* **14**, 461–465.

9.14. De Koning, A.U. (1981) 'A simple crack closure model for prediction of fatigue crack growth rates under variable-amplitude loading', in *Fatigue Mechanics – Thirteenth Conference, ASTM STP* **743**, American Society for Testing and Materials, Philadelphia, pp. 63–85.

Chapter 10

Micromechanics of Fracture

10.1. Introduction

The phenomenon of fracture of solids may be approached from different viewpoints depending on the scale of observation. At one extreme is the atomic approach where the phenomena take place in the material within distances of the order of 10^{-7} cm; at the other extreme is the continuum approach which models material behavior at distances greater than 10^{-2} cm. In the atomic approach, the problem is studied using the concepts of quantum mechanics; the continuum approach uses the theories of continuum mechanics and classical thermodynamics. A different approach should be used to explain the phenomena that take place in the material between these two extreme scales: movement of dislocations; formation of subgrain boundary precipitates, slip bands, grain inclusions and voids. The complex nature of the phenomenon of fracture prohibits a unified treatment of the problem, and the existing theories deal with the subject either from the microscopic or the macroscopic point of view. Attempts have been made to bridge the gap between these two approaches.

In the previous chapters the phenomenon of fracture was studied within the context of continuum mechanics and most attention was paid to the problem of separation of a structural member due to the propagation of a dominant macrocrack. It is the objective of this final chapter to outline the basic mechanisms of fracture which take place in metals at the microscopic level and at distances which vary from atomic spacing up to grain size. Study of the phenomenon of fracture at the microscopic scale is greatly facilitated by examining the surfaces of the fragments. The discipline that is concerned with this study is called fractography and its basic tool is the electron microscope. Since the penetrating power of an electron beam is limited, a replica of the fracture surface is made which allows transmission of the electron beam. Substantial progress was made by the introduction of the scanning electron microscope; the main advantage of this is that examination of the fracture surface is made directly, without the need of the replica. Understanding the mechanism of fracture at the microscopic level is of importance, not only because it provides a basis for predetermining fundamental fracture parameters, but because it provides a

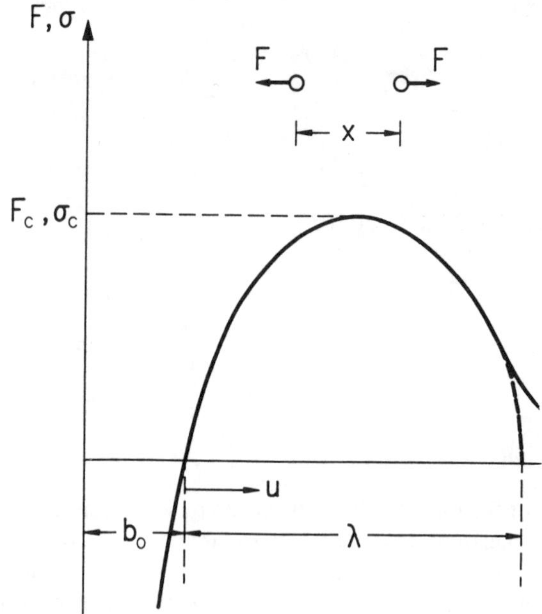

Fig. 10.1. Simplified force or stress versus interatomic distance relationship.

means of specifying the quantities which control the toughness of many materials by changing their microstructure.

This chapter starts with a simplified model for the estimation of the maximum theoretical cohesive strength of solids. The basic characteristic features of cleavage and fibrous fractures are presented, and some models for nucleation and growth of voids are reviewed. The chapter concludes with a brief description of the most widely used nondestructive testing methods for defect defection.

10.2. Cohesive strength of solids

An atomic approach to crack propagation in terms of intrinsic bond ruptures is important for understanding the crack nucleation process in perfectly brittle solids. In this respect a simplified model for the estimation of the maximum cohesive strength of solids is presented.

To begin, consider the force reactions between two free atoms as the interatomic distance changes. In all cases repulsive and attractive effects occur and their combination gives the form of curve shown in Figure 10.1. At equilibrium in the absence of applied forces let b_0 be the spacing between the atoms. Figure 10.1 shows that to decrease the distance b_0 by applying compression, a significant repulsive force must be overcome. To increase b_0 by applying tension requires force F; F can be increased until it reaches F_c and the bonds are broken. Further increase of the distance between the atoms requires a decreasing force.

The force-separation curve in the attractive region can be approximated by a sine curve having wavelength 2λ, namely

$$\begin{aligned} F &= F_c \sin \frac{\pi u}{\lambda}, \qquad 0 < u < \lambda \\ F &= 0 \qquad\qquad\quad u > \lambda \end{aligned} \tag{10.1}$$

where $u = x - b_0$ represents the displacement of an atom from its equilibrium position and λ is a range parameter of the bond.

If we define the stress as

$$\sigma = NF \tag{10.2}$$

where N is the number of atomic bonds intersecting a unit area, we can put Equation (10.1) in the form

$$\begin{aligned} \sigma &= \sigma_c \sin \frac{\pi u}{\lambda}, \qquad 0 < u < \lambda \\ \sigma &= 0 \qquad\qquad\quad u > \lambda \end{aligned} \tag{10.3}$$

where σ_c represents the cohesive strength of the solid.

At small displacements u, the sine function can be approximated by argument ($\sin x \simeq x$) so that

$$\sigma = \sigma_c \frac{\pi u}{\lambda}. \tag{10.4}$$

We assume now that displacements u obey Hooke's law, so that

$$\sigma = E\varepsilon = E \frac{u}{b_0} \tag{10.5}$$

where E is the modulus of elasticity. Then Equation (10.4) becomes

$$\sigma_c = \frac{\lambda}{\pi} \frac{E}{b_0}. \tag{10.6}$$

The surface energy 2γ (the factor 2 refers to the two surfaces which are created by material separation) which is the work required for a total separation of the lattice planes, is equal to the area under the force-extension curve. We have

$$2\gamma = \int_0^\infty \sigma \, du \tag{10.7}$$

and on account of Equation (10.3) we obtain

$$2\gamma = \int_0^\lambda \sigma_c \sin \frac{\pi u}{\lambda} \, du = \frac{2\lambda}{\pi} \sigma_c \, . \tag{10.8}$$

From Equations (10.6) and (10.8) we obtain the cohesive strength

$$\sigma_c = \sqrt{\frac{E\gamma}{b_0}} \, . \tag{10.9}$$

This equation indicates that high values of σ_c correspond to large values of modulus of elasticity E and surface energy γ and small distances b_0 of atomic planes.

For many materials γ is of the order of magnitude $Eb_0/100$. Then Equation (10.9) gives

$$\sigma_c \simeq \frac{E}{10} \, . \tag{10.10}$$

Other calculations of cohesive strength based on more precise force-separation laws give values of σ_c in the range $(1/4 - 1/3)E$. Such values of strength in solids can be achieved only when the material is tested in very thin fibers or whiskers. However, the actual strength of solids is much lower than σ_c, typically by three to four orders of magnitude. This is generally attributed to the existence of microstructural and other defects in the solid.

To obtain a rough estimate of the actual strength σ_u of the solid, let us assume that a microstructural defect exists in the solid in the form of an elliptic hole with major axis $2a$ and radius of curvature ρ. The maximum stress that can be developed at the end of the hole cannot exceed the cohesive strength σ_c. We obtain from Equation (1.7) with $\sigma_{\max} = \sigma_c$ and $\sigma = \sigma_u$

$$\sigma_c = 2\sigma_u \sqrt{\frac{a}{\rho}} \tag{10.11}$$

which on account of Equation (10.9) gives

$$\sigma_u = \sqrt{\frac{\pi\rho}{8b_0} \frac{2E\gamma}{\pi a}} \, . \tag{10.12}$$

For $\rho = 8b_0/\pi$ Equation (10.11) becomes identical to Griffith's Equation (4.14).

This model has been oversimplified. Some more sophisticated atomic theories have been proposed for the estimation of the cohesive strength of solids. Although atomic theories provide physical insight into the explanation of the fracture process, they involve great complication in analysis and fail to provide a satisfactory working tool for engineering applications.

10.3. Cleavage fracture

The term "cleavage fracture" means material separation that takes place by breaking atomic bonds along certain crystallographic planes. The fracture is transgranular and

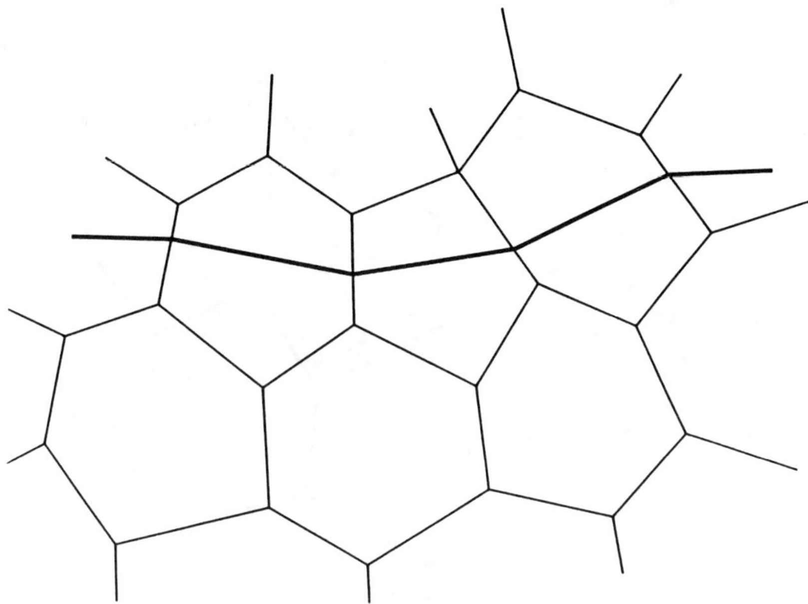

Fig. 10.2. Transgranular fracture.

the cleavage planes are those with the fewer bonds and the greater distances between the planes. This type of fracture is possible in body centered cubic crystals, like iron or low-carbon steel, that cleavage along the planes (1 0 0), and in hexagonal crystals like magnesium. Fracture takes place with the lowest expenditure of energy and the overall deformation is small.

In polycrystalline materials the grains are crystallographically disoriented with respect to each other, and a cleavage fracture changes its orientation each time it meets a grain boundary (Figure 10.2). The fracture planes in each grain are highly reflective, given a shiny appearance on the overall cleavage fracture surfaces. Cleavage fracture is promoted by low temperatures and high strain rates.

A number of micromechanical models have been developed for nucleation of cleavage fracture. For an analysis of these models the reader is referred to the book by Knott [10.1]. The model advanced by Smith [10.2] is of particular significance as it incorporates the important microstructural features of grain boundary carbides. The model considers stress concentration due to a dislocation pile-up at a grain boundary carbide. The fracture criterion takes the form

$$\left(\frac{c_0}{d}\right)\sigma_f^2 + \tau_{\text{eff}}^2\left[1 + \frac{4}{\pi}\left(\frac{c_0}{d}\right)^{1/2}\frac{\tau_i}{\tau_{\text{eff}}}\right]^2 \geq \frac{4E\gamma_p}{\pi(1-\nu^2)\,d} \tag{10.13}$$

where d is the grain diameter, c_0 the carbide thickness, γ_c the surface energy of the carbide, τ_i the friction stress, τ_{eff} the maximum stress that can be attained before yielding occurs and σ_f is the remote stress for fracture. Note that when the second

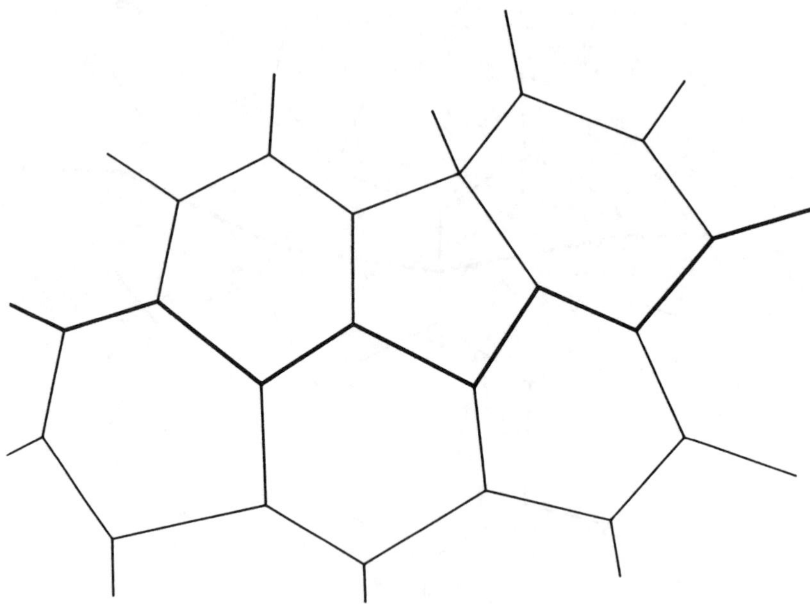

Fig. 10.3. Intergranular fracture.

term on the left-hand of Equation (10.13) is removed this equation reduces to Griffith's equation for a grain boundary microcrack. This term represents the contribution of dislocation to cleavage initiation. Equation (10.13) shows that larger carbides lower the fracture stress.

10.4. Intergranular fracture

Fracture along grain boundaries, so-called intergranular or intercrystalline fracture (Figure 10.3), is the exception rather than the rule in most metals under monotonic loading. Typically metals fail by transgranular fracture, since the fracture resistance along characteristic crystallographic planes in the grain is lower than along the grain boundaries. Intergranular fractures are the result of weak bonding between the grains, due to the segregation of embrittling particles and precipitates (for instance carbides, sulfides and oxides) to the grain boundaries. Cracks grow along grain boundaries leaving the grains intact. Little energy is consumed, and the fracture is brittle, as in the case of transgranular fracture. Intergranular fracture is usually promoted by aggressive environment and high temperatures. Hydrogen embrittlement, environment assisted cracking, intergranular corrosion and cracking at high temperatures lead to intergranular fractures.

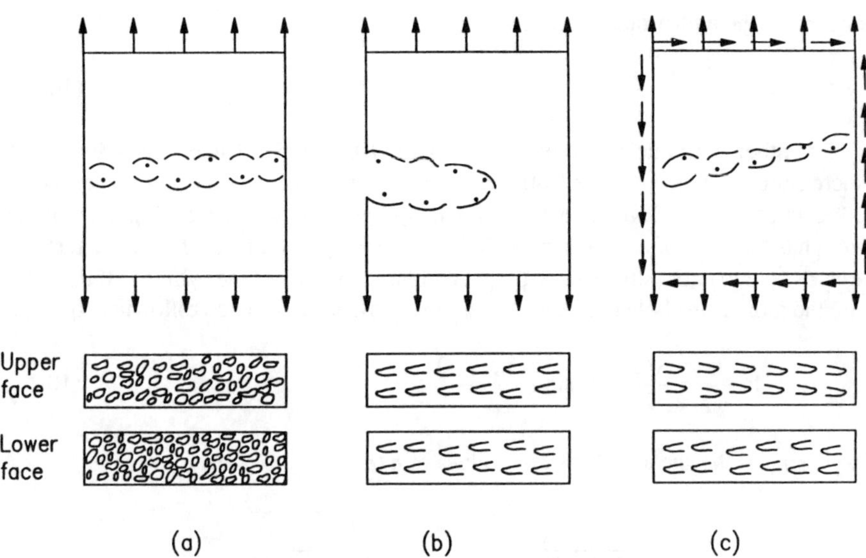

Fig. 10.4. Dimple patterns for macrocrack growth under uniaxial tensile and combined loading.

10.5. Ductile fracture

Crystalline materials always contain second phase particles or inclusions within the grains or at the grain boundaries. The size of the particles varies in the range from 10^{-8} m to 10^{-5} m and they may be added purposely to improve certain properties of the material. Under the influence of an applied load the particles or inclusions may fracture or debond, producing pores or voids. Under such circumstances a fracture surface is formed with a characteristic fibrous or dimpled appearance. Fracture of the material is a consequence of three distinct processes: void nucleation, growth and coalescence. The shape of voids is closely associated with the existing state of stress. Under uniaxial tensile load, the voids grow in a plane normal to the loading axis; when the fracture surfaces are observed in the electron microscope they consist of equiaxed dimples (Figure 10.4a). When the deformation process continues, the dimples become elongated and point in the same direction on the upper and lower faces of the fractured surface (Figure 10.4b). Under shear loading the dimples are elongated and point in opposite directions (Figure 10.4c). Fibrous fractures consume more energy than cleavage fractures and give rise to macroscopic ductile fractures.

A number of void nucleation models have appeared in the literature. Argon *et al.* [10.3] in their model, suggest that the decohesion stress σ_c is

$$\sigma_c = \sigma_e + \sigma_m \tag{10.14}$$

where σ_c is the effective stress given by

$$\sigma_e = \frac{1}{\sqrt{2}}\left[(\sigma_1 - \sigma_2)^2 + (\sigma_2 - \sigma_3)^2 + (\sigma_3 - \sigma_1)^2\right]^{1/2} \tag{10.15}$$

and σ_m is the hydrostatic stress given by

$$\sigma_m = \frac{\sigma_1 + \sigma_2 + \sigma_3}{3} \,. \tag{10.16}$$

For other void nucleation models refer to [10.4, 10.5]. Under the influence of increasing loading void nucleation is followed by void growth and coalescence. McClintock [10.6] developed a simplified void growth model. He assumed that the void has the form of a cylindrical hole in an infinite plate subjected to an axial strain rate $\dot{\varepsilon}_z$, and to equal stresses $\bar{\sigma}_\infty$ perpendicular to the axis of the cylinder. If a_0 and a are the radii of the hole before and after void growth we have the following equation

$$\varepsilon_a = \ln \frac{a}{a_0} = \frac{\sqrt{3}}{2} |\varepsilon_z| \sinh \frac{\bar{\sigma}_\infty}{\tau_Y} - \frac{1}{2} \varepsilon_z \tag{10.17}$$

where τ_Y is the shear yield stress. Values of a/a_0 for $\varepsilon_z = 1$ are given as

$\bar{\sigma}_\infty/\tau_Y$	1	2	3	4
a/a_0	1.7	14	3,500	10^{10}

High values of σ_∞/τ_Y are possible only if σ_z is also high. The increased values of a/a_0 as σ_∞/τ_Y increases demonstrate the strong effect of hydrostatic tension on void growth.

Rice and Tracey [10.7] studied a single void in an infinite plate subjected to principal stress $\sigma_{i\infty}$ $(i = 1, 2, 3)$. They obtained the following equation for the void extension rate $\dot{\varepsilon}_{ia} = \dot{a}_i/a$ in the direction of principal stress $\sigma_{i\infty}$:

$$\dot{\varepsilon}_{ia} = 2\dot{\varepsilon}_{i\infty} + 0.56\dot{\varepsilon}_{e\infty} \sinh \frac{\sqrt{3}}{2} \frac{\bar{\sigma}_\infty}{\tau_Y} \tag{10.18}$$

where $\bar{\sigma}_\infty$ is the mean stress, $\varepsilon_{e\infty}$ is the effective strain rate and $\dot{\varepsilon}_{i\infty}$ the strain rate in the direction of the principal stress σ_i. The effective strain rate is given by

$$\dot{\varepsilon}_e = \left(\frac{2}{3} \dot{\varepsilon}_{ij} \dot{\varepsilon}_{ij} \right)^{1/2} \,. \tag{10.19}$$

These models refer to a single void in an infinite plate and do not take into account void interaction. Gurson [10.8] advanced a model that studies plastic flow in a porous medium. In his model the yield condition is

$$\frac{3}{2} \frac{s_{ij} s_{ij}}{\sigma_Y^2} + 2f \cosh \left(\frac{3}{2} \frac{\sigma_m}{\sigma_Y} \right) - (1 + f^2) = 0 \tag{10.20}$$

where s_{ij} is the deviatoric stress defined by

$$s_{ij} = \sigma_{ij} - \sigma_m \delta_{ij} \tag{10.21}$$

and f is the void volume fraction. Note that Equation (10.20) for $f = 0$ reduces to the von Mises yield criterion.

Tvergaard [10.9] modified Gurson's equation by introducing two parameters q_1 and q_2; his equation is

$$\frac{3}{2}\frac{s_{ij}s_{ij}}{\sigma_Y^2} + 2q_1 f \cosh\left(\frac{3}{2}\frac{q_2\sigma_m}{\sigma_Y}\right) - [1 + (q_1 f)^2] = 0\,. \qquad (10.22)$$

Using experimental results he concluded that good predictions are obtained when $q_1 = 2$ and $q_2 = 1$.

10.6. Crack detection methods

(a) Introductory remarks

The fracture of machine or structural parts involves crack initiation, subcritical growth and final termination. Fracture mechanics is based on the realistic assumption that all materials contain initial defects which constitute the nuclei of fracture initiation. Design for the prevention of crack initiation is physically unrealistic. Initial defects appear in a material due to its composition, or they can be introduced in a structure during fabrication or service life. The detection of defects in structures plays an essential role in design using fracture mechanics. A number of nondestructive testing (NDT) methods for the detection, location and sizing of defects have been developed. Our ability to use fracture mechanics in design is largely due to the reliability of the NDT methods. At the production or service inspection stage, parts containing flaws larger than those determined according to fracture mechanics design must be rejected or replaced.

Six NDT methods that are widely used for defect detection will be briefly described below. These are dye penetration, magnetic particles, eddy currents, radiography, ultrasonics and acoustic emission. Each of these methods possesses advantages and disadvantages depending on the application. For further details on these methods the reader should consult references [10.10–10.15].

(b) Dye penetration

This technique is commonly used for detecting surface flaws. It involves application of a colored or fluorescent dye onto a cleaned surface of the component. After allowing sufficient time for penetration, the excess penetrant is washed off and the surface is dusted with a post-penetrant material (developer) such as chalk. The developer acts as a blotter and the defects show up as colored lines. The reliability of the method depends on the surface preparation of the component. The method is widely used, and can detect small cracks. It has the advantage of fast inspection at low cost. It applies, however, only to surface flaws.

(c) Magnetic particles

This method is based on the principle that flaws in a magnetic material produce a distortion to an induced magnetic field. Measuring this distortion provides information on the existing defects. The magnetic field is induced by passing a current through the component or using permanent or electromagnets. For detecting the distortion of the magnetic field the surface under inspection is coated with a fluorescent liquid that contains magnetic particles in suspension. The method is easy to apply, is speedy and economical. As in the dye penetration method, it can be used only for detecting cracks on or near the surface.

(d) Eddy currents

A coil carrying alternating current placed near a conducting surface induces eddy currents in the surface. The eddy currents create a magnetic field that affects the coil; its impedance changes when a defect is present. By measuring this change we can find information about the defect. The induced eddy currents concentrate near the surface of the conductor; this is the so-called "skin effect". The penetration depth is influenced by the frequency of the current, the magnetic permeability and electrical conductivity of the conductor, and the coil and conductor geometry. In a ferritic conductor, the penetration depth is smaller than 1 mm at most frequencies, while in nonmagnetic conductors it may be several millimetres. The sensitivity of the method is high for defects near the surface but decreases with increasing depth. Problems in the method arise from the difficulty of relating the defect size to the change in impedance, and the influence of a number of other factors on the impedance. These include: the relative position of the coil and the conductor; the presence of structural variations; material inhomogeneities. Measurement of defect size is made by comparing its effect to that observed from a standard defect.

(e) Radiography

Radiography is one of the oldest NDT methods for detecting subsurface defects. A source of X- or γ-rays is transmitted through the specimen. If the specimen has variations in thickness or density due – for example, to the presence of defects – the emerging radiation will not be of uniform intensity. Since defects absorb less X-rays than the surrounding material they can be detected by using a sensitive photographic film, on which they appear as dark lines. The method is particularly suitable for finding volumetric defects. The method may be used to detect cracks, but in order to obtain substantial differential absorption between rays passing through the crack and those passing through the surrounding material they should be oriented parallel to the plane of the incident radiation. Thus, the method is insensitive to cracks unless their orientations are known beforehand. This may involve a number of exposures at different positions of the X-rays.

(f) Ultrasonics

This method is based on the transmission of ultrasonic waves into the material by a transducer containing a piezoelectric crystal. Metallurgical defects and/or surface boundaries reflect the incident pulse which is monitored on an oscilloscope. The distance between the first pulse and the reflection gives the position of the crack. The size of the crack can also be estimated. The method is characterized by: high sensitivity for detection of cracks, at all positions; ability to measure crack position and size; fast response for rapid inspection; economy; applicability to thick material sections; portability of equipment for in-situ inspection. The application of the method is, however, limited by unfavorable specimen geometry and the difficulty of distinguishing between cracks and other types of defects, such as inclusions. The method is also characterized by the subjective interpretation of echoes by the operator.

(g) Acoustic emission

The method involves the use of a sensing transducer and sophisticated electronic equipment to detect sounds and stress waves emitted inside the material during the process of cracking. The detected emissions are amplified, filtered and interpreted. The method is capable of locating flaws without resorting to a point-by-point search over the entire surface of interest. It can be used to detect crack initiation and growth. A disadvantage of the method lies in the difficulty of interpreting the signals which are obtained.

References

10.1. Knott, J.F. (1981) *Fundamentals of Fracture Mechanics*, Butterworths, pp. 188–203.

10.2. Smith, E. (1966) 'The nucleation and growth of cleavage microcracks in mild steel', *Proc. Conference on Physical Basis of Yield and Fracture*, Institute of Physics and Physical Society, Oxford, pp. 36–46.

10.3. Argon, A.S., Im, J., and Safogly, R. (1975) 'Cavity formation from inclusions in ductile fracture', *Metallurgical Transactions* **6A**, 825–837.

10.4. Beremin, F.M. (1981) 'Cavity formation from inclusions in ductile fracture of A 508 steel', *Metallurgical Transactions* **12A**, 723–731.

10.5. Goods, S.H. and Brown, L.M. (1979) 'The nucleation of cavities by plastic deformation', *Acta Mechanica* **27**, 1–15.

10.6. McClintock, F.A. (1968) 'Ductile rupture by the growth of holes', *Journal of Applied Mechanics, Trans. ASME* **35**, 363–371.

10.7. Rice, J.R. and Tracey, D.M. (1969) 'On the ductile enlargement of voids in triaxial stress fields', *Journal of the Mechanics and Physics of Solids* **17**, 201–217.

10.8. Gurson, A.L. (1977) 'Continuum theory of ductile rupture by void nucleation and growth: Part 1 – Yield criteria and flow rules for porous ductile media', *Journal of Engineering Materials and Technology* **99**, 2–15.

10.9. Tvergaard, V. (1982) 'On localization in ductile materials containing spherical voids', *International Journal of Fracture* **18**, 237–252.

10.10. McGonnagle, W.J. (1971) 'Nondestructive testing', in *Fracture – An Advanced Treatise*, Vol. III, *Engineering Fundamentals and Environment Effects* (ed. H. Liebowitz), Academic Press, pp. 371–430.

10.11. Coffey, J.M. and Whittle, M.J. (1981) 'Non-destructive testing: its relation to fracture mechanics and component design', *Philosophical Transactions of the Royal Society of London* **A299**, 93–110.

10.12. Coffey, J.M. (1980) 'Ultrasonic measurement of crack dimensions in laboratory specimens', in *The Measurement of Crack Length and Shape During Fracture and Fatigue* (ed. C.J. Beevers), EMAS, pp. 345–385.

10.13. Catou, M.M., Flavenot, J.F., Flambard, C., Madelaine, A., and Anton, F. (1980) 'Automatic measurement of crack length during fatigue. Testing using ultrasonic surface waves', in *The Measurement of Crack Length and Shape During Fracture and Fatigue* (ed. C.J. Beevers), EMAS, pp. 387–392.

10.14. Richard, C.E. (1980) 'Some guidelines to the selection of techniques', in *The Measurement of Crack Length and Shape During Fracture and Fatigue* (ed. C.J. Beevers), EMAS, pp. 461–468.

10.15. Krautkramer, J. and Krautkramer, H. (1977) *Ultrasonic Testing of Materials* (2nd ed.), Springer-Verlag.

Index

Mechanics

SOLID MECHANICS AND ITS APPLICATIONS

Series Editor: G.M.L. Gladwell

Aims and Scope of the Series

The fundamental questions arising in mechanics are: *Why?, How?,* and *How much?* The aim of this series is to provide lucid accounts written by authoritative researchers giving vision and insight in answering these questions on the subject of mechanics as it relates to solids. The scope of the series covers the entire spectrum of solid mechanics. Thus it includes the foundation of mechanics; variational formulations; computational mechanics; statics, kinematics and dynamics of rigid and elastic bodies; vibrations of solids and structures; dynamical systems and chaos; the theories of elasticity, plasticity and viscoelasticity; composite materials; rods, beams, shells and membranes; structural control and stability; soils, rocks and geomechanics; fracture; tribology; experimental mechanics; biomechanics and machine design.

1. R.T. Haftka, Z. Gürdal and M.P. Kamat: *Elements of Structural Optimization*. 2nd rev.ed., 1990 ISBN 0-7923-0608-2
2. J.J. Kalker: *Three-Dimensional Elastic Bodies in Rolling Contact.* 1990 ISBN 0-7923-0712-7
3. P. Karasudhi: *Foundations of Solid Mechanics.* 1991 ISBN 0-7923-0772-0
4. N. Kikuchi: *Computational Methods in Contact Mechanics.* (forthcoming) ISBN 0-7923-0773-9
5. *Not published.*
6. J.F. Doyle: *Static and Dynamic Analysis of Structures.* With an Emphasis on Mechanics and Computer Matrix Methods. 1991 ISBN 0-7923-1124-8; Pb 0-7923-1208-2
7. O.O. Ochoa and J.N. Reddy: *Finite Element Analysis of Composite Laminates.* ISBN 0-7923-1125-6
8. M.H. Aliabadi and D.P. Rooke: *Numerical Fracture Mechanics.* ISBN 0-7923-1175-2
9. J. Angeles and C.S. López-Cajún: *Optimization of Cam Mechanisms.* 1991 ISBN 0-7923-1355-0
10. D.E. Grierson, A. Franchi and P. Riva: *Progress in Structural Engineering.* 1991 ISBN 0-7923-1396-8
11. R.T. Haftka and Z. Gürdal: *Elements of Structural Optimization.* 3rd rev. and exp. ed. 1992 ISBN 0-7923-1504-9; Pb 0-7923-1505-7
12. J.R. Barber: *Elasticity.* 1992 ISBN 0-7923-1609-6; Pb 0-7923-1610-X
13. H.S. Tzou and G.L. Anderson (eds.): *Intelligent Structural Systems.* 1992 ISBN 0-7923-1920-6
14. E.E. Gdoutos: *Fracture Mechanics.* An Introduction. 1993 ISBN 0-7923-1932-X
15. J.P. Ward: *Solid Mechanics.* An Introduction. 1992 ISBN 0-7923-1949-4
16. M. Farshad: *Design and Analysis of Shell Structures.* 1992 ISBN 0-7923-1950-8
17. H.S. Tzou and T. Fukuda (eds.): *Precision Sensors, Actuators and Systems.* 1992 ISBN 0-7923-2015-8
18. J.R. Vinson: *The Behavior of Shells Composed of Isotropic and Composite Materials.* 1993 ISBN 0-7923-2113-8

Kluwer Academic Publishers – Dordrecht / Boston / London

Mechanics

SOLID MECHANICS AND ITS APPLICATIONS

Series Editor: G.M.L. Gladwell

19. H.S. Tzou: *Piezoelectric Shells.* Distributed Sensing and Control of Continua. 1993
ISBN 0-7923-2186-3
20. W. Schiehlen: *Advanced Multibody System Dynamics.* Simulation and Software Tools. 1993
ISBN 0-7923-2192-8

Kluwer Academic Publishers – Dordrecht / Boston / London

Mechanics

FLUID MECHANICS AND ITS APPLICATIONS

Series Editor: R. Moreau

Aims and Scope of the Series

The purpose of this series is to focus on subjects in which fluid mechanics plays a fundamental role. As well as the more traditional applications of aeronautics, hydraulics, heat and mass transfer etc., books will be published dealing with topics which are currently in a state of rapid development, such as turbulence, suspensions and multiphase fluids, super and hypersonic flows and numerical modelling techniques. It is a widely held view that it is the interdisciplinary subjects that will receive intense scientific attention, bringing them to the forefront of technological advancement. Fluids have the ability to transport matter and its properties as well as transmit force, therefore fluid mechanics is a subject that is particularly open to cross fertilisation with other sciences and disciplines of engineering. The subject of fluid mechanics will be highly relevant in domains such as chemical, metallurgical, biological and ecological engineering. This series is particularly open to such new multidisciplinary domains.

1. M. Lesieur: *Turbulence in Fluids.* 2nd rev. ed., 1990 ISBN 0-7923-0645-7
2. O. Métais and M. Lesieur (eds.): *Turbulence and Coherent Structures.* 1991 ISBN 0-7923-0646-5
3. R. Moreau: *Magnetohydrodynamics.* 1990 ISBN 0-7923-0937-5
4. E. Coustols (ed.): *Turbulence Control by Passive Means.* 1990 ISBN 0-7923-1020-9
5. A.A. Borissov (ed.): *Dynamic Structure of Detonation in Gaseous and Dispersed Media.* 1991 ISBN 0-7923-1340-2
6. K.-S. Choi (ed.): *Recent Developments in Turbulence Management.* 1991 ISBN 0-7923-1477-8
7. E.P. Evans and B. Coulbeck (eds.): *Pipeline Systems.* 1992 ISBN 0-7923-1668-1
8. B. Nau (ed.): *Fluid Sealing.* 1992 ISBN 0-7923-1669-X
9. T.K.S. Murthy (ed.): *Computational Methods in Hypersonic Aerodynamics.* 1992 ISBN 0-7923-1673-8
10. R. King (ed.): *Fluid Mechanics of Mixing.* Modelling, Operations and Experimental Techniques. 1992 ISBN 0-7923-1720-3
11. Z. Han and X. Yin: *Shock Dynamics.* 1993 ISBN 0-7923-1746-7
12. L. Svarovsky and M.T. Thew (eds.): *Hydroclones.* Analysis and Applications. 1992 ISBN 0-7923-1876-5
13. A. Lichtarowicz (ed.): *Jet Cutting Technology.* 1992 ISBN 0-7923-1979-6
14. F.T.M. Nieuwstadt (ed.): *Flow Visualization and Image Analysis.* 1993 ISBN 0-7923-1994-X
15. A.J. Saul (ed.): *Floods and Flood Management.* 1992 ISBN 0-7923-2078-6
16. D.E. Ashpis, T.B. Gatski and R. Hirsh (eds.): *Instabilities and Turbulence in Engineering Flows.* 1993 ISBN 0-7923-2161-8
17. R.S. Azad: *The Atmospheric Boundary Layer for Engineers.* 1993 ISBN 0-7923-2187-1

Kluwer Academic Publishers – Dordrecht / Boston / London